Approximation of Nonlinear Evolution Systems

This is Volume 164 in
MATHEMATICS IN SCIENCE AND ENGINEERING
A Series of Monographs and Textbooks
Edited by RICHARD BELLMAN, *University of Southern California*

The complete listing of books in this series is available from the Publisher upon request.

Approximation of Nonlinear Evolution Systems

Joseph W. Jerome

Department of Mathematics
Northwestern University
Evanston, Illinois

1983

ACADEMIC PRESS

A Subsidiary of Harcourt Brace Jovanovich, Publishers

New York London
Paris San Diego San Francisco São Paulo Sydney Tokyo Toronto

ACADEMIC PRESS, INC.
111 Fifth Avenue, New York, New York 10003

United Kingdom Edition published by
ACADEMIC PRESS, INC. (LONDON) LTD.
24/28 Oval Road, London NW1 7DX

Library of Congress Cataloging in Publication Data

Jerome, W. Joseph.
Approximation of nonlinear evolution systems.

(Mathematics in science and engineering)
Includes bibliographical references and index.
1. Evolution equations, Nonlinear--Numerical solutions. 2. Approximation theory. I. Title. II. Series.
QA374.J47 1982 515.3'53 82-8808
ISBN 0-12-384680-3 AACR2

PRINTED IN THE UNITED STATES OF AMERICA

83 84 85 86 9 8 7 6 5 4 3 2 1

for my Parents

CONTENTS

II LOCAL SMOOTH SOLUTIONS

PREFACE

The evolution of complicated physical and biological systems is sufficiently subtle that it requires intricate analysis of the underlying mathematical models. This analysis must be focused so as to impart clear and original insights, yet it must be intrinsically generic, permitting examination of general questions such as existence, uniqueness, regularity, well-posedness, stability, and regularization, all related to the soundness of the models. The models themselves must possess sufficient novelty and importance to justify generality and suggest new directions for study. We believe that the models selected here accomplish this objective.

The book is directed primarily toward applied mathematical analysts who are concerned with the mathematical preliminaries to numerical computations in nonlinear partial differential equations, and toward physical, biological, and engineering scientists who are interested in phase transition, diffusion in porous media, diffusion with reaction, incompressible fluid flow, and general relativity. It is aimed, secondarily, at mathematicians with interests in partial differential equations, approximation theory, numerical analysis, convex analysis, or operator theory. Readers with some basic acquaintance with linear functional analysis and the notion of a weak solution of a partial differential equation will find parts of Chapters 1, 2, 4, 5, and 6 accessible. A stronger background in nonlinear functional analysis, vector measure and integration, partial differential equations, convex analysis, geometry, and operator theory is needed for Chapters 3 and 7, as well as some parts of Chapters 4, 5, and 6.

Part I of the book deals with the global existence in time of weak solutions of nonlinear evolution systems, and Part II deals with the local classical theory. The models are developed, at least in summary form, from basic principles, such as those of conservation. Where some subtleties arise, such as in the free-boundary problems associated with degenerate parabolic evolutions, distribution formulations are presented. Approximation via regularization is discussed in detail. Various levels of discrete analysis are presented, including semidiscrete maximum principles and invariant regions, an existence analysis based on semidiscretization in Parts I and II, and numerical estimates for finite-element approximations and regularizations. Some topics particularly worth mentioning are the "entropy" determination (specifically, n-widths) of function classes and the relation of this concept to the complexity of partial differential equations; the construction of the evolution operators for linear stable systems and their use in developing a local theory for quasi-linear equations in Banach spaces; the alternative stability analysis for quasi-linear equations via semidiscrete approximations and its application to invariant time interval determination in hydrodynamics; the evolution of systems that are not spatially gradient in nature but are nonetheless not far from such systems as measured by pseudo-monotone perturbations; the use of abstract integral operators, called lifting operators, in the recasting of the partial differential equations and their approximate formulations; and the use of regular Baire measures with duality to obtain convergence estimates.

This book shares with explicitly numerical texts the use of such tools as discrete analysis; with explicitly applied texts the introduction of physical and biological models; and with theoretical texts the use of various techniques to formulate and prove major existence, stability, and approximation theorems. These evolution systems encompass topics of interest to the applied and theoretical scientist.

ACKNOWLEDGMENTS

This book is based on lectures presented at Northwestern University, in which preliminary versions of six of the seven chapters were presented to small numbers of colleagues and students. Thanks are due to Stephen Fisher, Mark Pinsky, and Laurent Veron, who listened to some of these lectures and made many helpful comments. The policy of upper-level teaching seminars at Northwestern is a fruitful one, producing an atmosphere conducive to learning and research. Thanks are due also to T. Kato, J. E. Marsden, and J. R. Dorroh for their help in supplying appropriate literature and references. The author expresses his appreciation for the kind hospitality shown him by Leslie Fox during a leave at Oxford University; he also thanks Gunter Meyer, whose understanding of the quantitative aspects of the Stefan problem greatly assisted in the subsequent development of this volume.

Gratitude is also due those who have helped in certain technical matters related to the book. These include Garrett Birkhoff, David Fox, Richard Graff, Klaus Höllig, C. Ionescu-Tulcea, Ken Jackson, Mitchell Luskin, Ridgway Scott, Vidar Thomée, and Lars Wahlbin. Avner Friedman acted as a helpful conduit to the analytical aspects of the subject in his regular P. D. E. seminar and provided useful general information. Finally, the author expresses his appreciation to Miss Antoinette Trembinska for her very careful reading of the manuscript, and to Mrs. Alice Wagner for her excellent preparation of the manuscript.

LIST OF SYMBOLS AND DEFINITIONS

Term	Symbol or Definition
Arguments	
Part I	$(\mathbf{x}, t)$ $(\mathbf{x} \in \Omega \subset \mathbb{R}^n, t \in (0, T_0))$
Part II	$(t, \mathbf{x}, \mathbf{p})$ $(t \in J = [0, T_0], \mathbf{x} \in \mathbb{R}^n, \mathbf{p} \in \Omega \subset \mathbb{R}^{m(n+2)})$
Boundary regularity for bounded sets Ω in Part I	Restricted cone condition (see Ref. [4.1] p. 11) or the equivalent strong Lipschitz condition (see Ref. [4.25] p. 72) assumed throughout, unless replaced by stronger specific assumptions such as star-shaped, smooth, etc.
Closure (of G)	$\mathrm{Cl}(G)$, $\bar{G}$
Coercive	
Bilinear form	Usage as in (3.2.17b)
Functional	Usage as in (3.2.6) or (3.2.29) (for weaker alternate, see Remark 3.1.1)
Commutator (of operators A and S)	$[\mathrm{A}, \mathrm{S}] := \mathrm{AS} - \mathrm{SA}$
Connection coefficients	Γ^k_{ij}
Convergence	Sequential
Ordinary	$\rightarrow$
Weak	$\rightharpoonup$
Weak-*	$\rightharpoonup^*$

(*continued*)

Term	Symbol or Definition
Convolution	
Functions	*
Operators	(see (6.3.4))
Covariant derivative (in direction X)	D_X
Curvature	
Einstein tensor	G_{ij}
Operator	R_{XY}
Ricci tensor	R_{ij}
Riemann tensor	$R^i_{jk\ell}$
Scalar	R
Dispersion (of **B** from $\mathcal{M}$ in **X**)	$E_{\mathbf{X}}(\mathbf{B}, \mathcal{M})$
Distribution	
Differentiation	$D^{\boldsymbol{\alpha}} = \prod_{i=1}^{n} \left(\frac{\partial}{\partial x_i}\right)^{\alpha_i}, \boldsymbol{\alpha} = (\alpha_1, \ldots, \alpha_n)$
	$\left(\frac{\partial}{\partial \mathbf{x}}\right)^{\boldsymbol{\alpha}} := D^{\boldsymbol{\alpha}}$
Functional space	$\mathcal{D}'$
Pairing	$\langle \cdot, \cdot \rangle$
Test space	$\mathcal{D} = \mathbf{C}_0^\infty$
Domain (of mappings U)	$\mathbf{D}_{\mathrm{U}}$, $\mathbf{Dom}(\mathrm{U})$
Duality	
Pairing	$\langle \cdot, \cdot \rangle$
Symbol for space	*
Function/Measure spaces on Euclidean domain $\mathcal{O}$ (or $\overline{\mathcal{O}}$) with range in normed linear space **X**	Real-valued. In this list, $\mathcal{O}$ is a generic symbol: special choices occur in the text
Baire[†]	$\mathbf{M}(\overline{\mathcal{O}}) := \{\mu : \mu$ is measurable on Baire subsets of $\overline{\mathcal{O}}$ and $\int_{\overline{\mathcal{O}}} \lvert d\mu \rvert < \infty\}$
($k \geqslant 0$) Graded smooth	$\mathbf{C}^k(\mathcal{O}; \mathbf{X})$, $\mathbf{C}_b^k(\mathcal{O}; \mathbf{X})$ (see Def. 7.3.1 in the case $\mathcal{O} = \mathbb{R}^n \times \Omega$)
Lebesgue	$\mathbf{L}^p(\mathcal{O}; \mathbf{X}) := \left\{ f : \int_{\mathcal{O}} \lVert f \rVert_{\mathbf{X}}^p < \infty \right\} (1 \leqslant p < \infty)$
	$\mathbf{L}^\infty(\mathcal{O}; \mathbf{X}) := \{ f : \mathrm{ess}_{\mathcal{O}} \sup \lVert f \rVert_{\mathbf{X}} < \infty \}$
Lipschitz	$\mathbf{Lip}(\overline{\mathcal{O}}; X)$
Local Lebesgue	$\mathbf{L}^p_{\mathrm{loc}}(\mathcal{O}; \mathbf{X})$
($k \geqslant 0$) Sobolev[‡]	$\mathbf{W}^{k,p}(\mathcal{O}; \mathbf{X}) := \{ f : D^{\boldsymbol{\alpha}} f \in \mathbf{L}^p(\mathcal{O}; \mathbf{X}), \lvert \boldsymbol{\alpha} \rvert \leqslant k \}$

[†] The application is to a closed subset.

[‡] Applications are occasionally made to closed sets, such as the cube. In this case, interpret as the cube interior, etc.

Term	Symbol or Definition
$(k \geqslant 0)$	$\mathbf{W}^{k,p}_{\pi}$ (see Remark 4.2.8)
$(k \geqslant 0)$	$\mathbf{H}^k(\mathcal{O};\mathbf{X}) := \mathbf{W}^{k,2}(\mathcal{O};\mathbf{X})$
(nonintegral s, $k < s < k+1$)	$\mathbf{H}^s(\mathcal{O};\mathbf{X}) :=$ real interpolation space $(\mathbf{H}^k(\mathcal{O};\mathbf{X}), \mathbf{H}^{k+1}(\mathcal{O};\mathbf{X}))_{s-k,2}$
Uniformly local Lebesgue and Sobolev	$\mathbf{L}^p_{u\ell}(\mathcal{O};\mathbf{X}), \mathbf{W}^{k,p}_{u\ell}(\mathcal{O};\mathbf{X})$ (see Definition 7.3.3)
Functions	
Point action	$f: p \to f(p)$ or $p \mapsto f(p)$
Set action	$f: \mathrm{Dom}(f) \to K(\supset \mathrm{Range}(f))$
Interior (of G)	$\mathrm{int}(G)$, G^0
Lie bracket	$[\cdot,\cdot]$
Matrix dot product	$A \cdot B = \sum_{i,j} a_{ij} b_{ij}$
Measure (of Ω)	$\lvert\Omega\rvert$, $m(\Omega)$
Modular operation	$/$
Monomial	$\mathbf{x}^{\boldsymbol{\alpha}} := \prod_{i=1}^{n} x_i^{\alpha_i}$
Multinomial coefficient	$\binom{k}{k_1,\ldots,k_n} = \dfrac{k!}{k_1!\cdots k_n!} \left(k = \sum_{i=1}^{n} k_i\right)$
N-width (of $\mathscr{A}$ in $\mathbf{X}$)	$d_N(\mathscr{A},\mathbf{X})$
Negative	
Infinitesimal generator class	$G(\mathbf{X})$
Part of f	$f^- = \inf(0,f) \leqslant 0$
Norms	
$\lVert\mu\rVert_{\mathbf{M}(\bar{\Omega})}$	Variation of μ
$\lVert u\rVert_{\mathbf{C}^k_b(\mathcal{O};\mathbf{X})}$	$\sup_{\mathcal{O},\{\boldsymbol{\alpha}\}}$ {$\mathbf{X}$ norm of derivatives of order $\lvert\boldsymbol{\alpha}\rvert \leqslant k$ taken with respect to spatial variables comprising a fixed hyperplane in span $\mathcal{O}$}
$\lVert u\rVert_{\mathbf{L}^p(\mathcal{O};\mathbf{X})}$	$\left\{\int_{\mathcal{O}} \lVert u\rVert^p_{\mathbf{X}}\right\}^{1/p}$, $1 \leqslant p < \infty$
$\lVert u\rVert_{\mathbf{W}^{k,p}(\mathcal{O};\mathbf{X})}$	$\left\{\sum_{\lvert\boldsymbol{\alpha}\rvert\leqslant k} \lVert\mathbf{D}^{\boldsymbol{\alpha}}u\rVert^p_{\mathbf{L}^p(\mathcal{O};\mathbf{X})}\right\}^{1/p}$, $1 \leqslant p < \infty$
(Alternatively)	
$\lVert u\rVert_{\mathbf{W}^{k,2}(\mathcal{O};\mathbf{X})}$	$\left\{\sum_{\lvert\boldsymbol{\alpha}\rvert\leqslant k} C_{\boldsymbol{\alpha}} \lVert\mathbf{D}^{\boldsymbol{\alpha}}u\rVert^2_{\mathbf{L}^2(\mathcal{O};\mathbf{X})}\right\}^{1/2}$ with $C_{\boldsymbol{\alpha}} = \binom{\lvert\boldsymbol{\alpha}\rvert}{\alpha_1,\ldots,\alpha_n}$ (see Theorem 4.2.14)

(continued)

Term	Symbol or Definition
$\|u\|_{\mathbf{H}^s(\mathcal{O};\,\mathbf{X})}$	Either, interpolation space norm or equivalent Bessel potential norm defined via Fourier transform if $\mathcal{O} = \mathbb{R}^n$
$\|u\|_{\mathbf{L}^\infty(\mathcal{O};\,\mathbf{X})}$	$\operatorname{ess}_{\mathcal{O}} \sup \|u\|_{\mathbf{X}}$
$\|u\|_{\mathbf{W}^{k,\infty}(\mathcal{O};\,\mathbf{X})}$	$\max_{\|\alpha\| \leqslant k} \|D^\alpha u\|_{\mathbf{L}^\infty(\mathcal{O};\,\mathbf{X})}$
$\|\cdot\|_{\mathbf{L}^p_{u\ell}(\mathcal{O};\,\mathbf{X})}$, $\|\cdot\|_{\mathbf{W}^{k,p}_{u\ell}(\mathcal{O};\,\mathbf{X})}$	(see Definition 7.3.3)
$\|\cdot\|_{1,\mathbf{X}}$, $\|\cdot\|_{\infty,\mathbf{X}}$	(see Definition 6.3,1)
$\|f\|_{\mathbf{Lip}}$	$\inf\{$Lipschitz constants of $f\}$
Numbering format	
Definitions	Separate category
Equations	Separate category
Lemmas, Propositions, Corollaries and Theorems	As a category
Remarks	Separate category
Operators	
Bounded linear operators from **X** into **Y**	$\mathbf{B}(\mathbf{X}, \mathbf{Y})$ (with norm $\|\cdot\|_{\mathbf{X},\mathbf{Y}}$)
Bounded linear operators on **X**	$\mathbf{B}[\mathbf{X}]$ (with norm $\|\cdot\|_{\mathbf{X}}$)
Evolution	$\mathrm{U}(t, s)$ $(0 \leqslant s \leqslant t \leqslant T_0)$
Extension	$\mathrm{A} \supset \mathrm{B}$
Format	Roman
Inversion†	
Dirichlet	D_0
Neumann	N_σ
Robin	R_ω
Projection	
Energy space	E_h
Pivot space	P_h
Solenoidal (generalized)	P
Structure	
Diagonal	Scalar operator "multiplying" the identity matrix.
Order	
Big O	
Little o	
Partition	
Length	$\|\mathscr{P}\| := \min_k (t_k - t_{k-1})$ $(\mathscr{P} = \{t_k\}_0^M)$
Sequence	$\{\mathscr{P}^N\}$
Positive part (of f)	$f^+ = \sup(0, f)$ $(\geqslant 0)$

† Also called lifting or Riesz mappings.

Term	Symbol or Definition
Product	
Cartesian (of sets)	$A \times B$
Direct (of operators)	$\mathrm{A} \otimes \mathrm{B}$
Proximity mapping (induced by convex ϕ)	$\mathrm{prox}_\phi(\cdot)$
Range (of mapping U)	$\mathbf{R}_\mathrm{U}$, $\mathbf{Range}(\mathrm{U})$[†]
Real abstract space	Assumed unless indicated otherwise
Referencing	
$[n], \ldots,$	Refers to $\cdots$ in Reference n of the references of the current chapter
Chapter $m[n], \ldots,$	Refers to $\cdots$ in Reference n of the references of Chapter m
Resolvent	
Operator (U)	$\mathrm{R}(\lambda, \mathrm{U})$
Set	$\rho(\mathrm{U})$
Riesz	
Mapping	Isomorphism $\mathrm{T}: \mathbf{G}^* \to \mathbf{G}$ satisfying $\langle \ell, v \rangle = (\mathrm{T}\ell, v)_\mathbf{G}$
Potential	I_s
Schwartz class	$\mathscr{S}$
Semi-norm	$\lvert u \rvert_{\mathbf{W}^{k,p}(\mathcal{O};\mathbf{X})} = \left\{ \sum_{\lvert\boldsymbol{\alpha}\rvert = k} \lVert \mathrm{D}^{\boldsymbol{\alpha}} u \rVert^p_{\mathbf{L}^p(\mathcal{O};\mathbf{X})} \right\}^{1/p}, \ 1 \leqslant p < \infty$
	$\lvert u \rvert_{\mathbf{W}^{k,\infty}(\mathcal{O};\mathbf{X})} = \max_{\{\lvert\boldsymbol{\alpha}\rvert = k\}} \{ \lVert \mathrm{D}^{\boldsymbol{\alpha}} u \rVert_{\mathbf{L}^\infty(\mathcal{O};\mathbf{X})} \}$
Sequence	
Piecewise linear (in time)	$\{\tilde{U}^N\}$
Step function (in time)	$\{U^N\}$
Set complementation	$\backslash$
Signum function	$\mathrm{sgn}(x) := \begin{cases} 1, & x > 0, \\ 0, & x = 0, \\ -1, & x < 0 \end{cases}$
Spaces	
Format	Boldface (for abstract and function spaces and subsets)
Special symbols	
Del	$\boldsymbol{\nabla} := \left(\frac{\partial}{\partial x_1}, \ldots, \frac{\partial}{\partial x_n} \right)$
Identity function	id
Laplacian	$\Delta := \boldsymbol{\nabla} \cdot \boldsymbol{\nabla}$
Polynomials, degree $\leqslant \ell - 1$	$\mathscr{P}_\ell$

(continued)

[†] Not boldface if subset of Euclidean space.

Term	Symbol or Definition
Sobolev projection	P_ℓ
Triangle $\{0 \leqslant s \leqslant t \leqslant T_0\}$	Δ
Unit ball (closed) in $\mathbf{W}^{\ell,p}$	$\mathbf{U}^{\ell,p}$
Spectrum (of U)	$\sigma(\mathrm{U})$
Strictly	
Coercive (bilinear form)	$B(u,u) \geqslant C\|u\|^2$
Convex (norm)	A convex combination of distinct unit vectors has norm less than one.
Sum	
Algebraic	$\mathbf{M}_1 + \mathbf{M}_2$
Direct	
Operator	$\mathrm{A} \oplus \mathrm{B}$
Subspace	$\mathbf{M}_1 \oplus \mathbf{M}_2$
Tensor	Displayed with subscripts to indicate rank
Translation symbols	
f_σ	$f - \sigma\,\mathrm{id}$
$\tau_{\boldsymbol{\alpha}}$	$\mathrm{Q} \to \mathrm{Q}_{\boldsymbol{\alpha}}$, $\tau_{\boldsymbol{\alpha}}(\mathbf{y}) = \mathbf{x}_{\boldsymbol{\alpha}} + h\mathbf{y}$ (Definition 4.1.2)
$\sigma_{\boldsymbol{\alpha}} \circ u$	$u \circ \tau_\alpha^{-1}$
Triangulation	
Elements	e
Maximal element diameter	h
Symbol	$\Delta_h := \{e\}$
Uniform partition length	δt^N, Δt
Uniqueness	
(for) Multi-valued mapping H (of) Selection $H(u)$	Induced by formulation or sequential convergence
Upper Lebesgue integral (of f over J)	$\int_J^* f$
Vector	
Factorial	$\boldsymbol{\alpha}! = \prod_{i=1}^{n} \alpha_i!$
Format	Boldface
Length	$\|\boldsymbol{\alpha}\| = \sum_{i=1}^{n} \alpha_i$ (nonnegative integer entries)
Weakly asymptotic	$\approx$ (identical O orders)

INTRODUCTION

This book is devoted to the study of the initial-value and the mixed initial/boundary-value problems. It is divided into two parts, which superficially appear to be entirely unrelated: Part I, a weak solution theory, global in time, for nonlinear and degenerate parabolic evolutions, and Part II, a classical solution theory, local in time, for quasi-linear parabolic and hyperbolic evolutions. These theories are, in turn, motivated by models from the physical and biological sciences, whose formulations are described and succinctly developed, particularly in Chapter 1. These include phase transition through heat conduction, fluid filtration in porous media, reaction–diffusion systems, incompressible fluid dynamics, and general relativity.

The basic question of the determination of existence–uniqueness classes for the solutions of these models leads to a consideration of both types of solutions as presented in Parts I and II. For example, in the case of incompressible fluid dynamics, in spite of persistent current research efforts, such an existence–uniqueness class has not yet been determined for global solutions, though both properties have been proved for separate classes for viscous, incompressible fluids. One is then led to local solutions for the resolution of such basic well-posedness. On the other hand, very complicated analytical models such as the geometry of space–time have not yet been analyzed in any globally satisfactory way for the general initial-value problem. In fact, only fragmentary global theories exist for nonlinear hyperbolic problems in general. Some of the results obtained correlate the size of the initial datum, in some norm, to the length of the time interval on which existence of solutions is demonstrated. Such theories are clearly more closely related to local theories, in spirit, and are presented in Part II.

No attempt is made in this book to discuss recent global theories derived from applications of geometric measure theory to systems satisfying hyperbolic conservation laws, or applications of Nash–Moser† iteration to nonlinear wave equations. Nor do we discuss qualitative behavior of nonlinear evolutions such as shock formation, steady-state bifurcation, traveling wave solutions, finite support propagation in nonlinear "parabolic" equations, regularity of moving boundaries, or asymptotic analysis. In fact, the spirit of the book is closest to that of modern numerical analysis and approximation, although only one chapter is devoted to numerical methods as such, while approximation methods are somewhat more visible. This bias is accounted for by the method of proof. In Part I (see especially Chapter 5) existence is established by proving that certain consistent semidiscretizations in time lead to stable, and hence convergent, approximation schemes determined by the method of horizontal lines, introduced originally by Rothe (Chapter 5, Ref. [38]). This *modus operandi* is, of course, reminiscent of the Lax equivalence theorem in numerical analysis. In Part II, existence is established by the contraction mapping principle, applied to a linear theory of evolution operators, resting upon a delicate product integration construction. Resolvent stability (or quasi-stability) plays a decisive role here. An alternative horizontal line analysis is presented as a by-product in Section 7.5. No attempt is made, however, to formulate a general global theory based on accretive operators, for example.

In the writer's viewpoint, this book represents a first, but a necessary, step in the quantitative analysis of complex nonlinear evolution models. Ironically, this perspective is revealed here in the introduction, although it represents the writer's vantage point at the conclusion of this undertaking. What is needed is a reliable methodology, consisting of convergent algorithms for nonlinear problems in infinite-dimensional spaces, such that, upon interface with models such as those presented in this book, some semblance of optimality accrues for these algorithms with respect to known and possible computing processes. Indeed, there are at least two discernible levels at which this analysis can take place. One level is that of the reduction of complexity through a process roughly comparable to entropy determination. For example, one can fully discretize the two-phase Stefan problem (see

† Not to be confused with Moser iteration in elliptic regularity theory.

Section 1.1 for elaboration of this model), with a distribution equation describing the rate of change of (discontinuous) enthalpy

$$\frac{\partial H(u)}{\partial t} - \Delta u + f(u) = 0 \tag{1}$$

on a space–time domain $\mathscr{D}$, via a fully implicit method in time and a piecewise-linear, finite-element, Galerkin method in space and obtain a recursively generated sequence of well-posed finite-dimensional problems. While the first level of analysis computes the convergence of this scheme to the solution u, or solution pair $[u, H(u)]$ and compares this rate to the optimal rate in finite-dimensional approximation of *any* finite-energy function u on $\mathscr{D}$ satisfying an energy norm bound, it stops short of the second level, actually devising an algorithmic strategy for the solution of the finite-dimensional problems. The latter are usually solved on modern parallel-processing computers by some combination of a constrained Newton method, in conjunction with iterative and sparse direct matrix routines for linear problems, for example multigrid methods. While this second level of algorithm formulation is not discussed in the sequel, we shall comment on both levels now in order to focus the ideas. It can be shown (see Chapter 2, Ref. [9]) that, under reasonable hypotheses, the fully discrete scheme in space and time described above gives rise to a sequence of finite-dimensional approximations in $\mathbf{L}^2(\mathscr{D})$, convergent to u, $O(h|\ln(1/h)|)$ when $\Delta t = ch^2$; here h is the maximal finite-element cell diameter. The significance of this result, when h is related to the Euclidean dimension $(n + 1)$ of $\mathscr{D}$ and the dimension N of the approximating $\mathbf{L}^2(\mathscr{D})$ subspace according to the weakly asymptotic† formula $N \approx Ch^{-(n+2)}$, is that $O(N^{-(n+1)})$ is the N-width of the unit ball of $\mathbf{H}^1(\mathscr{D})$ as measured in $\mathbf{L}^2(\mathscr{D})$. Thus, the complexity of the fully discrete method may be directly related to that of computing, via finite-dimensional approximations from N-dimensional subspaces, an arbitrary element from the unit ball of $\mathbf{H}^1(\mathscr{D})$. The logarithm factor appears unavoidable; although $u \in \mathbf{H}^1(\mathscr{D})$, $\partial H(u)/\partial t$ is essentially bounded in t as a regular Baire measure in x, and not necessarily a square-integrable function in (x, t), for both the Stefan problem and the porous-medium equation (for this measure property, see Proposition 5.1.5 and Theorem 5.2.1). We address these questions in Chapter 4. Proposition 4.3.4 derives the convergence result for the continuous-time, Faedo–Galerkin method in the context of a general class of degenerate parabolic equations. The N-width results are developed in Chapter 4, Section 2 (see also Chapter 4 Introduction). Additional material is included involving mixed norms and also the relationship of the Weinstein–Aronszajn theory of intermediate eigenvalue problems to ellipsoidal N-widths.

† Two expressions are weakly asymptotic if they are of the same order (O).

We comment briefly now on algorithm formulation and, to be consistent with the general theme of the book, we confine our attention to infinite-dimensional spaces. The nonlinear elliptic theory developed in Chapter 3 for the steady-state and semidiscrete equations does not include the construction of solutions.[†] It is not enough, as emphasized earlier, simply to introduce complexity-reducing approximations, such as those defined by a Galerkin method. In lieu of this, we refer to the extremely promising work of Bank and Rose [3] on quasi-Newton methods for the solution of nonlinear elliptic boundary-value problems and operator equations more generally. This work extracts the essential features of gradient maps, for which Newton's method is globally convergent, and produces, on an explicitly defined set **S** containing the solution and starting values, a globally convergent quasi-Newton method with controlled residuals. The basic hypotheses are a form of coerciveness, Lipschitz continuity of the Frèchét derivative, and uniform boundedness of the derivative inverses on a set slightly larger than **S**. In certain cases, the last hypothesis simply reduces to the statement that the derivative (linear) maps A are injective. Interestingly, this last hypothesis applied to I-A is the major hypothesis employed by Sermange [8] in his construction of iterative methods, generalizing Picard successive-approximation convergence, as embodied in Kitchen's theorem [6]. Whereas the latter theorem requires the spectral radius of the compact complexification $\bar{\mathrm{A}}$ of the derivative of a given continuous operator T to be less than one at a prescribed fixed point of T, Sermange is able to construct a quadratic polynomial $p(\mathrm{T})$, with fixed points containing those of T, such that $p(\mathrm{T})$ satisfies the hypotheses of Kitchen's theorem, although T need not. While the derivative hypothesis, in principle, need hold only at the fixed point, the latter is undetermined, so that it is essentially a neighborhood hypothesis. Moreover, the convergence result is local, not global, and, hence, must be deemed somewhat weaker computationally than the result of [3]. Still, these results indicate that iterative methods, in conjunction with quasi-Newton methods, can yield effective algorithms down to the linear level where a new, highly nontrivial analysis must begin. Probably none of this will surprise the informed reader, who may realize that the proof of the implicit-function theorem in Banach space makes fundamental use of the contraction mapping principle, and that the generalization, embodied in Nash–Moser iteration, uses a quasi-Newton method involving approximate right inverses. Of course, all of this discussion has been begging a fundamental question: What is the effectiveness of these algorithms for degenerate problems, such as the two-phase Stefan problem, where differentiability requirements are not satisfied? While regularization, discussed in some detail in

[†] See, however, the contractive method of Hartman and Stampacchia (Ref. [15] in Chapter 3).

Chapter 2, can prove helpful here, it may also be true that adjustments of standard methods, applied to the nonregularized problem, may be superior. Some evidence that this is the case is furnished by the remarkable success of Wheeler (Chapter 1 [25]) in his use of constrained-Newton methods for solving the finite-element equations arising in modeling the melting of permafrost prior to the construction of the trans-Alaska pipeline.

Let us turn now to a synopsis of the key elements of the book. Many of the key ideas of Part II are due primarily to Kato and these results are taken from his papers (Chapter 6 [11, 12], and Chapter 7 [13, 17]), written either individually or with others. The reader will notice that some hypotheses have been strengthened for ease of presentation. The delicate part of this theory is the linear theory, or, more properly, the construction of the linear evolution operators and their invariance on a certain smooth space **Y**. Fundamental simplifications of the arguments by Dorroh are incorporated into the text. One of the most delicate aspects of Kato's construction surrounds his notion of quasi-stability of the class $\{A(t)\}$ of negative generators of strongly-continuous semigroups. This is suggested by the applications to quasi-linear systems, where (suppressing the nonlinear dependence) a similarity transformation on $\{A(t)\}$ gives rise to a perturbed family $\{A(t) + B(t)\}$. In general, it may be known only that $\|B(\cdot)\|_{\mathbf{X}}$ is upper Lebesgue integrable, leading to the notion of quasi-stability for $\{A(t) + B(t)\}$ on the ground space **X**; this, in turn, leads to quasi-stability for $\{A(t)\}$ on the smooth space **Y**. However, the quasi-linear systems considered in this book satisfy stability properties on **X** and **Y** appreciably stronger than the aforementioned quasi-stability. Although we present the details of the evolution operators' construction in Chapter 6, through semigroup splicing in the case of stability, we refer the reader to Kato's paper (Chapter 6 [12]) for the modifications of the details in the quasi-stable case. Here, uniform partitions of the time interval are no longer possible and delicate measure-theoretic questions enter, especially regarding approximation. Thus, Part II makes no pretense of being selfcontained with respect to proofs. We ask the reader's indulgence in the genuine interest of brevity particularly since the more general result is not required for the application. Proofs are also omitted for some other results, such as the underlying lemma (see Lemma 7.3.4) for the commutator estimate required to obtain a uniform norm bound on the perturbations $\{B(t)\}$, in the applications to quasi-linear hyperbolic systems (see Section 7.3). This lemma is a variant of an earlier result of Calderón (Chapter 6 [4]), and plays a decisive role in both Sections 7.3 and 7.5. In the latter section, it forms the basis for an explicit estimate, derived by the author (Chapter 7 [14]), for an invariant time interval, on which the viscosity method produces convergent Navier–Stokes approximations to a solution of the Euler equations for an ideal fluid. Section 7.5 contains the only results in Part II due to the author. The other physical application of Part II involves the verification that consistent initial data can be evolved forward in time, subject to the vanishing of the Ricci curvature tensor, by the use of harmonic coordinates. These equations represent the vacuum-field equations of general relativity.

The material on fluid dynamics in Part I is essentially well known, and is included for completeness and to serve as a reference point for the local result mentioned in the preceding paragraph. The analysis of the reaction–diffusion systems in Part I makes fundamental use of the idea of invariant regions with "inward-pointing" vector fields; however, while citing some of the fundamental literature here, we develop an independent approach based on semidiscretization. The analysis here also discriminates between concentration-type and potential-type variables, and the invariant regions are defined, via the confinement of the concentration variables, to a slab. This geometry is, of course, highly specialized.

It is now known that bounded solutions of the two-phase Stefan problem and the porous-medium equation are continuous in the interior of $\mathscr{D}$. However, $u \in \mathbf{H}^1(\mathscr{D})$ is all that can be expected, in the sense of gross Sobolev space regularity for weak solutions, and this provided $u_0 \in \mathbf{H}^1(\Omega)$ for the space domain Ω, due to the degeneracies induced by the function H in Equation (1); H is discontinuous at zero in the Stefan problem, and may have an unbounded derivative near zero in the case of the porous-medium equation. This $\mathbf{H}^1$ regularity is embedded in our formulations of both models in Chapter 1. Although we consider each separately in Chapters 1 and 2, for the good reason that the (smoothing) regularization-convergence theories are distinct with respect to rates of convergence, the technique of pointwise *a priori* estimates permits a unified numerical analysis in Section 4.3, and a unified stability and existence analysis in Sections 5.1 and 5.2. Central to the analysis of these degenerate-parabolic evolutions are the equivalent abstract integral equations, holding pointwise on D, which are obtained through lifting the weak differential version from the dual of some subspace $\mathbf{G}$ of $\mathbf{H}^1(\Omega)$ to $\mathbf{G}$, itself. The lifting, itself, is far from a novel idea, but perhaps our applications of this idea are not entirely evident. Among these we cite the following: (1) a simplified $\mathbf{L}^2$ stability analysis in Section 5.1; (2) a simplified uniqueness analysis in Section 1.5; (3) the semidiscrete error analysis in Section 4.4, in which the property that $\mathrm{T}H(\mathrm{u})$ is in $\mathbf{H}^2(\mathscr{D})$ is decisive; (4) the regularization error analysis in Chapter 2; and, finally, (5) the quadrature formulas of Section 3.3, via the analogy with initial-value problems for ordinary differential equations. Properties (2), (3), and (4) are introduced in the author's paper with Rose (Chapter 2 [9]). In his dissertation 4 [31], Rose made what was probably the first rigorous finite-element error analysis of a degenerate–parabolic model in his application of the lifting operators to the porous-medium equation. At that point in the late 1970s, the systematic use of these operators in the numerical analysis of linear and nonlinear parabolic evolutions was becoming widespread through the work of Bramble and his co-workers ([4], Chapter 2 [3]), as related to (5) above. However, the insight of (3) apparently originated in Chapter 2 [9], where it is applied in the context of fully-discrete approximations; the idea of (1) apparently originated in this text. Finally, if the lifting mapping is denoted by T, the estimate (for almost all t)

$$\left\| (\mathrm{T} - \mathrm{T}_h) \frac{\partial H(u)}{\partial t} \right\|_{\mathbf{M}(\Omega)} \leqslant Ch^2 \left[\ln\left(\frac{1}{h}\right) \right]^2 \left\| \frac{\partial H(u)}{\partial t} \right\|_{\mathbf{M}(\Omega)}, \tag{2}$$

where $\mathrm{T}_h = \mathrm{E}_h \circ \mathrm{T}$ is the composition of the piecewise-linear finite-element energy projection E_h with T, makes possible the continuous-time Galerkin estimates of Proposition 4.3.4†; here, C is independent of t. Note that (2) uses the essential boundedness of $\partial H(u)/\partial t$ as a regular Baire measure in conjunction with the natural duality estimate, making use of recent sharp $\mathbf{L}^\infty$ estimates in the finite-element method. These latter estimates are discussed in Remark 4.3.10. For completeness, we also present the local approximation theory in Section 4.1, derived from the Sobolev integral representation theorem, which provides appropriate upper bounds for $\mathbf{L}^2$ finite-element estimates, and for mixed norm estimates, as well. The operators T also play roles, here, in terms of the Aubin–Nitsche duality "trick" of Proposition 4.3.1.

The methods by which the stability estimates of Section 5.1 are derived, except for the $\mathbf{L}^1$ estimates of Proposition 5.1.5, depend upon an application of the discrete Gronwall inequality to the semidiscrete forms of these evolutions. We present several versions of the Gronwall inequality in Section 2.2, one of which was suggested by Veron [9]. The fundamental way in which the semidiscrete equations enter motivated the material presented in Chapter 3 on elliptic equations and inequalities. These are defined by mappings which are the direct sum of subgradient mappings and pseudomonotone mappings; in some of the applications the (multivalued) function H induces the appropriate subgradient mapping, whereas the pseudomonotone mapping is a convenient perturbation of the negative Laplace operator. In Chapter 3, we have tried to offer a reasonably compact presentation, adequately mixing the theory and applications; some proofs are omitted if they are readily reproduced elsewhere, or if they develop results not crucial to the sequel. The theory here was developed by many mathematicians during the 1960s, but the synthesis, which is preferred here, is due primarily to Brézis (Chapter 3 [6]) and proceeds from the proximity mapping developed by Moreau (Chapter 3 [24]). Some balance is given by the Tarski fixed-point theorem

† The estimates of this proposition are essentially best possible with respect to N-widths in $\mathbf{H}^1(\mathscr{D})$, in contrast to the fully discrete estimates.

(Chapter 3 [28]), which is presented with proof and applied to quasi-variational inequalities. The reader will notice that we have downplayed the role of variational inequalities, particularly because of the appearance of the book by Kinderlehrer and Stampacchia (Chapter 1 [10]).

We close the introduction with a brief description of the development of the N-width as a measure of optimality in numerical analysis. Following the author's doctoral dissertation [5], written as a generalization to several variables of Kolmogorov's original result on the N-widths of $\mathbf{L}^2$ classes defined by Sobolev seminorm bounds, two papers appeared simultaneously, that applied the asymptotic width computations to optimal Galerkin estimates. These were by Schultz [7] for the finite-element method and by the author (Chapter 4 [20]) for spectral approximations. Subsequently, Aubin, in his book [1], gave a systematic development of these ideas (see also Babuska [2]).

REFERENCES

[1] J. P. Aubin, "Approximation of Elliptic Boundary Value Problems." Wiley (Interscience), New York, 1972.

[2] I. Babuska, The rate of convergence for the finite element method, *SIAM J. Numer. Anal.* **8**, 304–315 (1971).

[3] R. E. Bank and D. J. Rose, Global approximate Newton methods, *Numer. Math.* **37**, 279–295 (1981).

[4] J. H. Bramble, A. H. Schatz, V. Thomée, and L. B. Wahlbin, Some convergence estimates for semi-discrete Galerkin type approximations for parabolic equations, *SIAM J. Numer. Anal.* **14**, 218–241 (1977).

[5] J. W. Jerome, On the L_2 n-Width of Certain Classes of Functions of Several Variables. Dissertation, Purdue Univ., Lafayette, Indiana (1966).

[6] J. W. Kitchen, Concerning the convergence of iterates to fixed points, *Studia Math.* **27**, 247–249 (1966).

[7] M. H. Schultz, Multivariate spline functions and elliptic problems, *in* "Approximations with Special Emphasis on Spline Functions" (I. J. Schoenberg, ed.), pp. 279–347. Academic Press, New York, 1969.

[8] M. Sermange, Une méthode numérique en bifurcation-application à un problème à frontière libre de la Physique des Plasmas, *Appl. Math. Optim.* **5**, 127–151 (1979).

[9] L. Veron, personal communication, (1980).

GLOBAL WEAK SOLUTIONS

I

1 PROBLEM FORMULATIONS AND UNIQUENESS FOR DISSIPATIVE PARABOLIC MODELS

1.0 INTRODUCTION

Weak solution formulations are introduced for the two-phase Stefan problem, for the porous medium equation, for reaction–diffusion systems of equations, and for the Navier–Stokes equations for incompressible fluids, by means of divergence conservation principles in distribution form, in Sections 1.1–1.4. The initial/boundary-value problems are considered for these models with homogeneous Neumann, Dirichlet, Robin, and Dirichlet boundary conditions, respectively. Uniqueness theorems for the first three models are derived in Section 1.5, and a brief discussion of this property is presented for the fourth model.

In the formulation of the reaction–diffusion systems, the variables are partitioned into those of concentration type and those of potential type, and the pointwise restriction of invariant region inclusion is imposed upon the concentration variables. Parts of the system are permitted to be degenerate with respect to diffusion. The Navier–Stokes system is placed in a generalized format of constrained systems, in which the usual transport mechanism $((\mathbf{u} \cdot \nabla)\mathbf{v}, \mathbf{v})_{\mathbf{L}^2} = 0$ for solenoidal velocities $\mathbf{u}$ and $\mathbf{v}$ is relaxed.

The uniqueness results, as well as several convergence results in the sequel, depend upon a lifting of the weak solution relations from the dual spaces, such as $[\mathbf{H}^1(\Omega)]^*$ to $\mathbf{H}^1(\Omega)$, where they are realized as pointwise relations. This is achieved by means of standard linear inversion operators, which are introduced and utilized, following the definition of solution, in Sections 1.1–1.3.

1.1 HEAT CONDUCTION WITH CHANGE OF PHASE: STEFAN PROBLEMS

Two-phase Stefan problems describe temperature variations in a substance undergoing a change of phase in a medium, in such a way that the temperature in both phases satisfies a diffusion equation. Strict conservation principles apply at the free boundary "interface", corresponding to the nominal change-of-phase temperature. We may consider as an illustration the thawing of permafrost, which is any soil or rock which has existed at a temperature colder than zero degrees Celsius for two or more years. It is known that pure water, dispersed in a fine grained soil, freezes over a finite-temperature interval, rather than at precisely zero degrees Celsius. In fact, a fraction of the dispersed water, far from any soil surface, freezes at zero degrees Celsius, and the remaining fraction freezes at temperatures less than zero degrees Celsius. The latter is, thus, the nominal freezing point, that is the highest temperature at which ice can exist in a given soil, and represents a singular temperature for the various thermal coefficients. This problem from geotechnical engineering embodies the essential features of the model we now develop.

Heat conduction with a change of state in an open region Ω can be formulated as the distribution equation for the temperature θ given by

$$\frac{\partial}{\partial t}(\alpha + s) - \nabla \cdot (k\nabla\theta) + q(\theta) = 0 \tag{1.1.1}$$

on $\mathscr{D} = \mathscr{D}_{T_0} = \Omega \times (0, T_0)$. Here $\alpha(\theta) = \int_0^\theta \beta(\xi)\,d\xi$, where β is the volumetric heat capacity, with units of energy per unit degree per unit volume; s is the latent energy content, with units of energy per unit volume and k is the thermal conductivity, with units chosen so that the flux, $-k\nabla\theta$, has units of energy per unit area per unit time. The quantity q represents a source or sink, and is negatively signed for a source and positively signed for a sink. The formulation (1.1.1) contains considerable information as a distribution relation, specifically (1.1.6) below.

To motivate (1.1.1) heuristically, start with the energy-balance relation, defined by an arbitrary open set $\mathscr{B} \subset \Omega$, with sufficiently smoothly oriented boundary $\partial\mathscr{B}$,

$$\frac{d}{dt}\int_{\mathscr{B}}(\alpha + s)\,d\mathbf{x} = \int_{\partial\mathscr{B}}(k\,\nabla\theta)\cdot \mathbf{v}\,d\sigma - \int_{\mathscr{B}} q(\theta)\,d\mathbf{x}, \tag{1.1.2}$$

where $\mathbf{v}$ denotes the outward normal to $\partial\mathscr{B}$. Equation (1.1.2) asserts that the net rate of heat flowing into $\mathscr{B}$, given by the right-hand side, increases the enthalpy, that is, the internal temperature and/or energy content, accordingly. We shall assume, for the present discussion, that β, s, and k are piecewise-smooth functions of θ only, with singularities restricted to the nominal change-of-phase temperature of zero; and that β and k are positive, with k bounded away from zero. Thus, except for special s, it is essential to interpret (1.1.2) as a distribution on $(0, T_0)$. We require further that, if $\{\theta = 0\}$ has positive measure in $\mathscr{D}$, then $\beta \circ \theta$, $s \circ \theta$, and $k \circ \theta$ can be extended to measurable functions on this set. If

$$\Sigma = \{(\mathbf{x}, t) \in \mathscr{D} : \theta(\mathbf{x}, t) = 0\}, \tag{1.1.3a}$$

we write $\Sigma = \Sigma_0 \cup \Sigma_+ \cup \Sigma_-$, where Σ_0 is the (possibly empty) interior of Σ and Σ_+, Σ_- are possibly overlapping sets given by

$$\Sigma_+ = \overline{\{\theta > 0\}} \cap \Sigma, \qquad \Sigma_- = \overline{\{\theta < 0\}} \cap \Sigma. \tag{1.1.3b}$$

The computations to follow assume that Σ_+ and Σ_- are sufficiently smoothly oriented hypersurfaces. We have the following form of (1.1.2)

$$\begin{aligned} -\int_0^{T_0}\int_{\mathscr{B}}(\alpha(\theta) + s(\theta))\,\frac{\partial\phi}{\partial t}\,d\mathbf{x}\,dt + \int_0^{T_0}\int_{\mathscr{B}} q(\theta)\phi\,d\mathbf{x}\,dt \\ = \int_0^{T_0}\int_{\partial\mathscr{B}} k(\theta)\,\nabla\theta\cdot\mathbf{v}\phi\,d\sigma\,dt, \end{aligned} \tag{1.1.4}$$

for $\phi \in \mathbf{C}_0^\infty(0, T_0)$. If $\theta = \theta^+ + \theta^-$ is the decomposition of θ into its positive and negative parts and $\mathscr{D}_{\mathscr{B}} = \mathscr{B} \times (0, T_0)$, we have, via integration by parts applied separately to the first and third terms of (1.1.4) in $\{\theta > 0\} \cap \mathscr{D}_{\mathscr{B}}$ and $\{\theta < 0\} \cap \mathscr{D}_{\mathscr{B}}$:

$$\begin{aligned} &\int_{\mathscr{D}_{\mathscr{B}}\setminus\Sigma}\left\{\frac{\partial}{\partial t}(\alpha(\theta) + s(\theta)) - \nabla\cdot(k(\theta)\,\nabla\theta) + q(\theta)\right\}\phi\,d\mathbf{x}\,dt \\ &\qquad - \int_{\mathscr{D}_{\mathscr{B}}\cap\Sigma_0}\left[s(\theta)\,\frac{\partial\phi}{\partial t} - q(\theta)\phi\right]d\mathbf{x}\,dt \\ &= -\int_{\bar{\mathscr{D}}_{\mathscr{B}}\cap\Sigma_-}\{s(\theta^-)\cos(\mathbf{v}_\sigma, \mathbf{1}_t) - k(\theta^-)\,\nabla\theta^-\cdot\pi\mathbf{v}_\sigma\}\phi\,d\sigma \\ &\qquad + \int_{\bar{\mathscr{D}}_{\mathscr{B}}\cap\Sigma_+}\{s(\theta^+)\cos(\mathbf{v}_\sigma, \mathbf{1}_t) - k(\theta^+)\,\nabla\theta^+\cdot\pi\mathbf{v}_\sigma\}\phi\,d\sigma \end{aligned} \tag{1.1.5}$$

where $\mathbf{v}_\sigma$ is the normal to Σ_+, Σ_-, respectively, outward for $\{\theta > 0\}$ and inward for $\{\theta < 0\}$; $\pi\mathbf{v}_\sigma$ is its projection into the plane of Ω; and $s(\theta^+)$, $k(\theta^+)$, $s(\theta^-)$, $k(\theta^-)$ are understood in the limiting sense as θ^+, $\theta^- \to 0$. Since $\mathscr{B}$ and ϕ are arbitrary, we conclude that the pointwise relations

$$\frac{\partial}{\partial t}(\alpha(\theta) + s(\theta)) - \mathbf{\nabla} \cdot (k(\theta)\mathbf{\nabla}\theta) + q(\theta) = 0, \qquad \text{in} \quad \mathscr{D}\backslash\Sigma, \tag{1.1.6a}$$

$$s(\theta^-)\cos(\mathbf{v}_\sigma, \mathbf{1}_t) - k(\theta^-)\,\mathbf{\nabla}\theta^- \cdot \pi\mathbf{v}_\sigma\Big|_{\Sigma_-} = s(\theta^+)\cos(\mathbf{v}_\sigma, \mathbf{1}_t) - k(\theta^+)\,\mathbf{\nabla}\theta^+ \cdot \pi\mathbf{v}_\sigma\Big|_{\Sigma_+} \tag{1.1.6b}$$

hold, and that the second term in (1.1.5) is zero for $\phi \in \mathbf{C}_0^\infty(0, T_0)$. However, none of the physical motivation nor any of the preceding calculations is affected if we include, in the test class for (1.1.4), functions in $\mathbf{C}_0^\infty(\Sigma_0)$, that is, the test class is the union $\mathbf{C}_0^\infty(0, T_0) \cup \mathbf{C}_0^\infty(\Sigma_0)$. Thus, we have

$$\frac{\partial s(\theta)}{\partial t} + q(\theta) = 0 \qquad \text{(distributionally) in} \quad \Sigma_0. \tag{1.1.6c}$$

Equation (1.1.1) is clearly a distributional form of (1.1.6), as can be seen by a set of calculations similar to (1.1.4) and (1.1.5). However, we wish to introduce a generalized temperature u, and define a weak solution of (1.1.1), which involves initial and boundary conditions. Thus, set

$$u = \int_0^\theta k(\xi)\,d\xi = K \circ \theta \tag{1.1.7a}$$

via a standard Kirchhoff transformation. If the standard enthalpy is defined by

$$Q(t) = \alpha(t) + s(t), \qquad t \neq 0, \tag{1.1.7b}$$

set $H = Q \circ K^{-1}$. The mathematical assumptions now follow after a brief remark concerning the physical properties of Q. There is some reason to believe (see [1] pp. 95–99) that Q can be absolutely continuous when thawed moisture is accounted for in s at temperatures lower than the nominal freezing point. A discontinuous Q results when some bulk freezing occurs, a situation expected away from soil surfaces. In the latter case, the apparent specific heat capacity dQ/dt is a distribution not represented by a function. In order to provide the maximum possible generality, we shall permit Q to be discontinuous. For simplicity, s is normalized by translation by a constant so that $s(0-) = 0$.

Remark 1.1.1. We *assume* that the function $H = Q \circ K^{-1}$, defined on $\mathbb{R}^1\backslash\{0\}$, is a monotone increasing function, $\mathbf{C}^1$ in $\mathbb{R}^1\backslash\{0\}$, with a jump discontinuity of height A at zero, and a derivative satisfying

$$0 < \lambda \leqslant H'(\xi) \leqslant \mu < \infty, \qquad \xi \neq 0, \tag{1.1.8a}$$

where $H'(0+)$ and $H'(0-)$ are assumed to exist. Normalization is chosen so that

$$H(0-) = 0, \tag{1.1.8b}$$

and the jump condition takes the form

$$H(0+) = A > 0. \tag{1.1.8c}$$

We also assume that the function $f = q \circ K^{-1}$ can be written

$$f = g + h, \tag{1.1.8d}$$

where g is a locally Lipschitz continuous monotone-increasing function vanishing at zero, and h is a Lipschitz continuous function. The distribution transform of (1.1.1) is now given by

$$\frac{\partial H(u)}{\partial t} - \Delta u + f(u) = 0. \tag{1.1.9}$$

In the example of permafrost thawing, the constant A represents the product of the heat of fusion B with that fraction of moisture content solidified at the nominal freezing point; at lower temperatures, $s - A$ is defined as the negatively-signed part of B specified by the ratio of unthawed moisture content to total moisture content. The specification of $H(u)|_{t=0}$ (on Ω) is an apt initial condition; $H(\cdot)$ is closely related to enthalpy, as we have seen. Since $J = H^{-1}$ exists (as a continuous function), specifying $H(u)|_{t=0}$ also specifies $u|_{t=0}$. Various physical boundary conditions are possible. Perhaps the most realistic allows for a time-dependent Dirichlet condition on a portion of $\partial\Omega$, and a time-dependent Neumann condition on the remainder of $\partial\Omega$.† We shall, however, select a homogeneous Neumann boundary condition on all of $\partial\Omega$, and compensate by including the source/sink term $f(\cdot)$, which easily can be modified to be position dependent, also. In the permafrost problem, where Ω typically is bounded by the atmosphere, by a (partially) insulated pipeline inducing a developing thaw bulb, and by (arbitrary) underground boundaries assumed virtually insulated, we are substituting for atmospheric and pipeline fluxes, appropriate sources/sinks.

Definition 1.1.1. Suppose the initial "enthalpy" $H(u)|_{t=0} = H(u_0) \in \mathbf{L}^\infty(\Omega)$ is specified. Then a function u in the class $\mathscr{C}$, defined by (1.1.11c), with $H(u)$

† As distinct from the Stefan–Signorini problem, where the liquid is removed and the flux condition is imposed on the moving boundary.

a selection satisfying

$$H(u(x)) \in [0, A] \qquad \text{if} \quad u(x) = 0, \tag{1.1.10a}$$

$$H(u) \in \mathbf{L}^{\infty}(\mathscr{D}) \cap \mathbf{H}^1([0, T_0]; [\mathbf{H}^1(\Omega)]^*), \tag{1.1.10b}$$

is said to be a weak solution of the initial/boundary-value problem for (1.1.9) with homogeneous Neumann condition, if $f(u) \in \mathbf{L}^{\infty}(\mathscr{D})$, and if the relation

$$\int_{\mathscr{D}} \left[H(u) \frac{\partial \psi}{\partial t} - \nabla u \cdot \nabla \psi - f(u)\psi \right] d\mathbf{x}\, dt$$

$$- \int_{\Omega \times \{T_0\}} H(u)\psi \, d\mathbf{x} + \int_{\Omega \times \{0\}} H(u_0)\psi \, d\mathbf{x} = 0 \tag{1.1.11a}$$

holds for all $\psi \in \mathscr{C}_0$, where

$$\mathscr{D} = \mathscr{D}_{T_0} = \Omega \times (0, T_0), \tag{1.1.11b}$$

and the solution class $\mathscr{C}$ and test class $\mathscr{C}_0$ are given by

$$\mathscr{C} = \mathbf{L}^{\infty}((0, T_0); \mathbf{H}^1(\Omega)) \cap \mathbf{H}^1([0, T_0]; \mathbf{L}^2(\Omega)) \cap \mathbf{L}^{\infty}(\mathscr{D}); \tag{1.1.11c}$$

$$\mathscr{C}_0 = \mathbf{C}([0, T_0]; \mathbf{H}^1(\Omega)) \cap \mathbf{H}^1([0, T_0]; \mathbf{L}^2(\Omega)). \tag{1.1.11d}$$

Here, Ω is a bounded, strongly Lipschitz, domain in $\mathbb{R}^n$ and, for consistency with the definition of $\mathscr{C}$, $u_0 \in \mathbf{H}^1(\Omega)$ is required. Moreover, the second integral in (1.1.11a) has the initial interpretation of a functional, since it is guaranteed that $H(u) \in \mathscr{C}([0, T_0]; [\mathbf{H}^1(\Omega)]^*)$ by the regularity of (1.1.10b), but it will be subsequently observed that $H(u)$ is weakly continuous from $[0, T_0]$ into $\mathbf{L}^2(\Omega)$ (see Definition 5.2.1 to follow).

Remark 1.1.2. A classical solution of the initial/boundary-value problem for (1.1.9) is a function on $\mathscr{D}$, satisfying (1.1.9) pointwise in $\{u > 0\}$ and $\{u < 0\}$, for which $\Sigma = \Sigma_+ = \Sigma_-$, and for which the energy conservation principle (see (1.1.5))

$$-\left[k \frac{\partial K^{-1}(u)}{\partial \pi \mathbf{v}_\sigma} \right] = s(0+) \cos(\mathbf{v}_\sigma, \mathbf{1}_t)^{\dagger} \tag{1.1.12}$$

holds across Σ. In addition, of course, the initial condition and boundary conditions are satisfied pointwise. Here $[\zeta]$ denotes the discontinuity of ζ across Σ directed from $\{\mathbf{z}: u > 0\}$ to $\{\mathbf{z}: u < 0\}$. It is a straightforward exercise in integration by parts to verify that every classical solution, satisfying the regularity requirements of Definition 1.1.1, is a weak solution, and every sufficiently regular weak solution is a classical solution. In the special case when Σ is a smooth surface, defined as the zero set of a smooth function ϕ,

† Differentiation with respect to $\pi \mathbf{v}_\sigma$ is interpreted as the gradient dot product with $\pi \mathbf{v}_\sigma$.

then multiplication of (1.1.12) by $|\nabla\phi|$ gives the usual relation, relating the front velocity and the discontinuity of $k\,\nabla\theta$ across Σ.

Remark 1.1.3. It is possible to lift the equation (1.1.11a) to obtain a pointwise relation on $\mathscr{D}$, via an appropriate inversion of $-\Delta$. We shall introduce this linear, Riesz mapping and carry out the lifting. In fact, we find it advantageous to introduce a one-parameter family N_σ of such mappings. We begin with N_0. The reader may view this class of mappings as inducing equivalent abstract integral equation relations.

Definition 1.1.2. Given $\ell \in \mathbf{F} := [\mathbf{H}^1(\Omega)]^*$, we define the element $w = \mathrm{N}_0\ell \in \mathbf{H}^1(\Omega)$ as the Riesz representer of ℓ, when the inner product

$$(v, w)_{[\mathbf{H}^1(\Omega)]_0} = \int_\Omega \nabla v \cdot \nabla w + \frac{1}{|\Omega|} j(v) j(w) \tag{1.1.13a}$$

is employed. Here,

$$j(v) = \int_\Omega v,^{\dagger} \tag{1.1.13b}$$

and we have

$$\langle \ell, v \rangle = \int_\Omega \nabla \mathrm{N}_0\ell \cdot \nabla v + \frac{1}{|\Omega|} j(v) j(\mathrm{N}_0\ell), \qquad v \in \mathbf{H}^1(\Omega). \tag{1.1.14}$$

The notation $\langle \cdot, \cdot \rangle$ represents the duality pairing on $\mathbf{F} \times \mathbf{H}^1(\Omega)$. For $\sigma > 0$,[‡] $w = \mathrm{N}_\sigma\ell$ is defined as the Riesz representer of ℓ, with respect to the inner product

$$(v, w)_{[\mathbf{H}^1(\Omega)]_\sigma} = \int_\Omega \nabla v \cdot \nabla w + \sigma \int_\Omega vw. \tag{1.1.15}$$

In this case,

$$\langle \ell, v \rangle = \int_\Omega \nabla \mathrm{N}_\sigma\ell \cdot \nabla v + \sigma \int_\Omega v \mathrm{N}_\sigma\ell. \tag{1.1.16}$$

Remark 1.1.4. The norm induced by (1.1.13) is equivalent to the standard norm on $\mathbf{H}^1(\Omega)$ (see Sobolev [22] or Section 4.1). The equivalence induced by (1.1.15) is, of course, trivial. We have writen $[\mathbf{H}^1(\Omega)]_\sigma$ to emphasize the explicit inner product chosen. The linear mapping N_σ, $\sigma \geqslant 0$, whose restriction to $\mathbf{L}^2(\Omega)$ is positive definite and symmetric by (1.1.13)–(1.1.16),

† Differential symbol suppressed; convention when no ambiguity is possible.

‡ We have used $d\sigma$, when appropriate to designate the surface differential. There should be no confusion with the present usage of σ.

induces the following (negative) norm on $\mathbf{F}$:

$$\|\ell\|_{[\mathbf{F}]_\sigma} = \langle \ell, \mathrm{N}_\sigma \ell \rangle^{1/2}. \tag{1.1.17}$$

The mappings N_σ are continuous from $[\mathbf{F}]_\sigma$ into $[\mathbf{H}^1(\Omega)]_\sigma$ and have closed range. These facts follow directly from the relation

$$\|\mathrm{N}_\sigma \ell\|_{[\mathbf{H}^1(\Omega)]_\sigma} = \|\ell\|_{[\mathbf{F}]_\sigma}. \tag{1.1.18}$$

Moreover, (1.1.18) may also be used in conjunction with (1.1.14) and (1.1.16) to deduce that

$$|||\ell|||_{\mathbf{F}} = \sup\{|\langle \ell, v \rangle| : \|v\|_{\mathbf{H}^1(\Omega)} \leqslant 1\} \tag{1.1.19}$$

is equivalent to (1.1.17) with constants dependent upon σ. The relation (1.1.19) defines the standard duality norm, of course. It follows immediately that the mappings $\mathrm{N}_\sigma|_{\mathbf{L}^2(\Omega)}$ are bounded into $\mathbf{H}^1(\Omega)$, and hence are self-adjoint, positive definite and compact as $\mathbf{L}^2$ operators. Because of the equivalence of the spaces $[\mathbf{H}^1(\Omega)]_\sigma$ and $\mathbf{H}^1(\Omega)$, and of $[\mathbf{F}]_\sigma$ and $\mathbf{F}$, we shall suppress the subscripts σ in the sequel.

Remark 1.1.5. An equivalent way of viewing $\mathrm{N}_\sigma : \mathbf{F} \to \mathbf{H}^1(\Omega)$ is as follows: $w = \mathrm{N}_0 \ell$ is the unique element satisfying the (weak form of the) linear elliptic Neumann problem

$$-\Delta w = \ell - \frac{1}{|\Omega|}\langle \ell, 1 \rangle, \qquad \text{in} \quad \Omega,^{\dagger} \tag{1.1.20a}$$

$$\frac{\partial w}{\partial \nu} = 0, \qquad \text{on} \quad \partial\Omega, \tag{1.1.20b}$$

$$\int_\Omega w = \langle \ell, 1 \rangle. \tag{1.1.20c}$$

For $\sigma > 0$, $w = \mathrm{N}_\sigma \ell$ satisfies

$$(-\Delta + \sigma) w = \ell, \qquad \text{in} \quad \Omega, \tag{1.1.21a}$$

$$\frac{\partial w}{\partial \nu} = 0, \qquad \text{on} \quad \partial\Omega. \tag{1.1.21b}$$

One may deduce directly from (1.1.20) and (1.1.21) that

$$-\Delta \mathrm{N}_0 \ell = \ell - \frac{1}{|\Omega|}\langle \ell, 1 \rangle, \qquad \mathrm{N}_0(-\Delta w) = w - \frac{1}{|\Omega|} j(w), \tag{1.1.22a}$$

$$(-\Delta + \sigma)\mathrm{N}_\sigma \ell = \ell, \qquad \mathrm{N}_\sigma(-\Delta + \sigma) w = w, \tag{1.1.22b}$$

for $\ell \in \mathbf{F}$ and $w \in \mathbf{Range}(\mathrm{N}_\sigma)$. It is easily seen that the latter is characterized as $\mathbf{H}^1(\Omega)$ for $\sigma > 0$, and as $\{w \in \mathbf{H}^1(\Omega) : j(w) = 0\}$ for $\sigma = 0$.

† Function identically one.

Remark 1.1.6. The mappings N_σ for $\sigma > 0$ possess the important property that

$$\ell \geqslant 0 \Rightarrow \mathrm{N}_\sigma \ell \geqslant 0, \tag{1.1.23}$$

where $\ell \geqslant 0$ is defined by $\langle \ell, v\rangle \geqslant 0$ for $v \geqslant 0$. The property (1.1.23) follows directly from setting

$$\mathrm{N}_\sigma \ell = (\mathrm{N}_\sigma \ell)^+ + (\mathrm{N}_\sigma \ell)^-, \qquad (\mathrm{N}_\sigma \ell)^+ = \sup(0, \mathrm{N}_\sigma \ell),$$

and making the choice $v = -(\mathrm{N}_\sigma \ell)^-$ in (1.1.16) to obtain $(\mathrm{N}_\sigma \ell)^- = 0$.

We now conclude this section with the pointwise relation(s) referred to earlier.

Proposition 1.1.1. Let u be a weak solution satisfying Definition 1.1.1. Then $(\partial/\partial t)\mathrm{N}_\sigma H(u) \in \mathbf{L}^2(\mathscr{D})$, $\sigma \geqslant 0$, and for all t, $0 < t < T_0$, the equations

$$\frac{\partial \mathrm{N}_0 H(u)}{\partial t} + u + \mathrm{N}_0 f(u) = \frac{1}{|\Omega|}\int_\Omega u, \tag{1.1.24}$$

$$\frac{\partial \mathrm{N}_\sigma H(u)}{\partial t} + u + \mathrm{N}_\sigma f_\sigma(u) = 0, \qquad \sigma > 0, \tag{1.1.25}$$

hold almost everywhere in Ω, where

$$f_\sigma(r) = f(r) - \sigma r, \qquad r \in \mathbb{R}^1. \tag{1.1.26}$$

Moreover, the initial condition

$$\mathrm{N}_\sigma H(u)\Big|_{t=0} = \mathrm{N}_\sigma H(u_0) \tag{1.1.27}$$

holds.

Proof: We give the proof explicitly only for $\sigma = 0$. For $\phi \in \mathbf{C}_0^\infty(\mathscr{D})$, let $\psi = \mathrm{N}_0\phi$ in (1.1.11a). Using the commutative relation $[(\partial/\partial t)\mathrm{N}_0 - \mathrm{N}_0(\partial/\partial t)]\phi = 0$ (see (1.1.14)), the self-adjointness of N_0, and (1.1.14) directly, we obtain

$$\int_{\mathscr{D}} \mathrm{N}_0 H(u)\frac{\partial \phi}{\partial t}\,d\mathbf{x}\,dt = \int_{\mathscr{D}}\left[u - \frac{1}{|\Omega|}j(u) + \mathrm{N}_0 f(u)\right]\phi\,d\mathbf{x}\,dt, \tag{1.1.28}$$

which shows that the distribution derivative $(\partial/\partial t)\mathrm{N}_0 H(u)$ is equal to $-u - \mathrm{N}_0 f(u) + (1/|\Omega|)j(u)$. Since this function is in $\mathbf{L}^2(\mathscr{D})$, it follows that $\mathrm{N}_0 H(u) \in \mathbf{H}^1(\mathscr{D})$ and, hence, (1.1.28) may be integrated by parts to obtain (1.1.24) for almost all t. To obtain (1.1.27), let $\psi \in \mathscr{C}_0$, $\psi(\cdot, T_0) = 0$. Then $\mathrm{N}_0\psi \in \mathscr{C}_0$, and this substitution into (1.1.11a), coupled with (1.1.24), yields, after an integration by parts in t,

$$\int_\Omega \mathrm{N}_0 H(u)\Big|_{t=0}\psi(\cdot, 0)\,d\mathbf{x} = \int_\Omega \mathrm{N}_0 H(u_0)\psi(\cdot, 0)\,d\mathbf{x},$$

from which (1.1.27) follows. Since $N_0 f(u) \in \mathbf{C}([0, T_0]; \mathbf{L}^2(\Omega))$, the final three, hence all, terms in (1.1.24) are in $\mathbf{C}([0, T_0]; \mathbf{L}^2(\Omega))$, which establishes the identity (1.1.24) for all $0 < t < T_0$. ■

Remark 1.1.7. The regularity of the initial datum u_0 plays a decisive role in the regularity properties of the solution. For example, in Section 5.2, it is shown that $(\partial H(u)/\partial t)$ is essentially bounded in t, as a regular Baire measure on $\bar{\Omega}$, if $u_0 \in \mathbf{W}^{2,1}(\Omega)$, and this property is exploited in Section 4.3 to obtain error estimates for the two-phase Stefan problem. In Section 2.2, it is shown that the semidiscretization defined there inherits pointwise stability, since $u_0 \in \mathbf{L}^\infty(\Omega)$, and this property is used in Section 5.2 to deduce that the solution is essentially bounded on its space–time domain in this case. Accordingly, it may be of interest to weaken the hypothesis on u_0 to $u_0 \in \mathbf{L}^2(\Omega)$. In this case, it is not to be expected that u_t† is an $\mathbf{L}^2(\mathscr{D})$ function, but a solution, in the sense of Definition 1.1.1 exists, nonetheless, if the regularity properties defining $\mathscr{C}$ are accordingly adjusted. This is discussed in Section 5.2, where the major existence theorems are deduced for (1.1.10) and (1.1.11) (see especially Corollary 5.2.7 and Proposition 5.2.12).

Remark 1.1.8. Analogous to the two-phase Stefan problem just described is the one-phase problem, in which diffusion processes are assumed to hold in only one of the phases; the temperature is assumed constant at the phase-change temperature in the other phase. There is an elementary change of variable,

$$u(\mathbf{x}, t) = \int_0^t \theta(\mathbf{x}, \tau)\, d\tau, \tag{1.1.29}$$

which converts the formulation corresponding to (1.1.1) into a parabolic variational inequality in this case. The reader is referred to Kinderlehrer and Stampacchia [10] for details. Solutions of such inequalities are considerably more regular than the corresponding solutions for the two-phase problem.

1.2 UNSATURATED FLUID INFILTRATION IN POROUS MEDIA‡

The density ρ of a gas expanding in a porous medium and the volumetric moisture content θ of a liquid of constant density infiltrating a porous medium

† $u_t = \dfrac{\partial u}{\partial t}$.

‡ The flow is assumed such that gravity can be neglected.

satisfy similar equations. In the case of the gas, the pressure serves as a potential; Darcy's law asserts that the velocity is given by

$$\mathbf{v} = -(k/\mu)\nabla p. \tag{1.2.1}$$

Here, k is the permeability, with units of area; μ is the viscosity, with units of mass per unit length per unit time; and p is the pressure, with units of force per unit area. In the case of the liquid, there is assumed a total potential, or piezometric head, Φ, which is measured in units of length. The analog of (1.2.1) simply is a flux defined for horizontal flow by

$$\mathbf{w} = -k\nabla\Phi. \tag{1.2.2}$$

Here k is the hydraulic conductivity, with units chosen so that $k\nabla\Phi$ has units of mass per unit length per unit time (the units of viscosity). Similar to the heat–balance equation (1.1.2), one sets up mass-balance equations of the forms

$$\frac{d}{dt}\int_{\mathscr{B}} \beta\rho\, d\mathbf{x} = -\int_{\partial\mathscr{B}} \rho\mathbf{v}\cdot\boldsymbol{\nu}\, d\sigma = \int_{\partial\mathscr{B}} \rho\frac{k}{\mu}\nabla p\cdot\boldsymbol{\nu}\, d\sigma, \tag{1.2.3}$$

and

$$\frac{d}{dt}\int_{\mathscr{B}} \theta\, d\mathbf{x} = -\int_{\partial\mathscr{B}} \mathbf{w}\cdot\boldsymbol{\nu}\, d\sigma = \int_{\partial\mathscr{B}} k\nabla\Phi\cdot\boldsymbol{\nu}\, d\sigma. \tag{1.2.4}$$

Here, (1.2.3) and (1.2.4) express the equality of the rate of mass crossing $\partial\mathscr{B}$ into $\mathscr{B}$, with the rate of increase of mass within $\mathscr{B}$ with convective effects assumed negligible. The quantity β is the dimensionless porosity of the medium, and the units of θ are expressed in mass per unit area. As in the previous section, the choice of the open set $\mathscr{B} \subset \Omega$ is essentially arbitrary and $\boldsymbol{\nu}$ is the outward normal to $\mathscr{B}$. Before the above equations can be utilized, so-called equations of state of the forms

$$\rho : p \to \rho(p), \qquad p \geqslant 0, \tag{1.2.5a}$$

$$\theta : \Phi \to \theta(\Phi), \qquad \Phi \geqslant 0, \tag{1.2.5b}$$

must be specified. In (1.2.3) and (1.2.4), β, k, and μ are assumed smooth and nonnegative; and, physically, ρ, $\theta \geqslant 0$. For example, a common form of (1.2.5a) is

$$\rho(p) = \rho_0 p^{1/(\gamma-1)} (\gamma > 1), \qquad p \geqslant 0, \tag{1.2.6}$$

where $\gamma = 2$ for an isothermal process, and $\gamma > 2$ for an adiabatic process. We may rewrite (1.2.3) and (1.2.4), respectively, via the compact expressions

$$\frac{d}{dt}\int_{\mathscr{B}} L(p)\, d\mathbf{x} = \int_{\partial\mathscr{B}} \nabla K \circ p\cdot\boldsymbol{\nu}\, d\sigma, \tag{1.2.7a}$$

where the functions K and L are defined by

$$K(p) = \int_0^p \rho(\xi) \frac{k(\xi)}{\mu(\xi)} \, d\xi, \qquad p \geqslant 0, \tag{1.2.7b}$$

$$L(p) = \beta(p)\rho(p), \qquad p \geqslant 0; \tag{1.2.7c}$$

and

$$\frac{d}{dt} \int_{\mathscr{B}} L(\Phi) \, d\mathbf{x} = \int_{\partial \mathscr{B}} \nabla K \circ \Phi \cdot \mathbf{v} \, d\sigma, \tag{1.2.8a}$$

where

$$K(\Phi) = \int_0^{\Phi} k(\xi) \, d\xi, \qquad \Phi \geqslant 0, \tag{1.2.8b}$$

$$L(\Phi) = \theta(\Phi), \qquad \Phi \geqslant 0. \tag{1.2.8c}$$

Under the assumption that K is an invertible function, (1.2.7) and (1.2.8) may be unified in the single statement

$$\frac{d}{dt} \int_{\mathscr{B}} H(u) \, d\mathbf{x} = \int_{\partial \mathscr{B}} \nabla u \cdot \mathbf{v} \, d\sigma, \tag{1.2.9}$$

where $u = K(\cdot)$ and $H = L \circ K^{-1}$. In the case of (1.2.6), if β, k, and μ are positive constants, we obtain (1.2.9), with

$$K(\lambda) = c\lambda^{\gamma/(\gamma-1)}, \qquad c = k\rho_0(\gamma - 1)/(\mu\gamma), \tag{1.2.10a}$$

$$H(\lambda) = \beta\rho_0 c^{-\gamma}\lambda^{1/\gamma}, \tag{1.2.10b}$$

for $\lambda \geqslant 0$. Just as in the interpretation of (1.1.2), it is necessary to interpret (1.2.9) as a distribution on $(0, T_0)$, since the free boundary,

$$\partial\{u > 0\} = \Sigma,$$

is a potential singularity set analogous to the set Σ of (1.1.3a). Under the assumption that Σ is a sufficiently smoothly oriented hypersurface, integration by parts applied to the distribution equation

$$-\int_{\mathscr{B} \times (0, T_0)} H(u) \frac{\partial \phi}{\partial t} \, d\mathbf{x} \, dt = \int_{\partial \mathscr{B} \times (0, T_0)} \nabla u \cdot \mathbf{v}\phi \, d\sigma \, dt, \qquad \phi \in \mathbf{C}_0^\infty(0, T_0), \tag{1.2.11}$$

yields, as in the previous section, since $\mathscr{B}$ and ϕ are arbitrary (see (1.1.6)),

$$\frac{\partial H(u)}{\partial t} - \Delta u = 0 \quad \text{in} \quad \{u > 0\}, \tag{1.2.12a}$$

$$\nabla u \cdot \pi \mathbf{v}_\sigma \Big|_\Sigma = 0, \tag{1.2.12b}$$

where $\mathbf{v}_\sigma$ is directed by the mass flux. Thus, if

$$\frac{\partial H(u)}{\partial t} - \Delta u = 0 \tag{1.2.13}$$

is interpreted in a distribution sense on $\mathscr{D} = \mathscr{D}_{T_0}$, (1.2.12) is generalized by (1.2.13). We shall now state the precise hypotheses on $H(\cdot)$. These are chosen so that we may consider the more delicate case, where H is not Lipschitz continuous. Other models are included in the format of Chapter 5.

Remark 1.2.1 We shall assume that H has properties which generalize (1.2.10b). Specifically, we assume

$$H \in \mathbf{C}(\mathbb{R}^1), \qquad H' \in \mathbf{C}(\mathbb{R}^1 \backslash \{0\}), \tag{1.2.14a}$$

$$H(-t) = -H(t), \qquad t \in \mathbb{R}^1, \tag{1.2.14b}$$

$$H'(t) > 0, \qquad t \neq 0, \tag{1.2.14c}$$

$$H \qquad \text{is concave on} \quad [0, \infty), \tag{1.2.14d}$$

$$c_1 t^{(1/\gamma)-1} \leqslant H'(t) \leqslant c_2 t^{(1/\gamma)-1}, \qquad \text{some} \quad \gamma > 1, \qquad \text{all} \quad t > 0, \tag{1.2.14e}$$

where c_1 and c_2 are positive constants. In particular, from $H(0) = 0$ and the concavity of H, we have that $H(t)/t$ is decreasing for $t > 0$, so that H is subadditive (see Lorentz [14] pp. 43–44), that is,

$$H(t_1 + t_2) \leqslant H(t_1) + H(t_2), \qquad t_1, t_2 \geqslant 0. \tag{1.2.14f}$$

We shall consider now the initial/boundary-value problem for (1.2.13) with a homogeneous Dirichlet boundary condition. Since the model involves unsaturated filtration, specifying $u = 0$ on $\partial\Omega$ amounts to the assumption that the liquid or gas diffuses from some initial configuration with compact support in Ω. We shall now define weak solutions for all time; the requirement on the support of the initial datum is somewhat relaxed.

Definition 1.2.1. Given $0 \leqslant u_0 \in \mathbf{L}^\infty(\Omega)$, a function $u \geqslant 0$ in the class $\mathscr{C}$ (see (1.2.16a)) is said to be a weak solution of the initial/boundary-value problem (1.2.13) with a homogeneous Dirichlet condition if, for every $T_0 > 0$,

$$\int_{\mathscr{D}_{T_0}} \left[H(u) \frac{\partial \psi}{\partial t} - \nabla u \cdot \nabla \psi \right] d\mathbf{x}\, dt - \int_{\Omega \times \{T_0\}} H(u)\psi\, d\mathbf{x} + \int_{\Omega \times \{0\}} H(u_0)\psi\, d\mathbf{x} = 0 \tag{1.2.15}$$

holds for all $\psi \in \mathscr{C}_0$, where the solution class $\mathscr{C}$ and the test class $\mathscr{C}_0$ are given by

$$\mathscr{C} = \mathbf{L}^2((0,\infty);\mathbf{H}_0^1(\Omega)) \cap \mathbf{H}^1([0,\infty);\mathbf{L}^2(\Omega)) \cap \cdots \tag{1.2.16a}$$
$$\mathbf{L}^\infty((0,\infty);\mathbf{H}_0^1(\Omega)) \cap \mathbf{L}^\infty((0,\infty);\mathbf{L}^\infty(\Omega)),$$

and

$$\mathscr{C}_0 = \mathbf{C}([0,T_0];\mathbf{H}_0^1(\Omega)) \cap \mathbf{H}^1([0,T_0];\mathbf{L}^2(\Omega)). \tag{1.2.16b}$$

Here Ω is a bounded, strongly Lipschitz domain in $\mathbb{R}^n$ and, for consistency with the definition of $\mathscr{C}$, $u_0 \in \mathbf{H}_0^1(\Omega)$ is required.

Remark 1.2.2. It is possible to lift (1.2.15), as in the previous section. The corresponding linear, Riesz mapping is somewhat different, since the homogeneous Neumann condition of the previous section has been replaced by a homogeneous Dirichlet condition. We shall close this section with such a description.

Definition 1.2.2. Given the element $\ell \in \mathbf{H}^{-1}(\Omega) = [\mathbf{H}_0^1(\Omega)]^*$, we define the element $w = \mathrm{D}_0\ell \in \mathbf{H}_0^1(\Omega)$ as the Riesz representer of ℓ, when the inner product

$$(v, w)_{\mathbf{H}_0^1(\Omega)} = \int_\Omega \nabla \mathbf{v} \cdot \nabla w \tag{1.2.17}$$

is employed. We, thus, have

$$\langle \ell, v\rangle = \int_\Omega \nabla \mathrm{D}_0\ell \cdot \nabla v. \tag{1.2.18}$$

Remark 1.2.3. The linear mapping D_0, which has the positivity property $\ell \geqslant 0 \Rightarrow \mathrm{D}_0\ell \geqslant 0$, and whose restriction to $\mathbf{L}^2(\Omega)$ is a symmetric mapping, induces the following (negative) norm on $\mathbf{H}^{-1}(\Omega)$:

$$\|\ell\|_{\mathbf{H}^{-1}(\Omega)} = \langle \ell, \mathrm{D}_0\ell\rangle^{1/2}. \tag{1.2.19}$$

This norm is equivalent to the standard $\mathbf{H}^{-1}(\Omega)$ norm and D_0 is bounded as a mapping of $\mathbf{H}^{-1}(\Omega)$ onto $\mathbf{H}_0^1(\Omega)$. In particular, $\mathrm{D}_0|_{\mathbf{L}^2(\Omega)}$ is self-adjoint, positive–definite, and compact.

Remark 1.2.4. A more graphic way of viewing $\mathrm{D}_0\colon \mathbf{H}^{-1}(\Omega) \to \mathbf{H}_0^1(\Omega)$ is as follows. The quantity $w = \mathrm{D}_0\ell$ is the unique element satisfying the weak form of the linear elliptic Dirichlet problem

$$-\Delta w = \ell \quad \text{in} \quad \Omega, \tag{1.2.20a}$$

$$w = 0 \quad \text{on} \quad \partial\Omega. \tag{1.2.20b}$$

Note that $-\Delta \mathrm{D}_0\ell = \ell$ and $\mathrm{D}_0(-\Delta)w = w$, that is D_0 is the inverse of $-\Delta$.

Proposition 1.2.1. Let u be a weak solution satisfying Definition 1.2.1. Then $\mathrm{D}_0 H(u)|_{\mathscr{D}_{T_0}} \in \mathbf{H}^1(\mathscr{D}_{T_0})$, $T_0 > 0$, and the equation

$$\frac{\partial \mathrm{D}_0 H(u)}{\partial t} + u = 0 \tag{1.2.21}$$

holds almost everywhere in Ω for each $0 < t < \infty$. Moreover, the initial condition

$$\mathrm{D}_0 H(u)\Big|_{t=0} = \mathrm{D}_0 H(u_0) \tag{1.2.22}$$

holds.

Proof: For $\phi \in \mathbf{C}_0^\infty(\mathscr{D}_{T_0})$, let $\psi = \mathrm{D}_0\phi$ in (1.2.15) and use the self-adjointness of D_0 and its commutation with $\partial/\partial t$ to conclude that

$$\int_{\mathscr{D}_{T_0}} \mathrm{D}_0 H(u) \frac{\partial \phi}{\partial t}\, d\mathbf{x}\, dt = \int_{\mathscr{D}_{T_0}} u\phi\, d\mathbf{x}\, dt, \tag{1.2.23}$$

which shows that the distribution derivative $(\partial/\partial t)\mathrm{D}_0 H(u)$ is equal to $-u$. Since $H(u) \in \mathbf{L}^2(\mathscr{D}_{T_0})$, it follows that $(\partial/\partial x_i)\mathrm{D}_0 H(u) \in \mathbf{L}^2(\mathscr{D}_{T_0})$, $i = 1, \ldots, n$, that is, $\mathrm{D}_0 H(u) \in \mathbf{H}^1(\mathscr{D}_{T_0})$. In particular, (1.2.23) can be integrated by parts to obtain (1.2.19) for $0 < t < T_0$, since ϕ is arbitrary and the left side of (1.2.21) is a member of $\mathbf{C}([0, T_0]; \mathbf{L}^2(\Omega))$. Of course, $T_0 < \infty$ is arbitrary. The verification of (1.2.22) parallels that of (1.1.27). ■

Remark 1.2.5. The reason $u_0 \in \mathbf{L}^\infty(\Omega)$ is required is that the convex inverse of H is of superlinear growth at infinity. Thus, to conclude that $u_t \in \mathbf{L}^2(\Omega \times (0, \infty))$, we require not only that $u_0 \in \mathbf{H}_0^1(\Omega)$, in analogy with the two-phase Stefan problem, but also require pointwise estimates on u over its space–time domain. These are deduced in Section 5.2, via the pointwise estimates of the semidiscrete schemes of Section 2.4. The existence result is presented in Corollary 5.2.8. Similar properties for $(\partial H(u))/(\partial t)$ also hold in this case as well (see Remark 1.1.7).

1.3 REACTION–DIFFUSION SYSTEMS

Reaction in conjunction with diffusion is found in chemical catalysis, biological transmission of nerve impulses, and ecological and genetic models, among many others. Usually such physical and biological systems are modeled by a simple continuity equation for the (vector) concentration of a physical or biological species, whose flux (determined by Fick's law analogously to (1.2.1)) and rate of generation describe the diffusion and reaction of the system. We shall briefly mention some examples.

Example 1.3.1. Chemically Reacting Systems

Chemical reactions are frequently expedited by catalysis. The catalytic agent, and possibly several promoting substances, are mixed and prepared in the form of a porous pellet. One or more chemical reactants are adsorbed at sites on the surface of the pellet, and react spontaneously and begin to diffuse within the pellet, where further reaction occurs. Though the medium is porous, it and the process are viewed as continuous. When transport effects are considered negligible, the conservation equations typically assume the forms

$$\varepsilon \frac{\partial c_i}{\partial t} = \sum_{k=1}^{m} \nabla \cdot (D_{ik} \nabla c_k) + \sum_{j=1}^{l} \beta_{ji} r_j, \qquad i = 1, \ldots, m, \tag{1.3.1a}$$

$$\beta \frac{\partial \theta}{\partial t} = \nabla \cdot (k \nabla \theta) + \sum_{j=1}^{l} h_j r_j, \tag{1.3.1b}$$

where c_i is the concentration of the ith reactant, with units of moles per unit volume, and θ denotes the system temperature. The first m equations represent the mass balance for each reactant, and the last equation is an enthalpy-balance equation. It is assumed that $l < m$ reactions occur, and r_j is the reaction rate of the jth reaction, with units of moles per unit time per unit volume. The β_{ji} terms are the dimensionless stoichiometric integer coefficients for the ith reaction, satisfying $\sum_{j=1}^{l} \beta_{ji} = 0$, $i = 1, \ldots, m$; β and k are the volumetric heat capacity and thermal conductivity, and ε is the total pore volume as a fraction of the total volume. Also, h_j is the heat of reaction of the jth reaction, and D_{ik} is the binary diffusion coefficient of the ith reactant in the jth, with units of area per unit time. We shall not consider in the sequel the cases when $D_{ik} \neq 0$.

Example 1.3.2. FitzHugh–Nagumo Equations

The Hodgkin–Huxley equations model the conduction of electrical impulses along nerve fibers, or axons. Potential differences are created by ionic activation, including sodium and potassium ions. The system involves four coupled equations, which we shall not display explicitly. For this, the reader is referred to the original article [8]. The first equation is a reaction–diffusion equation for the electromotive potential, and the remaining three are reaction equations for the sodium activation, sodium inactivation, and potassium activation, respectively. The FitzHugh–Nagumo model compresses the system to two linked equations in two dependent variables, which are in some sense equivalent, respectively, to the first two and the final two of the Hodgkin–Huxley variables. When the FitzHugh–Nagumo model is simulated by electric circuits, a so-called distributed line with interstage coupling resistances, to simulate diffusion, is employed in conjunction with a tunnel diode to simulate reaction. The two FitzHugh–Nagumo variables are essentially voltage and current in this simulated case. The model may be written explicitly as

$$\frac{\partial u_1}{\partial t} = \Delta u_1 - f(u_1) - k(u_1, u_2), \tag{1.3.2a}$$

$$\frac{\partial u_2}{\partial t} = h(u_1, u_2), \tag{1.3.2b}$$

where the Lipschitz functions h and k have been chosen typically as

$$h(u_1, u_2) = \sigma u_1 - \gamma u_2, \qquad k(u_1, u_2) = u_2, \qquad \sigma > 0, \quad \gamma \geqslant 0, \tag{1.3.2c}$$

and f is typically a cubic polynomial satisfying $f' \geqslant -c$. The explicit form of f depends on the number of stable equilibrium solutions which the model is expected to possess.

Example 1.3.3. Predator–Prey Models

If u_1 and u_2 denote population densities of prey and predator in a system including saturation and migration effects, their time evolution is governed by

$$\frac{\partial u_1}{\partial t} = D_{u_1} \Delta u_1 + h(u_1)(f(u_1) - a(u_2)), \tag{1.3.3a}$$

$$\frac{\partial u_2}{\partial t} = D_{u_2} \Delta u_2 + k(u_2)(-g(u_2) + b(u_1)). \tag{1.3.3b}$$

Here a, b, h, and k are positive functions, typically linear, and f and g are monotone decreasing and increasing, respectively. These latter functions describe saturation. Migration is described by the diffusion terms, particularly the diffusion coefficients D_{u_1} and D_{u_2}, assumed nonnegative.

Remark 1.3.1. Let $\mathbf{u} = (u_1, \ldots, u_m)$. The general reaction–diffusion model we shall consider is of the form

$$\frac{\partial \mathbf{u}}{\partial t} = \mathbf{D}' \cdot \Delta \mathbf{u} - \mathbf{f}(\mathbf{u}) \qquad \text{in} \quad \Omega, \tag{1.3.4}$$

where $\mathbf{D}' = (D_1, \ldots, D_m)$ is an m-tuple of nonnegative constants and $\mathbf{f}: \mathbb{R}^m \to \mathbb{R}^m$ is $\mathbf{C}^1$. The additional hypotheses upon $\mathbf{f}$ are motivated by the natural grouping of the variables u_i into those of concentration type, which are naturally nonnegative, possibly even further restricted, and those of potential type, which are unrestricted in sign. We now define these two classes of variables.

Definition 1.3.1. Let the integer i_0 satisfy $0 \leqslant i_0 \leqslant m$. If $i_0 > 0$, we say that the variables $u_1, \ldots, u_{i_0}$ are of concentration type if there exists a slab

$$\Sigma = \prod_{i=1}^{i_0} [a_i, b_i] \subset \mathbb{R}^{i_0}, \tag{1.3.5a}$$

$a_i \leqslant 0 \leqslant b_i$, $(b_i - a_i) > 0$, $i = 1, \ldots, i_0$ such that the vector field $(f_1, \ldots, f_{i_0})$ is bounded on $\mathbb{R}^m$ and satisfies certain properties on

$$Q = \Sigma \times \mathbb{R}^{m - i_0}. \tag{1.3.5b}$$

We describe these now. Define the faces

$$Q_{a_i} = \{\mathbf{v} \in Q : v_i = a_i\}, \qquad Q_{b_i} = \{\mathbf{v} \in Q : v_i = b_i\} \tag{1.3.6a}$$

of Q for $i = 1, \ldots, i_0$. We assume explicitly that

$$f_i(\mathbf{v}) \leqslant 0, \qquad \mathbf{v} \in Q_{a_i}, \qquad\qquad f_i(\mathbf{v}) \geqslant 0, \qquad \mathbf{v} \in Q_{b_i}, \tag{1.3.6b}$$

$$f_i(\mathbf{v}) \neq 0, \qquad \mathbf{v} \notin Q, \tag{1.3.6c}$$

$i = 1, \ldots, i_0$. Finally, we assume an ordering of sign regions property described by

$$f_i(v_1, \ldots, v_{i-1}, v_i', v_{i+1}, \ldots, v_m) \leqslant 0, \qquad f_i(v_1, \ldots, v_{i-1}, v_i'', v_{i+1}, \ldots, v_m) \geqslant 0,$$
$$\Rightarrow v_i' \leqslant v_i'' \qquad \text{for each fixed} \quad \{v_j\}_{j \neq i}. \tag{1.3.6d}$$

The remaining variables are assumed to be of potential type, that is

$$f_i(\mathbf{u}) = g_i(u_i) + h_i(\mathbf{u}), \qquad i = i_0 + 1, \ldots, m, \tag{1.3.7}$$

where h_i is a Lipschitz continuous function on Q; and g_i is monotone increasing in the variable u_i, such that

$$[g_i(u_i) - g_i(v_i)](u_i - v_i) \geqslant 0, \tag{1.3.8a}$$

and g_i vanishes at 0,

$$g_i(0) = 0. \tag{1.3.8b}$$

Remark 1.3.2. The set Q is termed invariant, since initial datum $\mathbf{u}_0$ with range in Q implies that the solution $\mathbf{u}$, of the appropriate initial/boundary-value problem corresponding to (1.3.4), also has range in Q. The most physically meaningful boundary conditions corresponding to Example 1.3.1 are Robin boundary conditions; these are also important for the remaining two examples. Thus, if $D_i > 0$ (note that $D_i = 0$ is permitted), a boundary condition of the form

$$\frac{\partial u_i}{\partial \nu} + \omega_i u_i = 0 \qquad \text{on} \quad \partial\Omega \quad (\boldsymbol{\nu} = \text{outward normal}) \tag{1.3.9}$$

is specified, where ω_i is a nonnegative, bounded function on $\partial\Omega$, satisfying

$$\int_{\partial\Omega} \omega_i > 0. \tag{1.3.10}$$

Moreover, in (1.3.11a) to follow, we set $\omega_i = 0$ if $D_i = 0$.

Definition 1.3.2. Given $\mathbf{u}_0 \in \mathbf{L}^\infty(\Omega; \mathbb{R}^m)$ with range in the set Q, defined by (1.3.5), and given, for each i, such that $D_i > 0$, functions $0 \leqslant \omega_i \in \mathbf{L}^\infty(\partial\Omega)$, then a function $u \in \mathscr{C}$ (see (1.3.11b)) is said to be a weak solution of the initial/boundary-value problem for (1.3.4) with Robin condition if the range of $\mathbf{u}$ is in Q; if $u|_{t=0} = \mathbf{u}_0$; and if, for each $i = 1, \ldots, m$,

$$\int_\Omega \left[\frac{\partial u_i}{\partial t} v_i + D_i \nabla u_i \cdot \nabla v_i + f_i(\mathbf{u}) v_i \right] d\mathbf{x} + D_i \int_{\partial\Omega} \omega_i u_i v_i \, d\sigma = 0 \tag{1.3.11a}$$

holds for all $0 < t < T_0$ and all $\mathbf{v} \in \mathbf{H}^1(\Omega; \mathbb{R}^m)$, where

$$\mathscr{C} = \mathbf{L}^\infty((0, T_0); \mathbf{H}^1(\Omega, \mathbb{R}^m)) \cap \mathbf{H}^1([0, T_0]; \mathbf{L}^2(\Omega; \mathbb{R}^m)) \cap \mathbf{L}^\infty(\mathscr{D}; \mathbb{R}^m). \tag{1.3.11b}$$

Here Ω is a bounded, strongly Lipschitz domain in $\mathbb{R}^n$ and, for consistency with the definition of $\mathscr{C}$ and Q, $\mathbf{u}_0 \in \mathbf{Lip}(\bar{\Omega}; \mathbb{R}^m)$ is required. The domain Ω is also required to satisfy the regularity conditions specified by (2.5.7) and Definition 1.3.3 to follow.

Remark 1.3.3. We shall briefly comment on the relation of the earlier models, introduced in Examples 1.3.1–1.3.3, to the general model of Definitions 1.3.1–1.3.2. In the FitzHugh–Nagumo model, it is feasible to select u_1 and u_2 as potential variables, with $D_1 = 1$, $D_2 = 0$, although one might wish the alternate choice of u_2 as a concentration variable if the initial datum permits this choice. The assumption $f' \geqslant -c$ in this model means that f differs from a monotone increasing function by a linear function. The model as described, with h and k Lipschitz continuous, is, thus, included, if $\mathbf{L}^\infty \cap \mathbf{C}^1$ initial datum is prescribed, with a Robin boundary condition for u_1.

In the case of the predator–prey model, the choice of u_1 and u_2 as concentration variables appears to be dictated in many cases of interest. The assumptions

$$h(a_1)[f(a_1)-a(b_2)]\geqslant 0, \qquad h(b_1)[f(b_1)-a(a_2)]\leqslant 0, \tag{1.3.12a}$$

$$k(a_2)[-g(a_2)+b(a_1)]\geqslant 0, \qquad k(b_2)[-g(b_2)+b(b_1)]\leqslant 0, \tag{1.3.12b}$$

guarantee (1.3.6b) if, in addition to the properties already stated, a and b are monotone increasing (note the sign of f in (1.3.4)). The monotonicity properties of f and g in (1.3.12), together with the sign properties of h and k (nonnegative), guarantee that (1.3.6d) holds. The presence and/or amount of diffusion will vary with the model.

The model of the chemically reacting systems is already general, and we shall not attempt a specific analysis except to make two observations. The first involves the r_j terms, which are typically nonnegative rational functions of $u_1, \ldots, u_j$. The second concerns the choice of variables. Here, $u_1, \ldots, u_m$ are typically concentration variables and u_{m+1} is a potential variable. This model demonstrates clearly why a partitioning of the variables is desirable and, indeed, necessary.

Given a Robin boundary condition specified by ω, we shall introduce the linear, Riesz lifting mapping R_ω in analogy with the mappings N_σ and D_0 introduced previously. We first make some preparatory remarks.

Remark 1.3.4. It is known that, under certain regularity conditions on $\partial\Omega$, the trace operator $\Gamma: \mathbf{H}^1(\Omega) \to \mathbf{L}^2(\partial\Omega)$ exists as a continuous linear mapping and coincides with the restriction mapping on $\mathbf{C}^\infty(\bar{\Omega})$, which is dense in $\mathbf{H}^1(\Omega)$ in this case. Moreover, the range of Γ is $\mathbf{H}^{1/2}(\partial\Omega)$ and, in fact, Γ is an isomorphism between $\mathbf{H}^1(\Omega)/\ker \Gamma$ and $\mathbf{H}^{1/2}(\partial\Omega)$. These facts are documented in Lions and Magenes (see [13]). Let $0 \leqslant \omega \in \mathbf{L}^\infty(\partial\Omega)$ be such that

$$\int_{\partial\Omega} \omega > 0. \tag{1.3.13}$$

Then the inner product

$$(v, w)_{[\mathbf{H}^1(\Omega)]_\omega} = \int_\Omega \nabla v \cdot \nabla w + \int_{\partial\Omega} \omega \Gamma v \Gamma w^\dagger \tag{1.3.14}$$

induces an equivalent Hilbert space norm on $\mathbf{H}^1(\Omega)$. To prove completeness, let $\{v_j\}$ be a Cauchy sequence. In particular, by the Cauchy–Schwarz in-

† Suppressed $d\sigma$.

equality, there is a $\rho \in \mathbb{R}^1$, such that

$$\int_{\partial\Omega} \omega \Gamma v_j \, d\sigma \to \rho, \tag{1.3.15}$$

and, by taking the quotient space of $\mathbf{H}^1(\Omega)$ with the constants, we deduce the existence of $v \in \mathbf{H}^1(\Omega)$ and $\{\sigma_j\} \subset \mathbb{R}^1$, such that

$$v_j + \sigma_j \to v \quad \text{in} \quad \mathbf{L}^2(\Omega), \tag{1.3.16a}$$

$$\frac{\partial v_j}{\partial x_i} \to \frac{\partial v}{\partial x_i} \quad \text{in} \quad \mathbf{L}^2(\Omega), \qquad i = 1, \ldots, n. \tag{1.3.16b}$$

From (1.3.16), we deduce that

$$\int_{\partial\Omega} |\Gamma v_j + \sigma_j - \Gamma v|^2 \to 0, \tag{1.3.17}$$

so that

$$\int_{\partial\Omega} \omega[\Gamma v_j + \sigma_j] d\sigma \to \int_{\partial\Omega} \omega \Gamma v \, d\sigma.$$

Combining (1.3.15) and (1.3.17), we deduce that

$$\sigma_j \to \sigma_0 = \left(\int_{\partial\Omega} \omega \Gamma v \, d\sigma - \rho \right) \Big/ \int_{\partial\Omega} \omega \, d\sigma.$$

It follows from (1.3.16) that $v_j \to v - \sigma_0$ in the standard $\mathbf{H}^1(\Omega)$ norm and, hence, in the norm induced by (1.3.14), since the standard norm dominates this norm up to a constant multiplier. A standard application of the open mapping theorem now gives the equivalence of norms on $\mathbf{H}^1(\Omega)$, and justifies the notation in (1.3.14).

Definition 1.3.3. Under the assumption of the existence of the linear continuous trace operator $\Gamma : \mathbf{H}^1(\Omega) \to \mathbf{L}^2(\partial\Omega)$ described above, let $0 \leqslant \omega \in \mathbf{L}^\infty(\Omega)$ be given satisfying (1.3.13). Given $\ell \in \mathbf{F} = [\mathbf{H}^1(\Omega)]^*$, we define the element $w = \mathbf{R}_\omega \ell \in \mathbf{H}^1(\Omega)$ as the Riesz representer of ℓ in the inner product (1.3.14). We have, then,

$$\langle \ell, v \rangle = \int_\Omega \nabla \mathbf{R}_\omega \ell \cdot \nabla v \, d\mathbf{x} + \int_{\partial\Omega} \omega \Gamma(\mathbf{R}_\omega \ell) \Gamma v \, d\sigma. \tag{1.3.18}$$

Remark 1.3.5. The restriction of $\mathbf{R}_\omega$ to $\mathbf{L}^2(\Omega)$ is symmetric by (1.3.14) and (1.3.18). The mapping $\mathbf{R}_\omega$ induces the following (negative) norm on $\mathbf{F}$:

$$\|\ell\|_{[\mathbf{F}]_\omega} = \langle \ell, \mathbf{R}_\omega \ell \rangle^{1/2}. \tag{1.3.19}$$

As in Section 1.1, we may conclude that the mappings $\mathbf{R}_\omega$ are continuous and have closed range, and that $[\mathbf{F}]_\omega$ is topologically equivalent to $\mathbf{F}$, with the standard duality norm. In particular, $R_\omega|_{L^2(\Omega)}$ is self-adjoint, positive–definite, and compact. The subscript ω will be frequently suppressed in the sequel.

Remark 1.3.6. An equivalent way of viewing $R_\omega : \mathbf{F} \to \mathbf{H}^1(\Omega)$ is as follows. The element $w = R_\omega \ell$ is the unique element satisfying the (weak form of the) linear elliptic Robin problem

$$-\Delta w = \ell \qquad \text{in } \Omega, \tag{1.3.20a}$$

$$\frac{\partial w}{\partial \nu} + \omega w = 0 \qquad \text{on } \partial\Omega. \tag{1.3.20b}$$

One may deduce directly that

$$-\Delta R_\omega \ell = \ell, \qquad R_\omega(-\Delta)w = w, \tag{1.3.21}$$

for $\ell \in \mathbf{F}$ and $w \in \mathbf{Range}(R_\omega)$. The latter is easily seen to be $\mathbf{H}^1(\Omega)$.

Remark 1.3.7. We may deduce the property

$$\ell \geqslant 0 \Rightarrow R_\omega \ell \geqslant 0 \tag{1.3.22}$$

directly from (1.3.18) by making the choice $v = -(R_\omega \ell)^-$, as in Remark 1.1.6.
We now close this section with the lifting relations.

Proposition 1.3.1. Let $\mathbf{u}$ be a weak solution satisfying Definition 1.3.2 and let i, $1 \leqslant i \leqslant m$, be an index for which $D_i > 0$. Then the equation

$$\frac{\partial}{\partial t} R_{\omega_i} u_i + D_i u_i + \mathbf{R}_{\omega_i} f_i(\mathbf{u}) = 0 \tag{1.3.23}$$

holds almost everywhere in Ω for $0 < t < T_0$.

Proof: For $0 < t < T_0$, we set $v_i = R_{\omega_i}\phi_i$ in (1.3.11a), for ϕ_i a component of a suitable test function in $\mathbf{C}^\infty(\bar{\Omega})$, apply (1.3.18) and the self-adjointness of R_{ω_i}, together with the commutation relation $R_{\omega_i}(\partial/\partial t) - (\partial/\partial t)R_{\omega_i} = 0$, to deduce (1.3.23). ■

1.4 INCOMPRESSIBLE, VISCOUS FLUID DYNAMICS AT CONSTANT TEMPERATURE: NAVIER–STOKES EQUATIONS AND GENERALIZATIONS

A moving fluid transferring heat is described by a system of five equations, characterized by conservation of mass, momentum, and energy, including the continuity equation of mass balance, the three equations of motion, and the energy-balance equation, augmented by an equation of state. The standard dependent variables are density, pressure, temperature, and the three velocity components of the fluid. For this analysis, temperature variations will be assumed negligible. For incompressible fluids, the equation of state in piezotropic flow, wherein density is a function of pressure, reduces to an expression of constant density ρ. In this case, the continuity equation

$$\frac{\partial \rho}{\partial t} + \nabla \cdot (\rho \mathbf{u}) = 0,$$

derived by a mass balance with respect to a motionless set $\mathscr{B} \subset \Omega$, reduces to the equation

$$\nabla \cdot \mathbf{u} = 0 \tag{1.4.1}$$

for the velocity $\mathbf{u}$. For a so-called ideal fluid, which neglects the irreversible processes created by internal friction, the set $\mathscr{B}$, in which the balance is computed, may be selected to move with the fluid to obtain the force law, in a distribution sense on $\mathscr{D} = \Omega \times (0, T_0)$, given by

$$\rho \frac{d\mathbf{u}}{dt} = -\nabla p,$$

where p denotes pressure and $d\mathbf{u}/dt$ is computed with respect to moving coordinates. With respect to stationary Cartesian coordinates,

$$\frac{d\mathbf{u}}{dt} = \frac{\partial \mathbf{u}}{\partial t} + (\mathbf{u} \cdot \nabla)\mathbf{u},$$

so that the equations of motion assume the form

$$\rho \left[\frac{\partial \mathbf{u}}{\partial t} + (\mathbf{u} \cdot \nabla)\mathbf{u} \right] + \nabla p = 0. \tag{1.4.2}$$

If a stationary set $\mathscr{B}$ is now employed to set up a momentum balance of the form

$$\frac{\partial}{\partial t} \int_{\mathscr{B}} \rho u_i \, dx = - \int_{\partial \mathscr{B}} (\Pi_{i1}, \Pi_{i2}, \Pi_{i3}) \cdot \nu \, d\sigma, \tag{1.4.3}$$

then (1.4.2) may be written according to the format, upon use of (1.4.1),

$$\frac{\partial}{\partial t}(\rho u_i) = - \sum_{k=1}^{3} \frac{\partial}{\partial x_k} \Pi_{ik}, \qquad i = 1, 2, 3, \tag{1.4.4a}$$

where the momentum flux tensor Π_{ik} has components

$$\Pi_{ik} = p\delta_{ik} + \rho u_i u_k. \tag{1.4.4b}$$

Thus, the equations of motion for an incompressible ideal fluid are given by (1.4.4). Usual derivations of the (Navier–Stokes) equations of motion for a *viscous* fluid employ the format of (1.4.3) and (1.4.4), and proceed by modifying (1.4.4b) to account for internal friction. One writes

$$\Pi_{ik} = p\delta_{ik} + \rho u_i u_k - \sigma'_{ik}, \tag{1.4.5}$$

where σ'_{ik} is called the viscosity stress tensor and, on physical grounds, is assumed to be proportional to the rate of (shear) deformation, that is

$$\sigma'_{ik} = \eta\left(\frac{\partial u_i}{\partial x_k} + \frac{\partial u_k}{\partial x_i}\right), \tag{1.4.6}$$

which measures the amount of nonuniform rotation of the fluid. A second (isotropic) component of σ'_{ik}, involving the divergence of $\mathbf{u}$, is missing according to (1.4.1). The relationship (1.4.6) is the analog of a stress–strain relationship in elasticity, and characterizes incompressible Newtonian fluids. Altogether, then, we have the equations

$$\frac{\partial \mathbf{u}}{\partial t} - \eta\,\Delta\mathbf{u} + (\mathbf{u}\cdot\nabla)\mathbf{u} + \frac{1}{\rho}\nabla p = \mathbf{0}, \tag{1.4.7}$$

when (1.4.5) and (1.4.6) are substituted into (1.4.4). Here, $\eta > 0$ is the (assumed constant) viscosity, and external forces, such as gravity, are neglected or assumed conservative.

Because of the assumption that the motion is isothermal, the energy-balance equation is unnecessary, and we are left with the Navier–Stokes system of (1.4.1) and (1.4.7) for the equations of motion of an incompressible fluid in a region Ω, with velocity $\mathbf{u}$, pressure p, and constant (positive) viscosity and density, η and ρ, respectively. When these equations are understood variationally, or distributionally, a significant reduction due to Leray[†] is possible. We shall describe this reduction in the case where the fluid adheres to its retaining boundary. Following Temam [23], we shall base the analysis on a distribution surjectivity property of the gradient mapping due to DeRham[‡], and a corresponding isomorphism property of this operator in Sobolev spaces. The following two results are discussed and proved in Temam ([23] Chapter 1, Sections 1 and 2).

Proposition 1.4.1. Let Ω be an open subset of $\mathbb{R}^n$ and let $f_i \in \mathscr{D}'(\Omega)$, $i = 1, \ldots, n$. A necessary and sufficient condition that

$$\mathbf{f} = (f_1, \ldots, f_n) = \nabla p \tag{1.4.8a}$$

for some $p \in \mathscr{D}'(\Omega)$ is that

$$0 = \langle \mathbf{f}, \boldsymbol{\phi} \rangle = \sum_{j=1}^{n} \langle f_j, \phi_j \rangle \tag{1.4.8b}$$

[†] *J. Math. Pures et Appl.* **XII** (1933), 1–82; *ibid.*, **XIII** (1934), 331–418.
[‡] "Variétés Différentiables." Paris, Hermann, 1960.

for all $\phi \in \mathscr{V}$, where

$$\mathscr{V} = \{\phi \in \mathscr{D}(\Omega; \mathbb{R}^n): \nabla \cdot \phi = 0\}. \tag{1.4.8c}$$

Moreover, if $f_i \in \mathbf{H}^{-1}(\Omega)$, $i = 1, \ldots, n$, then $p \in \mathbf{L}^2_{\text{loc}}(\Omega)$. The conclusion $p \in \mathbf{L}^2(\Omega)$ holds if Ω is a bounded, strongly Lipschitz domain and, in this case, the mapping $\mathbf{f} \to p$ is continuous from $\mathbf{H}^{-1}(\Omega; \mathbb{R}^n)$ to $\mathbf{L}^2(\Omega)/\mathbb{R}^1$.

The following corollary is then immediate.

Corollary 1.4.2. Suppose that

$$\mathbf{u}_t - \eta \, \Delta \mathbf{u} + (\mathbf{u} \cdot \nabla)\mathbf{u} \in \mathbf{L}^2((0, T_0); \mathscr{D}'(\Omega; \mathbb{R}^n)), \tag{1.4.9a}$$

and that, for almost all t, $0 < t < T_0$, the relation

$$\langle \mathbf{u}_t - \eta \, \Delta \mathbf{u} + (\mathbf{u} \cdot \nabla)\mathbf{u}, \phi \rangle = 0 \qquad \text{for all} \quad \phi \in \mathscr{V} \tag{1.4.9b}$$

holds distributionally. Then there exists $p \in \mathbf{L}^2((0, T_0); \mathscr{D}'(\Omega))$, such that

$$\langle \mathbf{u}_t - \eta \, \Delta \mathbf{u} + (\mathbf{u} \cdot \nabla)\mathbf{u} + \frac{1}{\rho} \nabla p, \phi \rangle = 0, \qquad \text{for all} \quad \phi \in \mathscr{D}(\Omega; \mathbb{R}^n). \tag{1.4.10}$$

Remark 1.4.1. The reduction alluded to earlier is that of (1.4.10) to (1.4.9), provided (1.4.9a) holds. However, (1.4.9a) is weaker than the regularity expected to hold; in fact, the left-hand side of (1.4.9a) is in $\mathbf{L}^2((0, T_0);$ $\mathbf{H}^{-1}(\Omega; \mathbb{R}^n))$ for $n \leqslant 4$, and $p \in \mathbf{L}^2(\mathscr{D}_{T_0})$ in this case. To obtain a (weak) variational formulation of maximum sharpness, we have the following proposition, also discussed in Temam ([23] Chapter 2, Section 1). For completeness, and because of modifications in presentation, we present a proof.

Proposition 1.4.3. Consider the trilinear form on $\mathscr{V} \times \mathscr{V} \times \mathscr{V}$ given by

$$b(\mathbf{u}, \mathbf{v}, \mathbf{w}) = ((\mathbf{u} \cdot \nabla)\mathbf{v}, \mathbf{w})_{\mathbf{L}^2(\Omega; \mathbb{R}^n)}. \tag{1.4.11}$$

Then there exists a positive constant c_1, such that

$$|b(\mathbf{u}, \mathbf{v}, \mathbf{w})| \leqslant c_1 \|\mathbf{u}\|_{\mathbf{H}_0^1(\Omega; \mathbb{R}^n)} \|\mathbf{v}\|_{\mathbf{H}_0^1(\Omega; \mathbb{R}^n)} \|\mathbf{w}\|_{\mathbf{L}^n(\Omega; \mathbb{R}^n) \cap \mathbf{H}_0^1(\Omega; \mathbb{R}^n)} \tag{1.4.12}$$

for all $n \geqslant 1$. In particular, for $s \geqslant (n/2) - 1$, there exists a positive constant c_2, such that

$$|b(\mathbf{u}, \mathbf{v}, \mathbf{w})| \leqslant c_2 \|\mathbf{u}\|_{\mathbf{H}_0^1(\Omega; \mathbb{R}^n)} \|\mathbf{v}\|_{\mathbf{H}_0^1(\Omega; \mathbb{R}^n)} \|\mathbf{w}\|_{\mathbf{H}^s(\Omega; \mathbb{R}^n) \cap \mathbf{H}_0^1(\Omega; \mathbb{R}^n)}. \tag{1.4.13}$$

If $n \geqslant 3$, the inequality

$$|b(\mathbf{u}, \mathbf{v}, \mathbf{w})| \leqslant c_3 \|\mathbf{u}\|_{\mathbf{L}^2(\Omega; \mathbb{R}^n)} \|\mathbf{v}\|_{\mathbf{H}_0^1(\Omega; \mathbb{R}^n)} \|\mathbf{w}\|_{\mathbf{H}^s(\Omega; \mathbb{R}^n)} \tag{1.4.14}$$

holds for $s \geqslant (n/2)$, and some positive constant c_3.

Proof: To prove (1.4.12), we first set $n \geqslant 3$. By Hölder's inequality, we have

$$\begin{aligned} |b(\mathbf{u},\mathbf{v},\mathbf{w})| &= \left| \sum_{i,j=1}^{n} \int_{\Omega} u_i \frac{\partial v_j}{\partial x_i} w_j \, d\mathbf{x} \right| \\ &\leqslant \sum_{i,j=1}^{n} \|u_i\|_{\mathbf{L}^{2n/(n-2)}(\Omega)} \left\| \frac{\partial v_j}{\partial x_i} \right\|_{\mathbf{L}^2(\Omega)} \|w_j\|_{\mathbf{L}^n(\Omega)}. \end{aligned} \tag{1.4.15}$$

Since $\mathbf{H}_0^1(\Omega) \subset \mathbf{L}^{2n/(n-2)}(\Omega)$ by Sobolev's inequality, (1.4.12) follows directly from (1.4.15). For $n = 2$, Hölder's inequality is applied to give

$$|b(\mathbf{u},\mathbf{v},\mathbf{w})| \leqslant \sum_{i,j=1}^{n} \|u_i\|_{\mathbf{L}^4(\Omega)} \left\| \frac{\partial v_j}{\partial x_i} \right\|_{\mathbf{L}^2(\Omega)} \|w_j\|_{\mathbf{L}^4(\Omega)},$$

and (1.4.12) is now immediate from $\mathbf{H}_0^1(\Omega) \subset \mathbf{L}^4(\Omega)$.

Now if $s \geqslant (n/2) - 1$, it follows from Sobolev's inequality that $\mathbf{H}^s(\Omega) \subset \mathbf{L}^n(\Omega)$, since $(1/n) \geqslant (1/2) - (s/n)$ in this case. This remark applied to (1.4.12) gives (1.4.13). Prior to the derivation of (1.4.14), we observe that

$$b(\mathbf{u},\mathbf{v},\mathbf{w}) = -b(\mathbf{u},\mathbf{w},\mathbf{v}) \tag{1.4.16}$$

on $\mathscr{V} \times \mathscr{V} \times \mathscr{V}$. This follows from

$$\begin{aligned} b(\mathbf{u},\mathbf{v},\mathbf{w}) &= \sum_{i,j=1}^{n} \int_{\Omega} u_i \frac{\partial v_j}{\partial x_i} w_j \, d\mathbf{x} = -\sum_{i,j=1}^{n} \int_{\Omega} v_j \frac{\partial}{\partial x_i}(u_i w_j)\, d\mathbf{x} \\ &= -\sum_{i,j=1}^{n} \int_{\Omega} u_i \frac{\partial w_j}{\partial x_i} v_j \, d\mathbf{x} = -b(\mathbf{u},\mathbf{w},\mathbf{v}), \end{aligned}$$

where we have used integration by parts and $\nabla \cdot \mathbf{u} = 0$. Now suppose $n \geqslant 3$ and $s \geqslant n/2$. By (1.4.16) and Hölder's inequality, we have

$$|b(\mathbf{u},\mathbf{v},\mathbf{w})| = |b(\mathbf{u},\mathbf{w},\mathbf{v})| \leqslant \sum_{i,j=1}^{n} \|u_i\|_{\mathbf{L}^2(\Omega)} \left\| \frac{\partial w_j}{\partial x_i} \right\|_{\mathbf{L}^n(\Omega)} \|v_j\|_{\mathbf{L}^{2n/(n-2)}(\Omega)}. \tag{1.4.17}$$

By Sobolev's inequality applied to the second and third terms in the product in (1.4.17), we have

$$|b(\mathbf{u},\mathbf{v},\mathbf{w})| \leqslant c \sum_{i,j=1}^{n} \|u_i\|_{\mathbf{L}^2(\Omega)} \left\| \frac{\partial w_j}{\partial x_i} \right\|_{\mathbf{H}^{s-1}(\Omega)} \|v_j\|_{\mathbf{H}_0^1(\Omega)},$$

from which (1.4.14) is immediate. Note that the application of Sobolev's inequality to the second term in the product in (1.4.17) used the inclusion $\mathbf{H}^{s-1}(\Omega) \subset \mathbf{L}^n(\Omega)$. The proof is now complete. ■

Remark 1.4.2. The result of the previous proposition may be used to define $b(\cdot,\cdot,\cdot)$ on the completion of $\mathscr{V} \times \mathscr{V} \times \mathscr{V}$ in

$$\mathbf{H}_0^1(\Omega;\mathbb{R}^n) \times \mathbf{H}_0^1(\Omega;\mathbb{R}^n) \times (\mathbf{H}^s(\Omega;\mathbb{R}^n) \cap \mathbf{H}_0^1(\Omega;\mathbb{R}^n))$$

for $s \geqslant n/2 - 1$. It follows from (1.4.16) that the extended form $b(\cdot,\cdot,\cdot)$ satisfies

$$b(\mathbf{u}, \mathbf{v}, \mathbf{v}) = 0 \tag{1.4.18}$$

on this completion, which we denote by $\mathbf{V} \times \mathbf{V} \times \mathbf{V}_s$. In fact, if Ω is a bounded strongly Lipschitz domain, then (see [23] Chapter 1, Section 1)

$$\begin{aligned} \mathbf{V} &= \{\mathbf{v} \in \mathbf{H}_0^1(\Omega; \mathbb{R}^n) : \nabla \cdot \mathbf{v} = 0\}, \\ \mathbf{V}_s &= \{\mathbf{v} \in \mathbf{H}^s(\Omega; \mathbb{R}^n) \cap \mathbf{H}_0^1(\Omega; \mathbb{R}^n) : \nabla \cdot \mathbf{v} = 0\}. \end{aligned} \tag{1.4.19}$$

In conclusion, we may sharpen (1.4.9b), by requiring that $\boldsymbol{\phi}$ belong to the extended test class $\mathbf{V}_s$, provided $\mathbf{u}$ is required to belong to $\mathbf{L}^2((0, T_0); \mathbf{V})$.

Remark 1.4.3. We shall now consider appropriate generalizations of (1.4.9b). Let P be an orthogonal projection in $\mathbf{L}^2(\Omega; \mathbb{R}^n)$, such that

$$\mathbf{V} = \text{closure} \quad \mathscr{V} \quad \text{in} \quad \mathbf{H}_0^1(\Omega; \mathbb{R}^n), \tag{1.4.20a}$$

where

$$\mathbf{V} := \{\mathbf{v} \in \mathbf{H}_0^1(\Omega; \mathbb{R}^n) : \mathrm{P}\mathbf{v} = 0\}, \tag{1.4.20b}$$

and

$$\mathscr{V} := \{\mathbf{v} \in \mathbf{C}_0^\infty(\Omega; \mathbb{R}^n) : \mathrm{P}\mathbf{v} = 0\}. \tag{1.4.20c}$$

For example, we have seen above that (1.4.20) is satisfied by the projection onto the divergence-free functions. Furthermore, let $a(\cdot,\cdot,\cdot)$ be a trilinear form on $\mathbf{V} \times \mathbf{V} \times \mathbf{V}_s$, where $s \geqslant 1$, and

$$\mathbf{V}_s = \{\mathbf{v} \in \mathbf{H}^s(\Omega; \mathbb{R}^n) \cap \mathbf{H}_0^1(\Omega; \mathbb{R}^n) : \mathrm{P}\mathbf{v} = 0\}, \tag{1.4.21}$$

such that the continuity relation

$$|a(\mathbf{u}, \mathbf{v}, \mathbf{w})| \leqslant c \|\mathbf{u}\|_{\mathbf{V}} \|\mathbf{v}\|_{\mathbf{V}} \|\mathbf{w}\|_{\mathbf{V}_s} \tag{1.4.22a}$$

holds on $\mathbf{V} \times \mathbf{V} \times \mathbf{V}_s$ for some constant c, as well as the dissipation relation

$$a(\mathbf{v}, \mathbf{v}, \mathbf{v}) \geqslant 0, \qquad \text{for all} \quad \mathbf{v} \in \mathbf{V}_s. \tag{1.4.22b}$$

Finally, set

$$\mathbf{H} = \{\mathbf{u} \in \mathbf{L}^2(\Omega; \mathbb{R}^n) : \mathrm{P}\mathbf{u} = 0\}. \tag{1.4.23}$$

We assume, explicitly, that

$$\begin{aligned} &\mathbf{u}_m \rightharpoonup \mathbf{u} \quad (\text{weakly in} \quad \mathbf{V}), \qquad \mathbf{u}_m \to \mathbf{u} \quad (\text{strongly in} \quad \mathbf{H}) \\ &\Rightarrow a(\mathbf{u}_m, \mathbf{u}_m, \mathbf{w}) \to a(\mathbf{u}, \mathbf{u}, \mathbf{w}) \qquad \text{for all} \quad \mathbf{w} \in \mathrm{V}_s. \end{aligned} \tag{1.4.24a}$$

This relation does, in fact, hold for the trilinear form defined by (1.4.11) (see Temam [23] pp. 165 and 166). We also require an analog of (1.4.24a) on

$\mathscr{D}$, which, in certain cases, is a corollary of (1.4.24a), but is assumed explicitly here.

$$\begin{aligned} \mathbf{u}_m \rightharpoonup \mathbf{u} \quad &(\text{weakly in} \quad \mathbf{L}^2((0, T_0); \mathbf{V})), \\ \mathbf{u}_m \to \mathbf{u} \quad &(\text{strongly in} \quad \mathbf{L}^2((0, T_0); \mathbf{H})) \\ \Rightarrow \int_0^{T_0} a(\mathbf{u}_m, \mathbf{u}_m, \mathbf{w})\, dt &\to \int_0^{T_0} a(\mathbf{u}, \mathbf{u}, \mathbf{w})\, dt, \end{aligned} \tag{1.4.24b}$$

for all $\mathbf{w} \in \mathbf{C}([0, T_0]; \mathbf{V}_s)$. Again, this relation holds for a as defined by (1.4.11) (see Temam [23] p. 289). As a complement of (1.4.24b), we state a further convergence result, to be used in the sequel, which follows directly from (1.4.22a).

$$\begin{aligned} \mathbf{u} \in \mathbf{B}(0, r) \subset \mathbf{V}, \qquad \mathbf{w}_m \to \mathbf{w} \qquad &(\text{strongly in} \quad \mathbf{L}^2((0, T_0); \mathbf{V}_s)) \\ \Rightarrow \int_0^{T_0} a(\mathbf{u}, \mathbf{u}, \mathbf{w}_m)\, dt \to \int_0^{T_0} a(\mathbf{u}, \mathbf{u}, \mathbf{w})\, dt & \end{aligned} \tag{1.4.24c}$$

uniformly in $\mathbf{u}$.

We now wish to consider the weak form of the initial/boundary-value problem for

$$\frac{\partial \mathbf{u}}{\partial t} - \eta\, \Delta \mathbf{u} + a(\mathbf{u}, \mathbf{u}, \cdot) = 0, \tag{1.4.25a}$$

$$\mathbf{P}\mathbf{u} = 0, \tag{1.4.25b}$$

constituting the generalized Navier–Stokes system.

For ease of expression in the following definition, we introduce the notation

$$A \cdot B = \sum_{i=1}^{n} \mathbf{a}^i \cdot \mathbf{b}^i, \tag{1.4.26}$$

if $\mathbf{a}^i$ and $\mathbf{b}^i$ are the rows of two $n \times n$ matrices A and B, for example, $A = \nabla \mathbf{u}$, $B = \nabla \boldsymbol{\phi}$. Here, $\mathbf{a} \cdot \mathbf{b}$ is the usual dot product; we shall frequently suppress the dot for such vectors.

Definition 1.4.1. Given initial datum $\mathbf{u}_0 \in \mathbf{V}$ and a trilinear form $a(\cdot, \cdot, \cdot)$: $\mathbf{V} \times \mathbf{V} \times \mathbf{V}_s \to \mathbb{R}^1$ satisfying (1.4.22) and (1.4.24), we say that $\mathbf{u} \in \mathscr{C}$ (see (1.4.28a)) is a weak solution of the constrained initial/boundary-value problem for (1.4.25), if

$$\begin{aligned} &\int_{\mathscr{D}} \left[\mathbf{u} \frac{\partial \boldsymbol{\phi}}{\partial t} - \eta\, \nabla \mathbf{u} \cdot \nabla \boldsymbol{\phi} \right] d\mathbf{x}\, dt - \int_0^{T_0} a(\mathbf{u}, \mathbf{u}, \boldsymbol{\phi})\, dt \\ &\quad - \int_{\Omega \times \{T_0\}} \mathbf{u}\boldsymbol{\phi}\, d\mathbf{x} + \int_{\Omega \times \{0\}} \mathbf{u}_0 \boldsymbol{\phi}\, d\mathbf{x} = 0 \end{aligned} \tag{1.4.27}$$

holds for all $\phi \in \mathscr{C}_0$. Here the solution space $\mathscr{C}$ and the test function space $\mathscr{C}_0$ are defined by

$$\mathscr{C} = \mathbf{L}^2((0, T_0); \mathbf{V}) \cap \mathbf{H}^1([0, T_0]; \mathbf{V}_s^*), \tag{1.4.28a}$$

and

$$\mathscr{C}_0 = \mathbf{C}([0, T_0]; \mathbf{V}_s) \cap \mathbf{H}^1([0, T_0]; \mathbf{L}^2(\Omega)). \tag{1.4.28b}$$

Moreover, the interpretation of the third integral is that of a functional; the regularity condition (1.4.28a) guarantees that $\mathbf{u} \in \mathbf{C}([0, T_0]; \mathbf{V}_s^*)$.

Remark 1.4.4. The case of compressible fluids requires a modification in the equations of motion to include a term of the form grad(div $\mathbf{u}$). In this case, the form $b(\cdot,\cdot,\cdot)$ of (1.4.11) may be perturbed by the term (div $\mathbf{v}$, div $\mathbf{w}$), which is insensitive to $\mathbf{u}$. Although our generalized format for the trilinear form can absorb this case, the reduction to (1.4.9b) is no longer valid and the continuity equation must be adjoined.

1.5 UNIQUENESS OF SOLUTIONS

We shall begin with a general uniqueness theorem for equations of the form

$$\frac{\partial \mathrm{T}H(u)}{\partial t} + u + \mathrm{T}g(u) + \mathrm{T}h(u) = 0 \qquad \text{almost everywhere in} \quad \Omega, \qquad 0 < t < T_0. \tag{1.5.1}$$

Definition 1.5.1. Let g be a monotone increasing locally Lipschitz continuous function, let h be a Lipschitz continuous function, and let H be strictly monotone and continuous, except possibly at zero, with $H(0-) = 0$. Let $\mathbf{G} = \mathbf{H}^1(\Omega)$ or $\mathbf{G} = \mathbf{H}_0^1(\Omega)$, respectively, and let $\mathrm{T}: \mathbf{G}^* \to \mathbf{G} \subset \mathbf{H}^1(\Omega)$ denote one of the operators N_σ, $\sigma > 0$, or D_0, introduced in Sections 1.1 and 1.2, respectively. The choice $\mathrm{T} = \mathrm{R}_\omega$ of Section 1.3 is also acceptable with $\mathbf{G} = \mathbf{H}^1(\Omega)$. By a solution of (1.5.1) is meant a pair $[u, H(u)]$ satisfying (1.5.1) almost everywhere in Ω for $0 < t < T_0$, where it is required that

$$u \in \mathbf{C}([0, T_0]; \mathbf{L}^2(\Omega)) \cap \mathbf{L}^2((0, T_0); \mathbf{G}), \tag{1.5.2a}$$

$$g(u), \mathrm{T}H(u) \in \mathbf{C}([0, T_0]; \mathbf{L}^2(\Omega)), \tag{1.5.2b}$$

$$H(u(\mathbf{x}, t)) \in [0, H(0+)] \quad \text{if} \quad u(\mathbf{x}, t) = 0, \qquad H(u) \in \mathbf{L}^2(\Omega), \tag{1.5.2c}$$

$$\frac{\partial}{\partial t}\mathrm{T}H(u) \in \mathbf{L}^2(\mathscr{D}_{T_0}), \tag{1.5.2d}$$

$$[H(t) - H(s)]/(t - s) \geqslant \lambda > 0 \qquad \text{if} \quad \|h\|_{\mathbf{Lip}} > 0. \tag{1.5.2e}$$

In (1.5.2e), $H(0)$ is understood as the set $[0, H(0+)]$.

Proposition 1.5.1. The solution pair of (1.5.1) is unique within the class described in Definition 1.5.1, provided $\mathrm{T}H(u)|_{t=0}$ is specified.

Proof: We recall that the operator T has the property of being pointwise nonnegative:

$$q \geqslant 0 \Rightarrow \mathrm{T}q \geqslant 0, \qquad q \in \mathbf{L}^2(\Omega). \tag{1.5.3}$$

Now let u, $H(u)$ and w, $H(w)$ be solution pairs satisfying $\mathrm{T}H(u)|_{t=0} = \mathrm{T}H(w)|_{t=0}$. Multiplying the relations (1.5.1) satisfied by u and w, respectively, by $H(u) - H(w)$, and subtracting, we find, after integration over Ω, that

$$\begin{aligned} \tfrac{1}{2}\frac{d}{dt}\|H(u) - H(w)\|_{\mathbf{G}^*}^2 &+ (u - w, H(u) - H(w))_{\mathbf{L}^2(\Omega)} \\ &+ (\mathrm{T}[g(u) - g(w)], H(u) - H(w))_{\mathbf{L}^2(\Omega)} \\ &+ (\mathrm{T}[h(u) - h(w)], H(u) - H(w))_{\mathbf{L}^2(\Omega)} = 0. \end{aligned} \tag{1.5.4}$$

Here, we have used the relation

$$\left(\frac{\partial}{\partial t}\mathrm{T}[H(u) - H(w)], H(u) - H(w)\right)_{\mathbf{L}^2(\Omega)} = \tfrac{1}{2}\frac{d}{dt}\|H(u) - H(w)\|_{\mathbf{G}^*}^2. \tag{1.5.5}$$

However, the second and third terms of (1.5.4) are nonnegative by the monotonicity of g and H, and by (1.5.3). By (1.5.2e), we have, thus,

$$\begin{aligned} \tfrac{1}{2}\|H(u) - H(w)\|_{\mathbf{G}^*}^2 &+ \lambda \operatorname{sgn}\|h\|_{\mathbf{Lip}} \int_0^t \|u - w\|_{\mathbf{L}^2(\Omega)}^2 \, d\tau \\ &\leqslant \int_0^t |(h(u) - h(w), \mathrm{T}[H(u) - H(w)])_{\mathbf{L}^2(\Omega)}| \, d\tau, \end{aligned} \tag{1.5.6}$$

after integration from $\tau = 0$ to $\tau = t$, for $0 < t < T_0$. By the Cauchy–Schwarz inequality in $\mathbf{G}^*$, the Lipschitz property of h, and the continuous injection of $\mathbf{L}^2(\Omega)$ into $\mathbf{G}$, we conclude that

$$\begin{aligned} |(h(u) - h(w), \mathrm{T}[H(u) - H(w)])_{\mathbf{L}^2(\Omega)}| &\leqslant \|h(u) - h(w)\|_{\mathbf{G}^*}\|H(u) - H(w)\|_{\mathbf{G}^*} \\ &\leqslant \tfrac{1}{2}\{\eta C\|u - w\|_{\mathbf{L}^2(\Omega)}^2 + \eta^{-1}\|H(u) - H(w)\|_{\mathbf{G}^*}^2\} \end{aligned} \tag{1.5.7}$$

holds for every $\eta > 0$, where C contains the explicit factor $\|h\|_{\mathbf{Lip}}$. If (1.5.7) is applied to (1.5.6), we obtain

$$\tfrac{1}{2}\|H(u) - H(w)\|_{\mathbf{G}^*}^2 + \tfrac{1}{2}\lambda\,\mathrm{sgn}\|h\|_{\mathbf{Lip}} \int_0^t \|u - w\|_{\mathbf{L}^2(\Omega)}^2\,d\tau$$

$$\leqslant \eta^{-1} \int_0^t \|H(u) - H(w)\|_{\mathbf{G}^*}^2\,d\tau \tag{1.5.8}$$

for $\eta C \leqslant \lambda$. By Gronwall's inequality applied to (1.5.8), we obtain

$$0 \leqslant \|H(u) - H(w)\|_{\mathbf{L}^\infty((0,T_0);\mathbf{G}^*)}^2 \leqslant 0, \tag{1.5.9}$$

so that $H(u) = H(w)$. By the invertibility of H, we conclude that $u = w$. ∎

Corollary 1.5.2. The solution u of (1.1.11) is unique within the class $\mathscr{C}$. In fact, if u and w are solutions of (1.1.11) in $\mathscr{C}$, then $u = w$ and $H(u) = H(w)$.

Proof: By Proposition 1.1.1, the initial condition (1.1.27) and the relation (1.1.25) hold for any (weak) solution of (1.1.11). The former is simply (1.5.1), with $\mathrm{T} = \mathrm{N}_\sigma$, any $\sigma > 0$, and h replaced by the Lipschitz function

$$h_\sigma(t) = h(t) - \sigma t. \tag{1.5.10}$$

The defining properties of $\mathscr{C}$, together with Proposition 1.1.1, show that such weak solutions are solutions of 1.5.1, in the sense of Definition 1.5.1. Note that (1.5.2e) holds via (1.1.8a). The result now follows from Proposition 1.5.1. ∎

Corollary 1.5.3. The solution u of (1.2.15) is unique within the class $\mathscr{C}$.

Proof: We use Proposition 1.2.1 in conjunction with Proposition 1.5.1, with $\mathrm{T} = \mathrm{D}_0$. Here $f = 0$ and (1.5.2e) is vacuous. ∎

Remark 1.5.1. The relations (1.1.25) and (1.2.21) hold without the condition $u_t \in \mathbf{L}^2(\mathscr{D}_{T_0})$, which may be replaced by $u \in \mathbf{C}([0, T_0];\mathbf{L}^2(\Omega))$. Thus, uniqueness holds over this broader class.

We pass now to uniqueness for reaction–diffusion systems.

Proposition 1.5.4. The solution of (1.3.11) is unique within the class $\mathscr{C}$.

Proof: Let $\mathbf{u}$ and $\mathbf{w}$ be solutions as described by Definition 1.3.2. Then $\mathbf{u}, \mathbf{w} \in \mathbf{C}([0, T_0];\ \mathbf{L}^\infty(\Omega))$, hence, have range in a bounded subset of $\mathbb{R}^m$. Thus, there is no loss of generality in assuming that f is Lipschitz, since it is assumed $\mathbf{C}^1$ and, hence, locally Lipschitz. Set $\mathbf{v}(\cdot, \tau) = \mathbf{u} - \mathbf{w}$ in (1.3.11).

For a fixed $i = 1, \ldots, m$, we have, after subtracting the two relations and integrating from $\tau = 0$ to $\tau = t$,

$$\tfrac{1}{2} \int_0^t \left\{ \frac{d}{dt} \|u_i - w_i\|^2_{\mathbf{L}^2(\Omega)} + D_i \| \nabla(u_i - w_i)\|^2_{\mathbf{L}^2(\Omega)} \right\} d\tau$$

$$+ D_i \int_0^t \|\sqrt{\omega_i}(u_i - w_i)\|^2_{\mathbf{L}^2(\partial\Omega)}\, d\tau = -\int_0^t (f_i(\mathbf{u}) - f_i(\mathbf{w}), u_i - w_i)_{\mathbf{L}^2(\Omega)}\, d\tau, \tag{1.5.11}$$

and (1.5.11) holds for all $0 < t < T_0$. If the Lipschitz property of f_i is used, and (1.5.11) is summed on i, we obtain, for each $0 < t < T_0$,

$$\tfrac{1}{2} \sum_{i=1}^{m} \|u_i - w_i\|^2_{\mathbf{L}^2(\Omega)} \leqslant C \int_0^t \sum_{i=1}^{m} \|u_i - w_i\|^2_{\mathbf{L}^2(\Omega)}\, d\tau. \tag{1.5.12}$$

Here we have used the pointwise Lipschitz estimate

$$|f_i(\mathbf{u}) - f_i(\mathbf{w})| \leqslant \frac{C}{m} |\mathbf{u} - \mathbf{w}| = \frac{C}{m} \left\{ \sum_{j=1}^{m} (u_j - w_j)^2 \right\}^{1/2}, \tag{1.5.13}$$

for some $C > 0$, and the nonnegativity of the second and third terms in (1.5.11). An application of Gronwall's inequality to (1.5.12) immediately yields $\sum_{i=1}^{m} \|u_i - w_i\|^2_{L^2(\Omega)} \equiv 0$ and, hence, $\mathbf{u} = \mathbf{w}$. ■

Remark 1.5.2. Uniqueness for the generalized model (1.4.27)–(1.4.28) and, indeed, for the Navier–Stokes equations, themselves, is a major open problem. For the latter, only the case $n = 2$ is satisfactorily understood. The reader is referred to the book of Ladyženskaja [11] for a detailed discussion. Roughly, the class within which global existence can be demonstrated is larger than the corresponding uniqueness class. Within this class, only local existence results can be demonstrated. This type of skewness of existence and uniqueness classes is not unusual in mathematical physics, and accounts for both parts of this book, that is, the local and global theory, are required for an understanding of this model.

1.6 BIBLIOGRAPHICAL REMARKS

Our format in (1.1.1) follows that of Wheeler [25], where the specific problem of thaw-bulb development, near a pipeline submerged in permafrost, is studied. This study was carried out prior to the construction of the trans-Alaska pipeline. Supporting physical discussion is given by Anderson

where

$$q_{0,\varepsilon} = \theta_0, \qquad q_{1,\varepsilon} = 2[3A - 2\varepsilon\theta_0 - \varepsilon H'(\varepsilon)]/\varepsilon^2,$$
$$q_{2,\varepsilon} = \{(H'(\varepsilon) - \theta_0)\varepsilon - 2[3A - 2\varepsilon\theta_0 - \varepsilon H'(\varepsilon)]\}/\varepsilon^3. \tag{2.1.6b}$$

Moreover, q_ε is *concave* on $[0, \varepsilon]$, that is, $q_{2,\varepsilon} \leqslant 0$. To see this, we argue as follows. Using the inequalities

$$\varepsilon \leqslant \varepsilon_* \leqslant \frac{2A}{6\theta_0 + 3\delta}, \qquad H'_\varepsilon(\varepsilon) \leqslant \theta_0 + 3\delta/2,$$

we conclude that

$$\varepsilon^2 q_{1,\varepsilon} - 4A \geqslant \varepsilon(6\theta_0 + 3\delta) - 4\varepsilon\theta_0 - 2\varepsilon(\theta_0 + 3\delta/2) \geqslant 0,$$

so that $\varepsilon q_{1,\varepsilon} \geqslant 4A/\varepsilon$. Thus,

$$\varepsilon^2 q_{2,\varepsilon} = [H'(\varepsilon) - \theta_0] - \varepsilon q_{1,\varepsilon} \leqslant \frac{3\delta}{2} - \frac{4A}{\varepsilon},$$

which, in conjunction with $\varepsilon \leqslant \varepsilon_* \leqslant 8A/(3\delta)$, yields

$$\varepsilon^2 q_{2,\varepsilon} \leqslant \frac{1}{\varepsilon}\left(\frac{3\delta\varepsilon}{2}\right) - \frac{1}{\varepsilon}(4A) \leqslant \frac{1}{\varepsilon}(4A - 4A) = 0.$$

This establishes the concavity of q_ε. In particular,

$$q_\varepsilon(t) \geqslant \min(\theta_0, H'(\varepsilon)) \geqslant \lambda > 0, \qquad 0 \leqslant t \leqslant \varepsilon, \tag{2.1.7}$$

where λ is the lower bound of H' assumed in Remark 1.1.1.

Proposition 2.1.1. H_ε and H'_ε converge uniformly to H and H', respectively, on compact subsets K of $\mathbb{R}^1\backslash\{0\}$. In fact, the estimate,

$$0 \leqslant H(t) - H_\varepsilon(t) \leqslant \mu\varepsilon \qquad \text{for} \quad t \geqslant \varepsilon \tag{2.1.8}$$

holds, where μ is given in (1.1.8a). Moreover, $H_\varepsilon(t)t \geqslant 0$ for $t \in \mathbb{R}^1$ and

$$\gamma/\varepsilon \geqslant H'_\varepsilon(t) \geqslant \lambda > 0, \qquad t \in \mathbb{R}^1, \tag{2.1.9}$$

for some positive constant γ not depending on ε. If $J = H^{-1}$ and $J_\varepsilon = H_\varepsilon^{-1}$, then J is a locally absolutely continuous, monotone increasing function, and

$$|J_\varepsilon(t) - J(t)| \leqslant (1 + \mu/\lambda)\varepsilon, \qquad t \in \mathbb{R}^1. \tag{2.1.10}$$

Proof: Since $H'_\varepsilon = H'$ and $H_\varepsilon = H$ on $(-\infty, 0)$ we may assume, for the convergence assertions concerning H'_ε and H_ε, that $K \subset (0, \infty)$. Now let ε

be given and set $t \geqslant \varepsilon$. We have

$$H(t) - H_\varepsilon(t) = \left[A + \int_0^t H'(s)\,ds\right] - \left[\int_0^\varepsilon q_\varepsilon(s)\,ds + \int_\varepsilon^t H'(s)\,ds\right]$$
$$= \int_0^\varepsilon H'(s)\,ds \leqslant \mu\varepsilon,$$

which verifies (2.1.8) and the assertion concerning K. The inequality $H'_\varepsilon \geqslant \lambda$ is immediate from (2.1.7), and the definition of H_ε, whereas $H'_\varepsilon = O(1/\varepsilon)$ follows from (2.1.6). That $H_\varepsilon(t)$ has the sign of t is immediate from (2.1.5). In order to verify (2.1.10), first note that $J(t) = 0$, $0 \leqslant t \leqslant A$, $J' = 1/(H' \circ J)$ on $\mathbb{R}^1 \backslash [0, A]$, hence, J is locally absolutely continuous. Also, (2.1.4) and (2.1.5) imply that $H_\varepsilon(\varepsilon) = A$, so that $J_\varepsilon(A) = \varepsilon$. Now $J(t) = J_\varepsilon(t)$ for $t < 0$ and, for $0 \leqslant t \leqslant A$, we have

$$J_\varepsilon(t) - J(t) = J_\varepsilon(t) \leqslant J_\varepsilon(A) = \varepsilon. \tag{2.1.11}$$

For $t > A$, there are two cases, depending upon whether $J(t) < \varepsilon$ or $J(t) \geqslant \varepsilon$. In the former case, we have

$$t < H(\varepsilon) = A + \int_0^\varepsilon H'(s)\,ds \leqslant A + \int_0^\varepsilon \mu\,ds = A + \varepsilon\mu,$$

so that $t - A \leqslant \mu\varepsilon$. This gives

$$0 \leqslant J_\varepsilon(t) - J_\varepsilon(A) = \int_A^t J'_\varepsilon(\tau)\,d\tau \leqslant \int_A^t (1/\lambda)\,d\tau = (t - A)/\lambda \leqslant \mu\varepsilon/\lambda,$$

so that

$$0 \leqslant J_\varepsilon(t) - J(t) \leqslant \mu\varepsilon/\lambda + \varepsilon - J(t) \leqslant \varepsilon(1 + \mu/\lambda), \tag{2.1.12}$$

if $t > A$ and $J(t) < \varepsilon$. For $J(t) \geqslant \varepsilon$, select s such that $J_\varepsilon(s) = J(t)$. We have, by the mean-value theorem,

$$J_\varepsilon(t) - J(t) = J_\varepsilon(t) - J_\varepsilon(s) = J'_\varepsilon(\xi)(t - s),$$

for ξ on the open interval determined by t and s. However,

$$|s - t| = |H_\varepsilon \circ J(t) - H \circ J(t)| \leqslant \mu\varepsilon,$$

by (2.1.8). Combining the two previous relations gives

$$|J_\varepsilon(t) - J(t)| \leqslant \mu\varepsilon/\lambda, \tag{2.1.13}$$

for $t > A$ and $J(t) \geqslant \varepsilon$. The relation (2.1.10) is now immediate from (2.1.11), (2.1.12), and (2.1.13). ■

Remark 2.1.2. It is natural to use the approximation H_ε to define regularized parabolic initial value problems with homogeneous Neumann boundary conditions. The next proposition describes the rate of convergence of the solutions u^ε of these regularized problems to the solution satisfying (1.1.11). However, we find it more economical to use the equivalent formulation, described by Proposition 1.1.1, in formulating these results. Recall that Proposition 1.5.1 guarantees uniqueness over the class

$$\mathscr{C}_1 = \mathbf{C}([0, T_0]; \mathbf{L}^2(\Omega)) \cap \mathbf{L}^2((0, T_0); \mathbf{H}^1(\Omega)).$$

Proposition 2.1.2. Let $\mathrm{T} = \mathrm{N}_\sigma$, $\sigma > 0$, be the Neumann inversion operator defined in Definition 1.1.2 and, if $\mathscr{C}_1$ denotes the class defined above, suppose that u, $u^\varepsilon \in \mathscr{C}_1$, respectively, satisfy (1.1.25) and

$$\frac{\partial \mathrm{T} H_\varepsilon(u^\varepsilon)}{\partial t} + u^\varepsilon + \mathrm{T} f_\sigma(u^\varepsilon) = 0 \quad \text{almost everywhere in } \Omega, \quad 0 < t < T_0, \tag{2.1.14}$$

for $0 < \varepsilon < \varepsilon_*$, where it is explicitly assumed that f_σ is defined by (1.1.26), and is Lipschitz continuous. If $\mathrm{T}H_\varepsilon(u^\varepsilon)|_{t=0} = \mathrm{T}H(u)|_{t=0}$, then the estimates

$$\|H_\varepsilon(u^\varepsilon) - H(u)\|_{\mathbf{L}^\infty((0,T_0);\mathbf{F})} \leqslant C\varepsilon^{1/2}, \tag{2.1.15a}$$

$$\|u^\varepsilon - u\|_{\mathbf{L}^2(\mathscr{D}_{T_0})} \leqslant C\varepsilon^{1/2} \tag{2.1.15b}$$

hold for some positive constant C, independent of ε, where $\mathbf{F} = [\mathbf{H}^1(\Omega)]^*$. The constant C depends upon $|\Omega|$ and T_0 explicitly.

Proof: Set $v = H(u)$ and $v^\varepsilon = H_\varepsilon(u^\varepsilon)$. After subtraction of (1.1.25) from (2.1.14), followed by multiplication by $(v^\varepsilon - v)$, and integration over Ω, we have

$$\begin{aligned}\tfrac{1}{2}\frac{d}{dt}\|v^\varepsilon - v\|_{\mathbf{F}}^2 &+ (J_\varepsilon(v^\varepsilon) - J_\varepsilon(v), v^\varepsilon - v)_{\mathbf{L}^2(\Omega)} \\ &= (J(v) - J_\varepsilon(v), v^\varepsilon - v)_{\mathbf{L}^2(\Omega)} \\ &\quad - (\mathrm{T}[f_\sigma \circ J_\varepsilon(v^\varepsilon) - f_\sigma \circ J(v)], v^\varepsilon - v)_{\mathbf{L}^2(\Omega)},\end{aligned} \tag{2.1.16}$$

where we have subtracted a term involving $J_\varepsilon(v)$ from both sides of (2.1.15), and have used (1.5.5). The second term on the right-hand side of (2.1.16) may be estimated in analogy with (1.5.7):

$$\begin{aligned}&|(f_\sigma \circ J_\varepsilon(v^\varepsilon) - f_\sigma \circ J(v), T(v^\varepsilon - v))_{\mathbf{L}^2(\Omega)}| \\ &\quad\leqslant \tfrac{1}{2}C[\eta\|J_\varepsilon(v^\varepsilon) - J(v)\|_{\mathbf{L}^2(\Omega)}^2 + \eta^{-1}\|v^\varepsilon - v\|_{\mathbf{F}}^2] \\ &\quad\leqslant C[\eta\|J_\varepsilon(v^\varepsilon) - J_\varepsilon(v)\|_{\mathbf{L}^2(\Omega)}^2 + \eta\|J_\varepsilon(v) - J(v)\|_{\mathbf{L}^2(\Omega)}^2 + \tfrac{1}{2}\eta^{-1}\|v^\varepsilon - v\|_{\mathbf{F}}^2],\end{aligned} \tag{2.1.17}$$

where η is an arbitrary positive constant, and $C = C_\sigma$. The first term on the right-hand side of (2.1.16) is estimated, in the standard way, by

$$|(J(v) - J_\varepsilon(v), v^\varepsilon - v)_{\mathbf{L}^2(\Omega)}| \leqslant \tfrac{1}{2}(\eta\|v^\varepsilon - v\|_{\mathbf{L}^2(\Omega)}^2 + \eta^{-1}\|J_\varepsilon(v) - J(v)\|_{\mathbf{L}^2(\Omega)}^2). \tag{2.1.18}$$

The key terms in these two inequalities are the first terms on the right, which we shall see can be absorbed on the left-hand side of (2.1.16), with appropriate choices of η. Indeed, the inequalities

$$(J_\varepsilon(v^\varepsilon) - J_\varepsilon(v), J_\varepsilon(v^\varepsilon) - J_\varepsilon(v))_{\mathbf{L}^2(\Omega)} \leqslant \frac{1}{\lambda}(J_\varepsilon(v^\varepsilon) - J_\varepsilon(v), v^\varepsilon - v)_{\mathbf{L}^2(\Omega)}, \quad (2.1.19a)$$

$$\begin{aligned}(v^\varepsilon - v, v^\varepsilon - v)_{\mathbf{L}^2(\Omega)} &= (H_\varepsilon \circ J_\varepsilon(v^\varepsilon) - H_\varepsilon \circ J_\varepsilon(v), v^\varepsilon - v)_{\mathbf{L}^2(\Omega)} \\ &\leqslant \frac{\gamma}{\varepsilon}(J_\varepsilon(v^\varepsilon) - J_\varepsilon(v), v^\varepsilon - v)_{\mathbf{L}^2(\Omega)}, \quad (2.1.19b)\end{aligned}$$

which follow from $J'_\varepsilon \leqslant 1/\lambda$ and $H'_\varepsilon \leqslant \gamma/\varepsilon$, respectively, lead to the reformulation of (2.1.16) as

$$\begin{aligned}&\tfrac{1}{2}\frac{d}{dt}\|v^\varepsilon - v\|_{\mathbf{F}}^2 + \tfrac{1}{2}(J_\varepsilon(v_\varepsilon) - J_\varepsilon(v), v^\varepsilon - v)_{\mathbf{L}^2(\Omega)} \\ &\quad \leqslant \left(\frac{\lambda}{4} + \frac{\gamma}{\varepsilon}\right)\|J_\varepsilon(v) - J(v)\|_{\mathbf{L}^2(\Omega)}^2 + \frac{2C^2}{\lambda}\|v^\varepsilon - v\|_{\mathbf{F}}^2, \quad (2.1.20)\end{aligned}$$

provided the choices $\eta = \lambda/(4C)$ and $\eta = \varepsilon/(2\gamma)$ are made in (2.1.17) and (2.1.18), respectively. If (2.1.10) is applied to (2.1.20), we obtain the inequality

$$\begin{aligned}&\tfrac{1}{2}\frac{d}{dt}\|v^\varepsilon - v\|_{\mathbf{F}}^2 + \tfrac{1}{2}(J_\varepsilon(v_\varepsilon) - J_\varepsilon(v), v^\varepsilon - v)_{\mathbf{L}^2(\Omega)} \\ &\quad \leqslant \frac{\lambda}{4}|\Omega|\left(1 + \frac{\mu}{\lambda}\right)^2 \varepsilon^2 + \gamma|\Omega|\left(1 + \frac{\mu}{\lambda}\right)^2 \varepsilon + \frac{2C^2}{\lambda}\|v^\varepsilon - v\|_{\mathbf{F}}^2. \quad (2.1.21)\end{aligned}$$

If (2.1.21) is integrated from $\tau = 0$ to $\tau = t$, one obtains, for obvious choices of C_1 and C_2, containing the factor $|\Omega| T_0$,

$$\begin{aligned}&\|v^\varepsilon - v\|_{\mathbf{F}}^2 + \int_0^t (J_\varepsilon(v_\varepsilon) - J_\varepsilon(v), v^\varepsilon - v)_{\mathbf{L}^2(\Omega)}\, d\tau \\ &\quad \leqslant C_1\varepsilon + C_2 \int_0^t \|v^\varepsilon - v\|_{\mathbf{F}}^2\, d\tau, \quad (2.1.22)\end{aligned}$$

for $0 < t < T_0$. By Gronwall's inequality applied to (2.1.22), we have

$$\begin{aligned}&\|v^\varepsilon - v\|_{L^\infty((0,T_0);\mathbf{F})}^2 + \int_0^{T_0} (J_\varepsilon(v^\varepsilon) - J_\varepsilon(v), v^\varepsilon - v)_{\mathbf{L}^2(\Omega)}\, d\tau \\ &\quad \leqslant C_1 \varepsilon e^{C_2 T_0}, \quad (2.1.23)\end{aligned}$$

which implies (2.1.15a). To obtain (2.1.15b), we use the triangle inequality, in conjunction with (2.1.10) and (2.1.19a), that is,

$$\begin{aligned}
&\|J_\varepsilon(v^\varepsilon) - J(v)\|^2_{\mathbf{L}^2(\mathscr{D})} \\
&\quad \leqslant 2\|J_\varepsilon(v^\varepsilon) - J_\varepsilon(v)\|^2_{\mathbf{L}^2(\mathscr{D})} + 2\|J_\varepsilon(v) - J(v)\|^2_{\mathbf{L}^2(\mathscr{D})} \\
&\quad \leqslant (2/\lambda)\int_0^{T_0} (J_\varepsilon(v^\varepsilon) - J_\varepsilon(v), v^\varepsilon - v)_{\mathbf{L}^2(\Omega)}\, d\tau + 2|\Omega| T_0 \left(1 + \frac{\mu}{\lambda}\right)^2 \varepsilon^2. \qquad (2.1.24)
\end{aligned}$$

Inequality (2.1.15b) follows from (2.1.23) and (2.1.24). ■

Remark 2.1.3. The smoothing introduced in this section attains its full power when $u_0 \notin \mathbf{H}^1(\Omega)$, in which case, the property $u_t \in \mathbf{L}^2(\mathscr{D})$ must be discarded. More precisely, the smoothing does not preserve $\mathbf{H}^1$ initial data, with the result that we do not expect the regularity condition $u_t^\varepsilon \in \mathbf{L}^2(\mathscr{D})$ to hold. The regularization introduced here, then, more properly represents a transferral of singularities. Note that the lifting formalism employed here substitutes for $u_t \in \mathbf{L}^2(\mathscr{D})$ the weaker property $(\mathrm{T}H(u))_t \in \mathbf{L}^2(\mathscr{D})$.

There is, of course, an obvious way to preserve $\mathbf{H}^1$ initial data, namely, to require $u_0^\varepsilon = u_0$, rather than $u_0^\varepsilon = J_\varepsilon \circ H(u_0)$ (see the initial condition in Proposition 2.1.2). Although this represents a more legitimate regularization, since $u^\varepsilon \in \mathscr{C}$ in this case, it introduces the additional term $\|H_\varepsilon(u_0) - H(u_0)\|^2_{\mathbf{F}}$ on the right-hand side of (2.1.22). As noted by Rose and the author [9], this term is of order ε, thus preserving the convergence rates of (2.1.15), provided the nondecreasing rearrangement Λ_{u_0} of u_0 satisfies on $\tilde{\Omega}$

$$\Lambda_{u_0}(t) \geqslant |\tilde{\Omega}| - c\varepsilon, \qquad 0 < t \leqslant \varepsilon,$$

for sufficiently small ε. Equivalently,

$$\|H_\varepsilon(u_0) - H(u_0)\|^2_{\mathbf{F}} \leqslant C\varepsilon, \qquad (2.1.25a)$$

provided on the set $\tilde{\Omega}$, where $u_0 \geqslant 0$,

$$\{\mathbf{x} : u_0(\mathbf{x}) \leqslant \varepsilon\} \leqslant c\varepsilon. \qquad (2.1.25b)$$

The condition (2.1.25b) already implies a certain degree of regularity, however, on the free boundary. Moreover, as stressed in the previous chapter, it is $H(u_0)$ which appears more physically significant, and it is this quantity, perhaps, which should dictate the initial datum for the regularization.

An existence theory for the initial/boundary-value problem corresponding to (2.1.14), with $\mathbf{L}^2$ initial datum, is presented in Ladyženskaja, Solonnikov, and Ural'ceva ([10] pp. 465–475). A change of variable $v_\varepsilon = H_\varepsilon(u^\varepsilon)$ is a necessary preliminary step to invoke the theory.

2.2 SEMIDISCRETE REGULARIZATION AND MAXIMUM PRINCIPLES IN THE STEFAN PROBLEM

We begin with discrete inequalities of Gronwall type, which will be used repeatedly in the sequel.

Proposition 2.2.1. Let $a = t_0 < t_1 < \cdots < t_M = b$ be a partition of $[a, b]$, and suppose that ϕ and ψ are nonnegative step functions, with values ϕ_k and ψ_k, respectively, on the intervals $[t_{k-1}, t_k)$, $k = 1, \ldots, M$. Suppose that, for some fixed $p \geqslant 1$, there exists $\sigma \geqslant 0$, such that

$$\phi_k^p \leqslant \sigma^p + \int_a^{t_{k-1}} \psi\phi \, dt = \sigma^p + \sum_{m=1}^{k-1} \psi_m \phi_m (t_m - t_{m-1}) \tag{2.2.1}$$

holds for each $k = 1, \ldots, M$. Then the following bounds are valid:

$$\phi^{p-1}(t) \leqslant \sigma^{p-1} + \left(1 - \frac{1}{p}\right) \int_a^t \psi(\tau)\, d\tau, \qquad a < t < b \quad (p > 1); \tag{2.2.2a}$$

$$\phi(t) \leqslant \sigma \exp\left(\int_a^t \psi(\tau)\, d\tau\right), \qquad a < t < b \quad (p = 1). \tag{2.2.2b}$$

In particular,

$$\phi_k^{p-1} \leqslant \sigma^{p-1} + \left(1 - \frac{1}{p}\right) \sum_{m=1}^{k-1} \psi_m (t_m - t_{m-1}), \qquad k = 1, \ldots, M, \tag{2.2.3a}$$

for $p > 1$, and

$$\phi_k \leqslant \sigma \exp\left\{\sum_{m=1}^{k-1} \psi_m (t_m - t_{m-1})\right\}, \qquad k = 1, \ldots, M, \tag{2.2.3b}$$

for $p = 1$.

Proof: Set

$$\omega(t) = \sigma^p + \int_a^t \psi(\tau)\phi(\tau)\, d\tau, \qquad a < t < b. \tag{2.2.4}$$

If $\omega \equiv 0$ on $[a, b]$, then $\phi \equiv 0$ by (2.2.1) and, thus, (2.2.2) clearly holds. Suppose then that $\omega(t) > 0$ for $t > t_i$, $0 \leqslant i < M$. By a change of variable $t' = t - t_i$, we may assume, without loss of generality, that $i = 0$, that is, $\omega(t) > 0$ for $a < t \leqslant b$. Note that ω is an absolutely continuous function of t, and its derivative exists and is equal to $\psi(t)\phi(t)$ at each point of continuity of $\psi\phi$, i.e., for $t \in (t_{k-1}, t_k)$, $k = 1, \ldots, M$. Thus, by (2.2.1),

$$\omega' = \psi\phi \leqslant \psi\omega^{1/p}, \tag{2.2.5}$$

and we obtain on (a, b) the formal differential inequality

$$\omega^{-1/p}\, d\omega \leqslant \psi\, dt. \tag{2.2.6}$$

Integration of (2.2.6) yields, for $p > 1$,

$$\omega^{1-(1/p)}(t) \leqslant \sigma^{p-1} + \left(1 - \frac{1}{p}\right)\int_a^t \psi(\tau)\, d\tau, \qquad a < t < b, \tag{2.2.7}$$

since $\omega(a) = \sigma^p$. Since $\phi^p \leqslant \omega$, we immediately obtain (2.2.2a) from (2.2.7). If $p = 1$, we apply a standard integrating factor $\exp\{-\int_a^t \psi(\tau)\, d\tau\}$ to (2.2.5), and obtain

$$\omega(t) \leqslant \sigma \exp\left(\int_a^t \psi(\tau)\, d\tau\right), \qquad a < t < b, \tag{2.2.8}$$

since $\omega(a) = \sigma$. Since $\phi \leqslant \omega$, we immediately obtain (2.2.2b) from (2.2.8). The relations (2.2.3) are immediate from (2.2.2), if we let $t \downarrow t_{k-1}$. ■

Remark 2.2.1. In the sequel, inequalities slightly different from (2.2.1) will also naturally arise, especially when implicit and explicit semidiscrete methods are mixed. Thus, for example, it will happen frequently that the inequality

$$\phi_k^p \leqslant \sigma^p + \sum_{m=1}^{k} \psi_m \phi_{m-1}(t_m - t_{m-1}), \qquad k = 1, \ldots, M \tag{2.2.9}$$

can be derived. If the new partition

$$\begin{gathered} 0 = t'_0 < t'_1 < \cdots < t'_M = T_0, \\ t'_i = t'_{i-1} + (t_{i+1} - t_i), \qquad i = 1, \ldots, M-1, \end{gathered} \tag{2.2.10}$$

is introduced, then (2.2.9) may be rewritten as

$$\phi_k^p \leqslant \sigma^p + \psi_1\phi_0(t_1 - t_0) + \sum_{l=1}^{k-1} \psi_{l+1}\phi_l(t'_l - t'_{l-1}),$$

and (2.2.3) assumes the form

$$\begin{aligned} \phi_k^{p-1} \leqslant{} & [\sigma^p + \psi_1\phi_0(t_1 - t_0)]^{(p-1)/p} \\ & + \left(1 - \frac{1}{p}\right)\sum_{l=1}^{k-1} \psi_{l+1}(t_{l+1} - t_l), \qquad k = 1, \ldots, M, \end{aligned} \tag{2.2.11a}$$

for $p > 1$, and

$$\phi_k \leqslant [\sigma + \psi_1\phi_0(t_1 - t_0)] \exp\left\{\sum_{l=1}^{k-1} \psi_{l+1}(t_{l+1} - t_l)\right\}, \qquad k = 1, \ldots, M, \tag{2.2.11b}$$

for $p = 1$. We summarize this result as

Corollary 2.2.2. If (2.2.9) holds for a given partition $\mathscr{P}$ of $[0, T_0]$ and some p, $1 \leqslant p < \infty$, then (2.2.11) is valid.

Remark 2.2.2. A little reflection shows that the preceding results are sharply formulated for $p = 1$, but not so for $p > 1$, in the following sense. It is possible to weaken hypothesis (2.2.1) to permit integration up to t_k, at the cost of slightly weakening conclusion (2.2.3a), with respect to the constant multiplier and, of course, subsequent summation to k in the bound. We illustrate this for the case $p = 2$. Suppose, then, that

$$\phi_k^2 \leqslant \sigma^2 + \sum_{m=1}^{k} \psi_m\phi_m(t_m - t_{m-1}), \qquad k = 1, \ldots, M. \tag{2.2.12a}$$

Then

$$\phi_k \leqslant \sigma + \sum_{m=1}^{k} \psi_m(t_m - t_{m-1}), \qquad k = 1, \ldots, M. \tag{2.2.12b}$$

This implication is a corollary of the following.

Proposition 2.2.3. Suppose $\{a_k\}$ and $\{b_k\}$ are two sequences of nonnegative numbers, such that

$$\tfrac{1}{2}a_k^2 \leqslant \tfrac{1}{2}a_0^2 + \sum_{j=1}^{k} a_j b_j \qquad \text{for all} \quad k \geqslant 0. \tag{2.2.13a}$$

Then

$$a_k \leqslant a_0 + 2\sum_{j=1}^{k} b_j \qquad \text{for all} \quad k \geqslant 0. \tag{2.2.13b}$$

Proof: Define the nonnegative sequence $\{\alpha_k\}$ by

$$\tfrac{1}{2}\alpha_k^2 = \tfrac{1}{2}a_0^2 + \sum_{j=1}^{k} \alpha_j b_j \qquad \text{for all} \quad k \geqslant 0, \tag{2.2.14}$$

and note that

$$a_k \leqslant \alpha_k, \qquad k \geqslant 0. \tag{2.2.15}$$

We deduce from (2.2.14) the quadratic recursion,

$$0 = \alpha_k^2 - 2\alpha_k b_k - \alpha_{k-1}^2,$$

so that

$$0 \leqslant \alpha_k = b_k + (b_k^2 + \alpha_{k-1}^2)^{1/2} \leqslant 2b_k + \alpha_{k-1}. \tag{2.2.16}$$

We obtain from the linear recursion (2.2.16),

$$\alpha_k \leqslant a_0 + 2 \sum_{j=1}^{k} b_j, \qquad k \geqslant 0,$$

from which (2.2.13b) follows via (2.2.15). ■

Remark 2.2.3. In the following chapter, we shall define a variety of semi-discrete approximations in time. However, in this chapter, we shall restrict attention to the fully-implicit, or backward-Euler method, combined with an explicit method, as applied to the Lipschitz part of f. The justification for singling out this special semidiscrete method at this time is that it readily permits estimates on the $\mathbf{L}^p$ norm of the approximate solutions at discrete time levels, in terms of the $\mathbf{L}^p$ norm of the initial datum. These estimates may be translated into $\mathbf{L}^\infty((0, T_0); \mathbf{L}^p(\Omega))$ estimates for the solution of the evolution equation in terms of $\mathbf{L}^p(\Omega)$ estimates of the initial datum. Our primary interest is in the cases $p = 2$ and $p = \infty$.

Definition 2.2.1. Let $0 = t_0^N < t_1^N < \cdots < t_{M(N)}^N = T_0$ denote an arbitrary sequence of partitions $\mathscr{P}^N$ of $[0, T_0]$. We introduce the semidiscrete version of (1.1.9), based on the backward-Euler method, applied on $\{\mathscr{P}^N\}$, as a recursively generated finite sequence $\{u_k^N, H(u_k^N)\}_{k=1}^{M(N)}$, for each $N = 1, 2, \ldots,$ satisfying

$$H(u_k^N(x)) \in [0, A] \qquad \text{if} \quad u_k^N(x) = 0, \tag{2.2.17a}$$

$$H(u_k^N) \in \mathbf{L}^2(\Omega), \tag{2.2.17b}$$

$$u_k^N \in \mathbf{H}^1(\Omega), \tag{2.2.17c}$$

and the variational condition

$$\begin{aligned}(t_k^N - t_{k-1}^N)^{-1}([H(u_k^N) - H(u_{k-1}^N)], v)_{\mathbf{L}^2(\Omega)} + (\nabla u_k^N, \nabla v)_{\mathbf{L}^2(\Omega)} \\ + (g(u_k^N) + h(u_{k-1}^N), v)_{\mathbf{L}^2(\Omega)} = 0 \qquad \text{for all} \quad v \in \mathbf{H}^1(\Omega),\end{aligned} \tag{2.2.18}$$

for $k = 1, \ldots, M(N)$. We require that

$$H(u_0^N) = H(u_0), \tag{2.2.19}$$

and that

$$f(u_k^N) \in \mathbf{L}^2(\Omega), \tag{2.2.20}$$

for each N and $1 \leqslant k \leqslant M(N)$. By the semidiscrete regularization of (2.2.17)–(2.2.20) is meant a recursively generated finite sequence $\{u_k^{\varepsilon,N}\}_{k=1}^{M(N)}$, for each $N = 1, 2, \ldots$, satisfying

$$u_k^{\varepsilon,N} \in \mathbf{H}^1(\Omega), \tag{2.2.21}$$

and the variational condition

$$\begin{aligned}(t_k^N - t_{k-1}^N)^{-1}([H_\varepsilon(u_k^{\varepsilon,N}) - H_\varepsilon(u_{k-1}^{\varepsilon,N})], v)_{\mathbf{L}^2(\Omega)} + (\nabla u_k^{\varepsilon,N}, \nabla v)_{\mathbf{L}^2(\Omega)} \\ + (g(u_k^{\varepsilon,N}) + h(u_{k-1}^{\varepsilon,N}), v)_{\mathbf{L}^2(\Omega)} = 0 \qquad \text{for all} \quad v \in \mathbf{H}^1(\Omega),\end{aligned} \tag{2.2.22}$$

for $k = 1, \ldots, M(N)$. As above, we require that

$$H_\varepsilon(u_0^{\varepsilon,N}) = H(u_0), \tag{2.2.23}$$

and

$$f(u_k^{\varepsilon,N}) \in \mathbf{L}^2(\Omega), \tag{2.2.24}$$

for each N and $1 \leqslant k \leqslant M(N)$.

Remark 2.2.4. Note that we have used the decomposition $f = g + h$ (see (1.1.8d)); current (implicit) values are used in the evaluation of g, whereas previous values are used for h. The existence and uniqueness of solutions are guaranteed by Proposition 3.3.1. We note this explicitly in Proposition 2.2.5. In the special case, the domain Ω possesses the property that

$$\mathbf{N}_\sigma \mathbf{L}^p(\Omega) \subset \mathbf{W}^{2,p}(\Omega), \qquad 1 \leqslant p < \infty, \tag{2.2.25}$$

then the solutions $\{u_k^N\}$ of (2.2.18) are progressively smoother, even if $H(u_0) \in \mathbf{L}^1(\Omega)$ is the assumption on the initial datum $H(u_0)$. In fact, an estimate can be given for k_0, such that $u_k^N \in \mathbf{L}^\infty(\Omega)$, $k \geqslant k_0$ (all N), via Sobolev's inequality. If we define the (finite) sequence

$$p_i = \frac{np_{i-1}}{n - 2p_{i-1}}, \qquad n - 2p_{i-1} > 0, \quad p_0 = 1, \tag{2.2.26}$$

for $i \geqslant 1$, then the sequence terminates at a value i_0, for which $n - 2p_{i_0} \leqslant 0$. Set

$$k_0 = \begin{cases} i_0, & n - 2p_{i_0} < 0, \\ i_0 + 1, & n - 2p_{i_0} = 0, \end{cases} \tag{2.2.27}$$

and in the latter case let $p_{k_0} > p_{i_0}$ be arbitrary. By Sobolev's inequality,

$$u_k^N \in \mathbf{W}^{2,p_{k_0}}(\Omega) \subset \mathbf{L}^\infty(\Omega), \qquad k \geqslant k_0. \tag{2.2.28}$$

We shall not attempt to exploit this fact to obtain the analog for evolution equations, which amounts to instantaneous smoothing, for $t > 0$, of $\mathbf{L}^1(\Omega)$ initial data. Note that modifications apply if (2.2.25) fails for $p = 1$.

Prior to the statement and proof of the next proposition, we shall introduce the truncation operators which provide a way of avoiding the hypothesis (2.2.25).

Definition 2.2.2. We define a sequence of bounded Lipschitz continuous, cutoff functions of the identity by

$$\theta_j(t) = \begin{cases} t, & |t| \leqslant j, \\ (\operatorname{sgn} t)j, & |t| > j, \end{cases} \tag{2.2.29a}$$

and the corresponding truncation operators $\Theta_j : \mathbf{L}^1(\Omega) \to \mathbf{L}^\infty(\Omega)$ by

$$\Theta_j z = \theta_j \circ z, \qquad z \in \mathbf{L}^1(\Omega). \tag{2.2.29b}$$

The "product" truncations $\Theta_{j,q} : \mathbf{L}^1(\Omega) \to \mathbf{L}^\infty(\Omega)$ are then given by

$$\Theta_{j,q} z = (\theta_j \circ z)^q, \qquad z \in \mathbf{L}^1(\Omega), \quad q \geqslant 1, \tag{2.2.30}$$

when this exponentiation is well defined. Here j is a positive integer.

Remark 2.2.5. It is readily seen that $\Theta_{j,q}$ maps $\mathbf{H}^1(\Omega)$ into itself. This "ring" property does not hold for $\mathbf{H}^1(\Omega)$ functions themselves except for $n = 1$.

Proposition 2.2.4. Suppose that the initial datum satisfies $H(u_0) \in \mathbf{L}^p(\Omega)$ for $2 \leqslant p \leqslant \infty$, and suppose that $\{u_k^{\varepsilon,N}\}$ is a semidiscrete regularization as in Definition 2.2.1. Then

$$\begin{aligned} &\|H_\varepsilon(u_k^{\varepsilon,N})\|_{\mathbf{L}^p(\Omega)} \\ &\quad \leqslant [|h(0)|\,|\Omega|^{1/p} T_0 + (1 + T_0\|h\|_{\mathbf{Lip}}/\lambda)\|H(u_0)\|_{\mathbf{L}^p(\Omega)}] \exp\{T_0\|h\|_{\mathbf{Lip}}/\lambda\}, \end{aligned} \tag{2.2.31}$$

$$\begin{aligned} &\|u_k^{\varepsilon,N}\|_{\mathbf{L}^p(\Omega)} \\ &\quad \leqslant \left(\frac{1}{\lambda}\right) [|h(0)|\,|\Omega|^{1/p} T_0 + (1 + T_0\|h\|_{\mathbf{Lip}}/\lambda)\|H(u_0)\|_{\mathbf{L}^p(\Omega)}] \exp\{T_0\|h\|_{\mathbf{Lip}}/\lambda\}, \end{aligned} \tag{2.2.32}$$

for all $N \geqslant 1$ and $k = 1, \ldots, M(N)$.

Proof: We shall obtain the estimates (2.2.31)–(2.2.32) for the cases $q = 2l/(2m - 1) \geqslant 2$, where l and m are positive integers. Since such rationals are dense in $[2, \infty)$ via tertiary expansions, for example, the result for general p follows by letting $q \uparrow p$. The essential property of such q, which is used in this proof, is the fact that

$$\theta_j^{q-2}(t) \geqslant 0, \qquad \theta_j^{q-1}(t)t \geqslant 0, \qquad t \in \mathbb{R}^1. \tag{2.2.33}$$

Now, set $v = \Theta_{j,q-1}(H_\varepsilon(u_k^{\varepsilon,N}))$ in (2.2.18), and assume, inductively, that $u_i^{\varepsilon,N} \in \mathbf{L}^q(\Omega)$ for $i < k$. Note that

$$u_0^{\varepsilon,N} := J_\varepsilon \circ H_\varepsilon(u_0^{\varepsilon,N}) = J_\varepsilon(H(u_0)) \tag{2.2.34}$$

has $\mathbf{L}^q$ norm $\leqslant (1/\lambda)\|H(u_0)\|_{\mathbf{L}^q(\Omega)}$.

Since g and $\theta_j^{q-1} \circ H_\varepsilon$ have the same sign as t, then

$$(g(u_k^{\varepsilon,N}), \Theta_{j,q-1}(H_\varepsilon(u_k^{\varepsilon,N})))_{\mathbf{L}^2(\Omega)} \geqslant 0, \tag{2.2.35}$$

and since

$$\nabla\Theta_{j,q-1}(H_\varepsilon(u_k^{\varepsilon,N})) = \begin{cases} [(q-1)\theta_j^{q-2} \circ H_\varepsilon(u_k^{\varepsilon,N})]H_\varepsilon'(u_k^{\varepsilon,N})\,\nabla u_k^{\varepsilon,N}, & |H_\varepsilon(u_k^{\varepsilon,N})| \leqslant j, \\ 0, \quad |H_\varepsilon(u_k^{\varepsilon,N})| > j, \end{cases} \tag{2.2.36a}$$

then, by the monotonicity of H_ε, and by (2.2.33),

$$(\nabla u_k^{\varepsilon,N}, \nabla\Theta_{j,q-1}(H_\varepsilon(u_k^{\varepsilon,N})))_{\mathbf{L}^2(\Omega)} \geqslant 0. \tag{2.2.36b}$$

Hence, we obtain from (2.2.18), after using the fact that the identity dominates θ_j, in the sense of absolute values,

$$\int_\Omega |\theta_j \circ H_\varepsilon(u_k^{\varepsilon,N})|^q \leqslant \frac{1}{r}\int_\Omega |\theta_j \circ H_\varepsilon(u_k^{\varepsilon,N})|^q + \frac{1}{q}\int_\Omega |H_\varepsilon(u_{k-1}^{\varepsilon,N}) - h(u_{k-1}^{\varepsilon,N})(t_k^N - t_{k-1}^N)|^q \tag{2.2.37}$$

where we have used the inequality

$$ab \leqslant \frac{1}{q}a^q + \frac{1}{r}b^r, \qquad a \geqslant 0, \quad b \geqslant 0, \quad \frac{1}{q} + \frac{1}{r} = 1, \tag{2.2.38}$$

with $a = |H_\varepsilon(u_{k-1}^{\varepsilon,N}) - h(u_{k-1}^{\varepsilon,N})(t_k^N - t_{k-1}^N)|$ and $b = |\theta_j \circ H_\varepsilon(u_k^{\varepsilon,N})|^{q-1}$. Note, here, that $(q-1)r = q$. Applying the equality $(1/q) + (1/r) = 1$ to (2.2.37), and taking the qth roots, we obtain

$$\begin{aligned} \|\theta_j \circ H_\varepsilon(u_k^{\varepsilon,N})\|_{\mathbf{L}^q(\Omega)} \leqslant{} & \|H_\varepsilon(u_{k-1}^{\varepsilon,N})\|_{\mathbf{L}^q(\Omega)} + \|h\|_{\mathbf{Lip}}\|u_{k-1}^{\varepsilon,N}\|_{\mathbf{L}^q(\Omega)}(t_k^N - t_{k-1}^N) \\ & + |h(0)||\Omega|^{1/q}(t_k^N - t_{k-1}^N) \end{aligned} \tag{2.2.39}$$

for all integers $j \geqslant 1$, so that

$$\begin{aligned} \|H_\varepsilon(u_k^{\varepsilon,N})\|_{\mathbf{L}^q(\Omega)} \leqslant{} & \|H_\varepsilon(u_{k-1}^{\varepsilon,N})\|_{\mathbf{L}^q(\Omega)} + \|h\|_{\mathbf{Lip}}\|u_{k-1}^{\varepsilon,N}\|_{\mathbf{L}^q(\Omega)}(t_k^N - t_{k-1}^N) \\ & + |h(0)||\Omega|^{1/q}(t_k^N - t_{k-1}^N). \end{aligned} \tag{2.2.40}$$

In particular, $u_k^{\varepsilon,N} = J_\varepsilon \circ H_\varepsilon(u_k^{\varepsilon,N}) \in \mathbf{L}^q(\Omega)$, and the estimate

$$\|H_\varepsilon(u_k^{\varepsilon,N})\|_{\mathbf{L}^q(\Omega)} \leqslant \|H_\varepsilon(u_0^{\varepsilon,N})\|_{\mathbf{L}^q(\Omega)} + |h(0)||\Omega|^{1/q}T_0 + (\|h\|_{\mathbf{Lip}}/\lambda)\sum_{i=1}^{k}\|H_\varepsilon(u_{i-1}^{\varepsilon,N})\|_{\mathbf{L}^q(\Omega)}(t_i^N - t_{i-1}^N)$$

holds for $k = 1, \ldots, M(N)$. By the version of the discrete Gronwall inequality contained in Corollary 2.2.2, we obtain (2.2.31). The result (2.2.32) is immediate from (2.2.31). ■

Proposition 2.2.5. The solution pair of (2.2.17)–(2.2.20) and that of (2.2.21)–(2.2.24) exist and are unique. If f is Lipschitz continuous, the estimates

$$\|H_\varepsilon(u_k^{\varepsilon,N}) - H(u_k^N)\|_{\mathbf{F}} \leqslant C\varepsilon^{1/2}, \tag{2.2.41a}$$

$$\left[\sum_{k=1}^{M(N)} \|u_k^{\varepsilon,N} - u_k^N\|_{\mathbf{L}^2(\Omega)}^2(t_k^N - t_{k-1}^N)\right]^{1/2} \leqslant C\varepsilon^{1/2} \tag{2.2.41b}$$

hold for some constant C, independent of ε, k and N, provided the local mesh ratios are bounded,

$$\max\left(\frac{t_k^N - t_{k-1}^N}{t_{k-1}^N - t_{k-2}^N}, \frac{t_{k-1}^N - t_{k-2}^N}{t_k^N - t_{k-1}^N}\right) \leqslant r_0,\ N \geqslant 1,\ k = 2, \ldots, M(N), \tag{2.2.41c}$$

and provided $\|\mathscr{P}^N\|$ is sufficiently small (cf. (2.2.47) below).

Proof: To establish the inequalities (2.2.41a, b), we begin with the equations

$$(t_k^N - t_{k-1}^N)^{-1}\mathrm{T}[H_\varepsilon(u_k^{\varepsilon,N}) - H_\varepsilon(u_{k-1}^{\varepsilon,N})] + u_k^{\varepsilon,N} + \mathrm{T}f_\sigma(u_{k-1}^{\varepsilon,N}) = 0, \tag{2.2.42a}$$

$$(t_k^N - t_{k-1}^N)^{-1}\mathrm{T}[H(u_k^N) - H(u_{k-1}^N)] + u_k^N + \mathrm{T}f_\sigma(u_{k-1}^N) = 0, \tag{2.2.42b}$$

where $\mathrm{T} = \mathrm{N}_\sigma$, $0 < \sigma < \lambda T_0^{-1}$ (see Definition 1.1.2) and f_σ is defined in (1.1.26). We find it advantageous to use the inverse formulation. Thus, set $J_\varepsilon = H_\varepsilon^{-1}$, $J = H^{-1}$, and

$$v_k^{\varepsilon,N} = H_\varepsilon(u_k^{\varepsilon,N}), \qquad v_k^N = H(u_k^N). \tag{2.2.43}$$

Subtracting (2.2.42b) from (2.2.42a), multiplying by $v_k^{\varepsilon,N} - v_k^N$, and integrating over Ω, we obtain

$$\begin{aligned}(t_k^N - t_{k-1}^N)^{-1}&[\|v_k^{\varepsilon,N} - v_k^N\|_{\mathbf{F}}^2 - (v_{k-1}^{\varepsilon,N} - v_{k-1}^N, v_k^{\varepsilon,N} - v_k^N)_{\mathbf{F}}] \\ &+ (J_\varepsilon(v_k^{\varepsilon,N}) - J_\varepsilon(v_k^N), v_k^{\varepsilon,N} - v_k^N)_{\mathbf{L}^2(\Omega)} \\ &= (J(v_k^N) - J_\varepsilon(v_k^N), v_k^{\varepsilon,N} - v_k^N)_{\mathbf{L}^2(\Omega)} \\ &- (f_\sigma \circ J_\varepsilon(v_{k-1}^{\varepsilon,N}) - f_\sigma \circ J(v_{k-1}^N), T[v_k^{\varepsilon,N} - v_k^N])_{\mathbf{L}^2(\Omega)}.\end{aligned} \tag{2.2.44}$$

The estimation of the right-hand side of (2.2.44) parallels that in Proposition 2.1.2 (see (2.1.17)–(2.1.19)), adjusted by the fact that f_σ is evaluated explicitly. If we estimate the left-hand side of (2.2.44), we obtain, altogether,

$$\begin{aligned}
&\tfrac{1}{2}(t_k^N - t_{k-1}^N)^{-1}[\|v_k^{\varepsilon,N} - v_k^N\|_{\mathbf{F}}^2 - \|v_{k-1}^{\varepsilon,N} - v_{k-1}^N\|_{\mathbf{F}}^2] \\
&\quad + (J_\varepsilon(v_k^{\varepsilon,N}) - J_\varepsilon(v_k^N), v_k^{\varepsilon,N} - v_k^N)_{\mathbf{L}^2(\Omega)} \\
&\leqslant \|J_\varepsilon(v_k^N) - J(v_k^N)\|_{\mathbf{L}^2(\Omega)}\|v_k^{\varepsilon,N} - v_k^N\|_{\mathbf{L}^2(\Omega)} \\
&\quad + C_0\|J_\varepsilon(v_{k-1}^{\varepsilon,N}) - J(v_{k-1}^N)\|_{\mathbf{L}^2(\Omega)}\|v_k^{\varepsilon,N} - v_k^N\|_{\mathbf{F}} \\
&\leqslant \left\{\frac{\gamma}{\varepsilon}\|J_\varepsilon(v_k^N) - J(v_k^N)\|_{\mathbf{L}^2(\Omega)}^2 + \frac{\varepsilon}{4\gamma}\|v_k^{\varepsilon,N} - v_k^N\|_{\mathbf{L}^2(\Omega)}^2\right\} \\
&\quad + \left\{\frac{\lambda}{4}\left(\frac{t_{k-1}^N - t_{k-2}^N}{t_k^N - t_{k-1}^N}\right)[\|J_\varepsilon(v_{k-1}^{\varepsilon,N}) - J_\varepsilon(v_{k-1}^N)\|_{\mathbf{L}^2(\Omega)}^2\right. \\
&\quad \left. + \|J_\varepsilon(v_{k-1}^N) - J(v_{k-1}^N)\|_{\mathbf{L}^2(\Omega)}^2] + \frac{2C_0^2}{\lambda}\left(\frac{t_k^N - t_{k-1}^N}{t_{k-1}^N - t_{k-2}^N}\right)\|v_k^{\varepsilon,N} - v_k^N\|_{\mathbf{F}}^2\right\},
\end{aligned} \tag{2.2.45}$$

so that, after multiplication of (2.2.45) by $t_k^N - t_{k-1}^N$, summation on $k = 1, \ldots, l$, and absorption of terms by the coercive resulting left-hand side, we obtain

$$\begin{aligned}
&\tfrac{1}{2}\|v_l^{\varepsilon,N} - v_l^N\|_{\mathbf{F}}^2 + \tfrac{1}{2}\sum_{k=1}^{l}(J_\varepsilon(v_k^{\varepsilon,N}) - J_\varepsilon(v_k^N), v_k^{\varepsilon,N} - v_k^N)_{\mathbf{L}^2(\Omega)}(t_k^N - t_{k-1}^N) \\
&\leqslant |\Omega|T_0\left(1 + \frac{\mu}{\lambda}\right)^2\left(\frac{\lambda\varepsilon^2}{4} + \gamma\varepsilon\right) + \frac{2C_0^2 r_0}{\lambda}\sum_{k=1}^{l}\|v_k^{\varepsilon,N} - v_k^N\|_{\mathbf{F}}^2(t_k^N - t_{k-1}^N),
\end{aligned} \tag{2.2.46}$$

where r_0 is given by (2.2.41c). The Gronwall inequality of Proposition 2.2.1 may be applied (to 2.2.46), provided

$$\|\mathscr{P}^N\| \leqslant \rho_0 < \lambda/(4C_0^2 r_0). \tag{2.2.47}$$

Here C_0 may be defined, from (2.2.45), as the product of the Lipschitz norm of f_σ and the norm of the injection $\mathbf{L}^2(\Omega) \to \mathbf{F}$. Under the assumption (2.2.47), the term $k = l$ in (2.2.46) may be absorbed in the first term on the left-hand side, with resulting constant $(1/2) - (2C_0^2 r_0 \rho_0)/\lambda$. We, thus, obtain (2.2.41a) for an appropriate choice of C. In fact, the Gronwall inequality also yields the estimate

$$\left[\sum_{k=1}^{t}(J_\varepsilon(v_k^{\varepsilon,N}) - J_\varepsilon(v_k^N), v_k^{\varepsilon,N} - v_k^N)_{\mathbf{L}^2(\Omega)}(t_k^N - t_{k-1}^N)\right]^{1/2} \leqslant C\varepsilon^{1/2}. \tag{2.2.48}$$

To obtain (2.2.41b), we use the triangle inequality in conjunction with (2.2.48), that is,

$$\begin{aligned}
\sum_{k=1}^{M} &\|J_\varepsilon(v_k^{\varepsilon,N}) - J(v_k^N)\|_{\mathbf{L}^2(\Omega)}^2 (t_k^N - t_{k-1}^N) \\
&\leqslant 2 \sum_{k=1}^{M} \|J_\varepsilon(v_k^{\varepsilon,N}) - J_\varepsilon(v_k^N)\|_{\mathbf{L}^2(\Omega)}^2 (t_k^N - t_{k-1}^N) \\
&\quad + 2 \sum_{k=1}^{M} \|J_\varepsilon(v_k^N) - J(v_k^N)\|_{\mathbf{L}^2(\Omega)}^2 (t_k^N - t_{k-1}^N) \\
&\leqslant (2/\lambda) \sum_{k=1}^{l} (J_\varepsilon(v_k^{\varepsilon,N}) - J_\varepsilon(v_k^N), v_k^{\varepsilon,N} - v_k^N)_{\mathbf{L}^2(\Omega)} (t_k^N - t_{k-1}^N) \\
&\quad + 2\varepsilon^2 \left(1 + \frac{\mu}{\lambda}\right)^2 |\Omega| T_0,
\end{aligned} \tag{2.2.49}$$

and (2.2.41b) follows from (2.2.48) and (2.2.49), with an adjustment in C. The existence and uniqueness are a consequence of Proposition 3.3.1. ■

Corollary 2.2.6. Suppose the initial datum $H(u_0)$ is in $\mathbf{L}^p(\Omega)$, $p \geqslant 2$, and suppose f is Lipschitz continuous. Then $\{H(u_k^N)\}_{k,N}$ satisfies the bound in (2.2.31), provided the partitions $\{\mathscr{P}^N\}$ satisfy (2.2.41c) and (2.2.47).

Proof: For $2 \leqslant p < \infty$ (resp $p = \infty$), there is a subsequence $\{H_{\varepsilon_j}(u_k^{\varepsilon_j,N})\}_j$ of $\{H_\varepsilon(u_k^{\varepsilon,N})\}_\varepsilon$ which is weakly (resp weak-*) convergent in $\mathbf{L}^p(\Omega)$ to a function χ_k^N as $j \to \infty$ and, by lower semicontinuity, χ_k^N satisfies the bound in (2.2.31). By composition of continuous embeddings, which preserve weak convergence, we conclude the weak convergence of this subsequence in $\mathbf{F}$ and, also, by Proposition 2.2.5, the convergence in $\mathbf{F}$ to $H(u_k^N)$. Uniqueness of weak limits yields $\chi_k^N = H(u_k^N)$. ■

Remark 2.2.6. Note that the Lipschitz continuity of f was used only in deducing the weak convergence of $H_{\varepsilon_j}(u_k^{\varepsilon_j,N})$ to $H(u_k^N)$ in $\mathbf{L}^p(\Omega)$ via Proposition 2.2.5. The actual *a priori* estimates for $H_\varepsilon(u_k^{\varepsilon,N})$, as contained in Proposition 2.2.4, do not require f to be Lipschitz continuous. However, we do observe that, if $p = \infty$ and f is locally Lipschitz, the result of Corollary 2.2.6 holds, since f can be replaced by an appropriate extension $\tilde{f}$ of the restriction of f to a set containing the range of u_k^N. This, of course, utilizes uniqueness properties.

In Chapter 5, we shall have cause to consider more general H and more general smoothings. For this more general class, the estimates (2.2.31) and

(2.2.32) will remain valid, since no properties beyond Definition 5.1.2 are required. However, Proposition 2.2.5 is expected to be valid only selectively for general classes of smoothings since it depends fundamentally upon (2.1.13). Finally, although we did not use both partition bounds, as defined by (2.2.41c), in the proof of Proposition 2.2.5, we shall have cause to make use of them in Chapter 5.

2.3 REGULARIZATION IN THE POROUS-MEDIUM EQUATION

The hypotheses on the function H of Section 1.2, which yields the moisture content or density in the case of the porous-medium equation, are given in Remark 1.2.1. In this section, we shall smooth H to obtain a strictly monotone increasing, continuously differentiable Hölder function H_ε, such that, on compact subsets of $\mathbb{R}^1\backslash\{0\}$, H_ε and H'_ε agree with H and H' for ε sufficiently small. As in Section 2.1, this will be followed by a regularization theorem.

Definition 2.3.1. Suppose $0 < \varepsilon \leqslant 1$. Define the linear function $\hat{\ell}_\varepsilon$ on $[0, \varepsilon]$ to be the uniquely determined function satisfying

$$\hat{\ell}_\varepsilon(\varepsilon) = H'(\varepsilon), \qquad \int_0^\varepsilon \hat{\ell}_\varepsilon(t)\, dt = H(\varepsilon). \tag{2.3.1}$$

The function $\hat{\ell}$ is given explicitly by

$$\hat{\ell}_\varepsilon(t) = \hat{\ell}_{0,\varepsilon} + \hat{\ell}_{1,\varepsilon} t, \qquad 0 \leqslant t \leqslant \varepsilon, \tag{2.3.2a}$$

where

$$\hat{\ell}_{0,\varepsilon} = [2H(\varepsilon) - \varepsilon H'(\varepsilon)]/\varepsilon, \qquad \hat{\ell}_{1,\varepsilon} = 2[\varepsilon H'(\varepsilon) - H(\varepsilon)]/\varepsilon^2. \tag{2.3.2b}$$

Note that $\hat{\ell}_{1,\varepsilon} \leqslant 0$ follows from the decrease of the difference quotients for the concave function H:

$$\frac{H(\varepsilon)}{\varepsilon} = \frac{H(\varepsilon) - H(0)}{\varepsilon - 0} \geqslant \frac{H(\varepsilon) - H(\delta)}{\varepsilon - \delta} \Leftarrow \left(1 - \frac{\varepsilon - \delta}{\varepsilon}\right) H(\varepsilon) + \frac{\varepsilon - \delta}{\varepsilon} H(0) \leqslant H(\delta), \tag{2.3.3}$$

for $0 < \delta < \varepsilon$. Letting $\delta \uparrow \varepsilon$ immediately gives

$$H(\varepsilon) \geqslant \varepsilon H'(\varepsilon). \tag{2.3.4}$$

We also conclude that $\hat{\ell}_{0,\varepsilon} \geqslant H(\varepsilon)/\varepsilon$. Let ℓ_ε denote the even extension of $\hat{\ell}_\varepsilon$ to $[-\varepsilon, \varepsilon]$, and set

$$h_\varepsilon(t) = \begin{cases} \ell_\varepsilon(t), & |t| \leqslant \varepsilon, \\ H'(t), & |t| \geqslant \varepsilon. \end{cases} \tag{2.3.5}$$

Finally, set

$$H_\varepsilon(t) = \int_0^t h_\varepsilon(s)\,ds, \qquad t \in \mathbb{R}^1. \tag{2.3.6}$$

Proposition 2.3.1. For any compact subset K of $\mathbb{R}^1\backslash\{0\}$, such that $(0, \varepsilon) \cap K = \varnothing, H_\varepsilon$ and H'_ε agree with H and H', respectively, on K. Moreover, $H_\varepsilon(t)t \geqslant 0$ for $t \in \mathbb{R}^1$ and, for some $c > 0$,

$$H'_\varepsilon(t) \leqslant c\varepsilon^{(1/\gamma)-1}, \qquad t \in \mathbb{R}^1. \tag{2.3.7}$$

In the notation of (1.2.14e), c is given explicitly by $c = 2c_2\gamma$. Moreover, the approximation

$$|H_\varepsilon(t) - H(t)| \leqslant c\varepsilon^{1/\gamma}, \qquad t \in \mathbb{R}^1, \tag{2.3.8a}$$

holds for the same c. If $J = H^{-1}$ and $J_\varepsilon = H_\varepsilon^{-1}$, then J, as well as J_ε, are continuously differentiable strictly monotone increasing functions and

$$|J_\varepsilon(t) - J(t)| \leqslant 2\varepsilon, \qquad t \in \mathbb{R}^1. \tag{2.3.8b}$$

Proof: By construction, H_ε and H'_ε agree with H and H', respectively, on $\mathbb{R}^1\backslash(-\varepsilon, \varepsilon)$. For $-\varepsilon \leqslant t \leqslant \varepsilon$, we have, from the definition of ℓ_ε,

$$H'_\varepsilon(t) = \ell_\varepsilon(t) \leqslant 2H(\varepsilon)/\varepsilon. \tag{2.3.9}$$

However,

$$H(\varepsilon) \leqslant c_2\gamma\varepsilon^{1/\gamma}, \tag{2.3.10}$$

from the second bound in (1.2.14e). From the same bound, we deduce

$$H'(t) \leqslant c_2 t^{(1/\gamma)-1} \leqslant c_2\varepsilon^{(1/\gamma)-1}, \qquad t \geqslant \varepsilon. \tag{2.3.11}$$

Altogether, from (2.3.9), (2.3.10), and (2.3.11), we deduce (2.3.7), with $c = 2c_2\gamma$. Now, (2.3.8a) is immediate if the difference is estimated crudely by the triangle inequality, where H_ε and H differ. Thus, since $H_\varepsilon(\varepsilon) = H(\varepsilon)$, by construction,

$$|H_\varepsilon(t) - H(t)| \leqslant H_\varepsilon(t) + H(t) \leqslant H_\varepsilon(\varepsilon) + H(\varepsilon) = 2H(\varepsilon) \leqslant 2c_2\gamma\varepsilon^{1/\gamma},$$

for $0 \leqslant t \leqslant \varepsilon$, hence, for $-\varepsilon \leqslant t \leqslant \varepsilon$, by the sign properties of odd functions. In a similar way, (2.3.8b) need only be estimated on $[-H(\varepsilon), H(\varepsilon)]$, and we find

$$|J_\varepsilon(t) - J(t)| \leqslant J_\varepsilon(t) + J(t) \leqslant J_\varepsilon(H(\varepsilon)) + J(H(\varepsilon))$$
$$= J_\varepsilon(H_\varepsilon(\varepsilon)) + J(H(\varepsilon)) = 2\varepsilon,$$

which is simply (2.3.8b). The remaining statements are clear.

Remark 2.3.1. As in Section 2.1, we may use H_ε to define regularized parabolic problems, which are defined more conveniently in terms of the lifting (1.2.18). The next proposition describes the rate of convergence induced by the regularization. We use the solution class

$$\mathscr{C}_1 = \mathbf{C}([0, T_0]; \mathbf{L}^2(\Omega)) \cap \mathbf{L}^2((0, T_0); \mathbf{H}_0^1(\Omega)).$$

Proposition 2.3.2. Let $\mathrm{T} = \mathrm{D}_0$ be the Dirichlet inversion operator of Definition 1.2.2. If $\mathscr{C}_1$ denotes the class defined above, suppose that $u, u^\varepsilon \in \mathscr{C}_1$, respectively, satisfy (1.2.21) and

$$\frac{\partial \mathrm{T} H_\varepsilon(u^\varepsilon)}{\partial t} + u^\varepsilon = 0, \tag{2.3.12}$$

for $0 < \varepsilon < 1$. If $\mathrm{T}H_\varepsilon(u^\varepsilon)|_{t=0} = \mathrm{T}H(u)|_{t=0}$, then the estimate

$$\|H_\varepsilon(u^\varepsilon) - H(u)\|_{\mathbf{L}^\infty((0,T_0);\mathbf{H}^{-1}(\Omega))} \leqslant C\varepsilon^{(1/2)(1+(1/\gamma))} \tag{2.3.13}$$

holds for some positive constant C. If u^ε and u are assumed nonnegative and to satisfy the weak maximum principle (see Sections 2.4 and 5.2)

$$\|u^\varepsilon\|_{\mathbf{L}^\infty(\mathscr{D})} \leqslant (1 + C_0)\|u_0\|_{\mathbf{L}^\infty(\Omega)}, \qquad \|u\|_{\mathbf{L}^\infty(\mathscr{D})} \leqslant \|u_0\|_{\mathbf{L}^\infty(\Omega)}, \qquad 0 < t < T_0, \tag{2.3.14}$$

then the additional estimate

$$\|u^\varepsilon - u\|_{\mathbf{L}^2((0,T_0);\mathbf{L}^2(\Omega))} \leqslant C\varepsilon^{1/\gamma} \tag{2.3.15}$$

holds. The constant C depends, in the first case, upon $|\Omega|$ and T_0 (see (2.3.18)), and, in the second case, on $|\Omega|$, T_0 and $\|u_0\|_{\mathbf{L}^\infty(\Omega)}$ (see (2.3.24)).

Proof: Setting $v = H(u)$ and $v^\varepsilon = H_\varepsilon(u^\varepsilon)$, we have, after subtraction of (1.2.21) from (2.3.12), multiplication by $v^\varepsilon - v$, and integration over Ω,

$$\tfrac{1}{2}\frac{d}{dt}\|v^\varepsilon - v\|^2_{\mathbf{H}^{-1}(\Omega)} + (J_\varepsilon(v^\varepsilon) - J_\varepsilon(v), v^\varepsilon - v)_{\mathbf{L}^2(\Omega)}$$
$$= (J(v) - J_\varepsilon(v), v^\varepsilon - v)_{\mathbf{L}^2(\Omega)}$$
$$\leqslant \tfrac{1}{2}c\varepsilon^{(1/\gamma)-1}\|J_\varepsilon(v) - J(v)\|^2_{\mathbf{L}^2(\Omega)} + \tfrac{1}{2}\varepsilon^{1-(1/\gamma)}c^{-1}\|v^\varepsilon - v\|^2_{\mathbf{L}^2(\Omega)}, \tag{2.3.16}$$

so that, for each $t > 0$,

$$\frac{1}{2}\frac{d}{dt}\|v^\varepsilon - v\|^2_{\mathbf{H}^{-1}(\Omega)} + \frac{1}{2}(J_\varepsilon(v^\varepsilon) - J_\varepsilon(v), v^\varepsilon - v)_{\mathbf{L}^2(\Omega)} \leqslant 2c|\Omega|\varepsilon^{1+(1/\gamma)}. \quad (2.3.17)$$

Here, we have used (2.3.8b), and the inequality

$$\begin{aligned}(v^\varepsilon - v, v^\varepsilon - v)_{\mathbf{L}^2(\Omega)} &= (H_\varepsilon \circ J_\varepsilon(v^\varepsilon) - H_\varepsilon \circ J_\varepsilon(v), v^\varepsilon - v)_{\mathbf{L}^2(\Omega)} \\ &\leqslant c\varepsilon^{(1/\gamma)-1}(J_\varepsilon(v^\varepsilon) - J_\varepsilon(v), v^\varepsilon - v)_{\mathbf{L}^2(\Omega)},\end{aligned}$$

which follows from (2.3.7). The inequality (2.3.13) follows from (2.3.17), upon integration from $\tau = 0$ to $\tau = t$, that is,

$$\begin{aligned}\|v^\varepsilon - v\|^2_{\mathbf{H}^{-1}(\Omega)} &+ \int_0^t (J_\varepsilon(v^\varepsilon) - J_\varepsilon(v), v^\varepsilon - v)_{\mathbf{L}^2(\Omega)}\,d\tau \\ &\leqslant 4c|\Omega|T_0\varepsilon^{1+(1/\gamma)}, \qquad 0 < t < T_0. \quad (2.3.18)\end{aligned}$$

To obtain (2.3.15), we repeat the steps which led to (2.3.16). This gives

$$\begin{aligned}\frac{1}{2}\frac{d}{dt}&\|H_\varepsilon(u^\varepsilon) - H(u)\|^2_{\mathbf{H}^{-1}(\Omega)} + (u^\varepsilon - u, H(u^\varepsilon) - H(u))_{\mathbf{L}^2(\Omega)} \\ &= (u^\varepsilon - u, H(u^\varepsilon) - H_\varepsilon(u^\varepsilon))_{\mathbf{L}^2(\Omega)} \\ &\leqslant \tfrac{1}{2}\{\eta\|u^\varepsilon - u\|^2_{\mathbf{L}^2(\Omega)} + \eta^{-1}\|H_\varepsilon(u^\varepsilon) - H(u^\varepsilon)\|^2_{\mathbf{L}^2(\Omega)}\}, \quad (2.3.19)\end{aligned}$$

where the choice of $\eta > 0$ is at our disposal. In order to utilize the coercive properties of the second term in (2.3.19), we require the inequality

$$\left(\frac{y^r - z^r}{y^s - z^s} \times \frac{s}{r}\right)^{1/(r-s)} \leqslant \max(y, z) \quad (2.3.20)$$

for $y \geqslant 0$, $z \geqslant 0$, $y \neq z$, and $r > 0$, $s > 0$, $r \neq s$ (see [12], p. 85). The identifications $r = 1$, $s = 1/\gamma$, $y = [H(u^\varepsilon)]^\gamma$, $z = [H(u)]^\gamma$ in (2.3.20), in conjunction with the assumed relations (2.3.14) and (1.2.14e) lead to, with $\kappa = (1 + C_0)^{1-(1/\gamma)}$,

$$\begin{aligned}\kappa\gamma|H(u^\varepsilon) - H(u)|&(c_2\gamma)^{\gamma-1}\|u_0\|^{1-(1/\gamma)}_{\mathbf{L}^\infty(\Omega)} \\ &\geqslant \gamma|H(u^\varepsilon) - H(u)|[\max([H(u^\varepsilon)]^\gamma, [H(u)]^\gamma)]^{1-(1/\gamma)} \\ &\geqslant |[H(u^\varepsilon)]^\gamma - [H(u)]^\gamma| \geqslant C|u^\varepsilon - u|, \quad (2.3.21)\end{aligned}$$

where the latter inequality, with $C = [c_1\gamma]^\gamma$, follows from the fact that the composite function

$$K(t) = [H(t)]^\gamma, \qquad t \geqslant 0 \quad (2.3.22a)$$

has derivative K', satisfying

$$K'(t) \geqslant [c_1\gamma]^\gamma, \qquad t \geqslant 0. \quad (2.3.22b)$$

If (2.3.21) is substituted into (2.3.19), we obtain, for $u_0 \neq 0$,

$$\tfrac{1}{2}\frac{d}{dt}\|H_\varepsilon(u^\varepsilon) - H(u)\|^2_{\mathbf{H}^{-1}(\Omega)} + \left(\frac{C}{\kappa\gamma}\right)(c_2\gamma)^{1-\gamma}\|u_0\|^{(1/\gamma)-1}_{\mathbf{L}^\infty(\Omega)}\|u^\varepsilon - u\|^2_{\mathbf{L}^2(\Omega)}$$
$$\leqslant \tfrac{1}{2}\eta\|u^\varepsilon - u\|^2_{\mathbf{L}^2(\Omega)} + \tfrac{1}{2}\eta^{-1}c^2|\Omega|\varepsilon^{2/\gamma}, \tag{2.3.23}$$

where we have used (2.3.8a). Multiplying (2.3.23) by $\|u_0\|^{1-(1/\gamma)}_{\mathbf{L}^\infty(\Omega)}$, we obtain, for sufficiently small η (say, $\eta = (C/(\kappa\gamma))(c_2\gamma)^{1-\gamma}\|u_0\|^{(1/\gamma)-1}_{\mathbf{L}^\infty(\Omega)}$),

$$\int_0^{T_0} \|u^\varepsilon - u\|^2_{\mathbf{L}^2(\Omega)}\,dt \leqslant C_1\|u_0\|^{2(1-(1/\gamma))}_{\mathbf{L}^\infty(\Omega)}\varepsilon^{2/\gamma}, \qquad 0 < t < T_0, \tag{2.3.24a}$$

where

$$C_1 = T_0\left(\frac{\kappa\gamma}{C}\right)(c_2\gamma)^{\gamma-1}c^2|\Omega|. \tag{2.3.24b}$$

The inequality (2.3.15) follows immediately from (2.3.24). ■

Remark 2.3.2. Uniqueness of solutions of (1.2.21) and (2.3.12) follows from Proposition 1.5.1. Although it is certainly possible to parallel the development of the previous section in developing semidiscrete regularizations, we shall consider the implicit method for the nonregularized problem only. A major reason for this is that it is possible to derive the maximum principles, and, in fact, the entire existence theory, directly from the nonregularized formulation.

The rate of convergence in (2.3.13) and (2.3.15) is preserved under the hypothesis (2.1.25b) if, instead of the specification $u_0^\varepsilon = J_\varepsilon \circ H(u_0)$, made in the regularization of Proposition 2.3.2, one specifies the initial datum $u_0^\varepsilon = u_0$. The latter choice of the initial datum leads to $u_t^\varepsilon \in \mathbf{L}^2(\Omega \times (0, \infty))$ if $u_0 \in \mathbf{H}_0^1(\Omega)$. In (2.1.25b), the zero set of u_0 may be ignored.

2.4 NONNEGATIVE SEMIDISCRETE SOLUTIONS OF THE POROUS-MEDIUM EQUATION AND MAXIMUM PRINCIPLES

We begin by defining the semidiscrete approximations.

Definition 2.4.1. Let $0 = t_0^N < t_1^N < \cdots < t_{M(N)}^N = T_0$ denote an arbitrary sequence of partitions $\mathscr{P}^N$ of $[0, T_0]$. By the implicit semidiscrete solution of the initial/homogeneous Dirichlet boundary-value problem corresponding to (1.2.13) is meant a recursively generated finite sequence $\{u_k^N\}_{k=1}^{M(N)}$, for each

$N = 1, 2, \ldots$, satisfying

$$u_k^N \in \mathbf{H}_0^1(\Omega), \tag{2.4.1}$$

and the variational conditions

$$(t_k^N - t_{k-1}^N)^{-1}([H(u_k^N) - H(u_{k-1}^N)], v)_{\mathbf{L}^2(\Omega)} + (\nabla u_k^N, \nabla v)_{\mathbf{L}^2(\Omega)} = 0, \tag{2.4.2}$$

for all $v \in \mathbf{H}_0^1(\Omega)$, for $k = 1, \ldots, M(N)$. We require that

$$u_0^N = u_0, \qquad \text{all} \quad N, \tag{2.4.3a}$$

and

$$u_0 \geqslant 0, \qquad u_0 \in \mathbf{L}^\infty(\Omega). \tag{2.4.3b}$$

Proposition 2.4.1. Solutions of (2.4.1)–(2.4.3) exist and are uniquely determined and nonnegative on Ω. They satisfy the weak maximum principle

$$\|u_k^N\|_{\mathbf{L}^\infty(\Omega)} \leqslant \|u_{k-1}^N\|_{\mathbf{L}^\infty(\Omega)} \leqslant \|u_0\|_{\mathbf{L}^\infty(\Omega)}, \tag{2.4.4}$$

for all $N \geqslant 1$, $k = 1, \ldots, M(N)$.

Proof: The existence and uniqueness of u_k^N follows from Proposition 3.3.2. Suppose, for $k \geqslant 1$, that $u_{k-1}^N \geqslant 0$. Using the decomposition of u_k^N into positive and negative parts, we obtain from (2.4.2), since $(u_k^N)^- \in \mathbf{H}_0^1(\Omega)$,

$$\begin{aligned}(t_k^N - t_{k-1}^N)^{-1} \int_\Omega H(u_k^N)(u_k^N)^-\, d\mathbf{x} + \int_\Omega \nabla u_k^N \cdot \nabla (u_k^N)^-\, d\mathbf{x} \\ = (t_k^N - t_{k-1}^N)^{-1} \int_\Omega H(u_{k-1}^N)(u_k^N)^-\, d\mathbf{x},\end{aligned} \tag{2.4.5}$$

so that

$$\begin{aligned}0 \leqslant \int_\Omega H(u_k^N)(u_k^N)^-\, d\mathbf{x} + (t_k^N - t_{k-1}^N) \int_\Omega |\nabla (u_k^N)^-|^2\, d\mathbf{x} \\ = \int_\Omega H(u_{k-1}^N)(u_k^N)^-\, d\mathbf{x} \leqslant 0,\end{aligned} \tag{2.4.6}$$

and it follows that $(u_k^N)^- = 0$. Thus, $u_k^N \geqslant 0$. The verification of (2.4.4) also proceeds by induction on k, and utilizes the truncation operators introduced in Definition 2.2.2. Thus, for $l \geqslant 1$, set

$$v = \Theta_{j,l\gamma}(H(u_k^N)) = \Theta_{j,l}([H(u_k^N)]^\gamma) \tag{2.4.7}$$

in (2.4.2). Since H^γ is a Lipschitz function vanishing at zero, it follows that $v \in \mathbf{H}_0^1(\Omega)$, and such a choice is permissible. In analogy with (2.2.36), we have

$$(\nabla u_k^N, \nabla v)_{\mathbf{L}^2(\Omega)} \geqslant 0, \tag{2.4.8}$$

where we have used the monotonicity of H and the nonnegativity of u_k^N. We, thus, conclude from (2.4.2), by the use of (2.4.8) and (2.2.38), that

$$\int_\Omega [\theta_j \circ H(u_k^N)]^q \leqslant \frac{1}{r}\int_\Omega [\theta_j \circ H(u_k^N)]^q + \frac{1}{q}\int_\Omega [H(u_{k-1}^N)]^q, \tag{2.4.9}$$

for $q = l\gamma + 1$ and $r = q/(q-1)$. Here, as in the derivation of (2.2.37), we have used the domination of θ_j by the identity, and we have set $a = H(u_{k-1}^N)$ and $b = [\theta_j \circ H(u_k^N)]^{q-1}$. The inductive proof is concluded by applying the equality $(1/q) + (1/r) = 1$, taking the qth roots and letting $l = (q-1)/\gamma$ tend to infinity. ■

Remark 2.4.1. Strictly speaking, the proof has shown that $\|H(u_k^N)\|_{\mathbf{L}^\infty(\Omega)} \leqslant \|H(u_{k-1}^N)\|_{\mathbf{L}^\infty(\Omega)} \leqslant \|H(u_0)\|_{\mathbf{L}^\infty(\Omega)}$. However, we obtain (2.4.4) directly by utilizing the continuity and monotonicity of H. Note that similar estimates are valid for the solutions u_k^ε of the semidiscrete regularized problems, which we have not formally introduced. One might expect a sharper version of (2.4.4) on the basis of recent decay estimates for the parabolic problem (see Aronson and Peletier†).

2.5 INVARIANT RECTANGLES AND MAXIMUM PRINCIPLES FOR REACTION–DIFFUSION SYSTEMS IN SEMIDISCRETE FORM

We begin by defining the implicit semidiscretization for the reaction–diffusion systems. We maintain the hypotheses of Section 1.3.

Definition 2.5.1. Given a sequence $\mathscr{P}^N$ of partitions of $[0, T_0]$, the semidiscrete solution of the initial/Robin boundary-value problem corresponding to (1.3.4) and (1.3.9) is a recursively generated finite sequence $\{\mathbf{u}_k^N\}_{k=1}^{M(N)}$, for each $N = 1, \ldots,$ satisfying the condition that

$$\mathbf{u}_k^N \in \mathbf{H}^1(\Omega; \mathbb{R}^m), \tag{2.5.1}$$

the variational conditions, for $\mathbf{v} \in \mathbf{H}^1(\Omega; \mathbb{R}^m)$,

$$\begin{aligned}(t_k^N - t_{k-1}^N)^{-1}((u_{k,i}^N - u_{k-1,i}^N), v_i)_{\mathbf{L}^2(\Omega)} &+ D_i(\nabla u_{k,i}^N, \nabla v_i)_{\mathbf{L}^2(\Omega)} \\ &+ (f_i(u_{k-1,1}^N, \ldots, u_{k-1,i-1}^N, u_{k,i}^N, u_{k-1,i+1}^N, \ldots, u_{k-1,m}^N), v_i)_{\mathbf{L}^2(\Omega)} \\ &+ D_i(\omega_i u_{k,i}^N, v_i)_{\mathbf{L}^2(\partial\Omega)} = 0,\end{aligned} \tag{2.5.2a}$$

† *J. Differential Eqs.* **39**, 378–412 (1981).

$i = 1, \ldots, i_0$, and

$$(t_k^N - t_{k-1}^N)^{-1}((u_{k,i}^N - u_{k-1,i}^N, v_i)_{\mathbf{L}^2(\Omega)} + D_i(\nabla u_{k,i}^N, \nabla v_i)_{\mathbf{L}^2(\Omega)}$$
$$+ (g_i(u_{k,i}^N), v_i)_{\mathbf{L}^2(\Omega)} + (h_i(u_{k-1}^N), v_i)_{\mathbf{L}^2(\Omega)} + D_i(\omega_i u_{k,i}^N, v_i)_{\mathbf{L}^2(\partial\Omega)} = 0, \tag{2.5.2b}$$

$i = i_0 + 1, \ldots, m$. We require, in addition,

$$\mathbf{u}_0^N = \mathbf{u}_0, \tag{2.5.3}$$

$$\mathbf{f}(\mathbf{u}_k^N) \in \mathbf{L}^2(\Omega; \mathbb{R}^m), \tag{2.5.4}$$

$$\text{Range}(\mathbf{u}_k^N) \subset Q, \tag{2.5.5}$$

all $N \geqslant 1$, $k = 1, \ldots, M(N)$. The partitions are required to satisfy (3.3.29) and (3.3.30) to follow.

Remark 2.5.1. Throughout this section, we shall assume that the region Q is actually contractive, that is, strict inequality of the vector field holds on the faces of Q. More precisely, (1.3.6b) is strengthened to

$$f_i(\mathbf{v}) < 0, \qquad \mathbf{v} \in Q_{a_i}, \qquad\qquad f_i(\mathbf{v}) > 0, \qquad \mathbf{v} \in Q_{b_i}, \tag{2.5.6}$$

$i = 1, \ldots, i_0$. This restriction will be removed in Section 5.3. We shall also assume that the Robin inversion operators R_{ω_i} possess the property that

$$\mathrm{R}_{\omega_i}\mathbf{L}^p(\Omega) \subset \mathbf{W}^{2,p}(\Omega), \qquad 1 < p < \infty, \tag{2.5.7}$$

which, of course, is a restriction upon Ω.

Definition 2.5.2. Let $\mathbf{e}^i$ denote the unit vector in $\mathbb{R}^m$, whose jth component is δ_{ij}, and set, for $1 \leqslant i \leqslant i_0$,

$$g_{k,i}^N(\mathbf{x}) = f_i(u_{k,i}^N(\mathbf{x})\mathbf{e}^i + \sum_{\substack{j \neq i \\ j=1}}^{m} u_{k-1,j}^N(\mathbf{x})\mathbf{e}^j), \qquad \mathbf{x} \in \Omega. \tag{2.5.8}$$

Now write the decomposition, for fixed k, i, N, as

$$\Omega = \Omega_+ \cup \Omega_- \cup \Omega_0, \tag{2.5.9a}$$

where

$$\Omega_+ = \{\mathbf{x} \in \Omega : g_{k,i}^N(\mathbf{x}) > 0\}, \qquad \Omega_- = \{x \in \Omega : g_{k,i}^N(\mathbf{x}) < 0\},$$
$$\Omega_0 = \{\mathbf{x} \in \Omega : g_{k,i}^N(\mathbf{x}) = 0\}. \tag{2.5.9b}$$

Finally, introduce the notation

$$\delta = \{i : 1 \leqslant i \leqslant m \qquad \text{and} \qquad D_i > 0\}. \tag{2.5.10}$$

Remark 2.5.2. Our analysis will proceed by showing that solutions $\mathbf{u}_k^N$ of (2.5.1)–(2.5.4) necessarily satisfy uniform bounds in $\mathbf{L}^\infty(\Omega;\mathbb{R}^m)$ and satisfy (2.5.5). We shall do this in stages. The verification of (2.5.5) is complicated by the fact that the components of Ω_+ and Ω_- need not have boundaries sufficiently regular for integration by parts to hold, thus necessitating component approximation. The correlative analysis makes fundamental use of (2.5.6).

Proposition 2.5.1. Unique solutions of (2.5.1)–(2.5.4) exist in $\mathbf{L}^\infty(\Omega;\mathbb{R}^m)$, and satisfy the estimates (see (2.5.14) and (2.5.15))

$$\|u_{k,i}^N\|_{\mathbf{L}^\infty(\Omega)} \leqslant \|u_{0,i}\|_{\mathbf{L}^\infty(\Omega)} + (\sup|f_i|)T_0, \qquad 1 \leqslant i \leqslant i_0, \tag{2.5.11a}$$

$$\sum_{i=1}^m \|u_{k,i}^N\|_{\mathbf{L}^\infty(\Omega)} \leqslant \Big[C_1 T_0 + (1 + C_2) \sum_{i=1}^m \|u_{0,i}\|_{\mathbf{L}^\infty(\Omega)}\Big] \exp(C_2 T_0), \tag{2.5.11b}$$

for $k = 1, \ldots, M(N)$, and $N \geqslant 1$. Moreover, for such k and N,

$$u_{k,i}^N \in \mathbf{W}^{2,p}(\Omega) \qquad \text{for} \quad 1 < p < \infty \qquad \text{and} \quad i \in \delta. \tag{2.5.12}$$

Proof: The existence and uniqueness of solutions of (2.5.1)–(2.5.4) follow from Proposition 3.3.3. The hypothesis (2.5.7), in conjunction with a bootstrapping sequence of applications of Sobolev's inequality, yields (2.5.12), via induction on k. To outline the verification of (2.5.11), we note that, in analogy with (2.2.39), we have, for a sequence of even integers q,

$$\|u_{k,i}^N\|_{\mathbf{L}^q(\Omega)} \leqslant \|u_{k-1,i}^N\|_{\mathbf{L}^q(\Omega)} + (\sup|f_i|)(t_k^N - t_{k-1}^N), \qquad 1 \leqslant i \leqslant i_0, \tag{2.5.13a}$$

$$\begin{aligned}\|u_{k,i}^N\|_{\mathbf{L}^q(\Omega)} \leqslant{}& \|u_{k-1,i}^N\|_{\mathbf{L}^q(\Omega)} + \|h_i\|_{\mathbf{Lip}} \sum_{l=1}^m \|u_{k-1,l}\|_{\mathbf{L}^q(\Omega)}(t_k^N - t_{k-1}^N)\\ &+ |h_i(0)|\,|\Omega|^{1/q}(t_k^N - t_{k-1}^N), \qquad i_0 + 1 \leqslant i \leqslant m.\end{aligned} \tag{2.5.13b}$$

Summing on $i = 1, \ldots, m$, we obtain

$$\begin{aligned}\sum_{i=1}^m \|u_{k,i}^N\|_{\mathbf{L}^q(\Omega)} \leqslant{}& \sum_{i=1}^m \{\|u_{0,i}\|_{\mathbf{L}^q(\Omega)} + c_i(q)T_0\}\\ &+ \left(\sum_{i=i_0+1}^m \|h_i\|_{\mathbf{Lip}}\right) \sum_{r=1}^k \sum_{l=1}^m \|u_{r-1,l}\|_{\mathbf{L}^q(\Omega)}(t_r^N - t_{r-1}^N),\end{aligned} \tag{2.5.14a}$$

for

$$c_i(q) = \begin{cases} \sup|f_i|, & 1 \leqslant i \leqslant i_0,\\ |h_i(0)|\,|\Omega|^{1/q}, & i_0 + 1 \leqslant i \leqslant m.\end{cases} \tag{2.5.14b}$$

By applying to (2.5.14) the version of the discrete Gronwall inequality contained in Corollary 2.2.2, and letting $q \to \infty$, we obtain (2.5.11b), with

$$C_1 = \left(\sum_{i=1}^{m} c_i(\infty) \right), \tag{2.5.15a}$$

and

$$C_2 = \sum_{i=i_0+1}^{m} \|h_i\|_{\mathbf{Lip}}. \tag{2.5.15b}$$

Note that (2.5.11a) is an immediate consequence of (2.5.13a). ■

Remark 2.5.3. The functions $u_{k,i}^N$ are continuous, hence, Ω_+ and Ω_- are open subsets of Ω with compact boundaries. For $i \in \delta$, the Lipschitz continuity (in fact, $C^{1,\lambda}$ property) follows from (2.5.12). For $i \notin \delta$, the Lipschitz continuity follows from the arguments preceding Proposition 3.3.3 and from the Lipschitz assumption on $\mathbf{u}_0$.

Proposition 2.5.2. For $i \in \delta$, $1 \leqslant i \leqslant i_0$, let Ω_c be a component of Ω_+ (respectively Ω_-). Then

$$\int_{\Omega_c} -\Delta u_{k,i}^N (u_{k,i}^N - b_i)^+ \, d\mathbf{x} \geqslant 0 \qquad \left(\text{respectively} \int_{\Omega_c} -\Delta u_{k,i}^N (u_{k,i}^N - a_i)^- \, d\mathbf{x} \geqslant 0 \right). \tag{2.5.16}$$

Proof: Each component Ω_c of Ω_+ and Ω_- may be approximated from within, to any desired accuracy in the Hausdorff metric, by an open set Ω_*, with boundary sufficiently smooth to apply the integration-by-parts formula. This may be seen by taking a finite subcover of balls of the compact boundary of Ω_c, and smoothing the piecewise spherical boundary of the resulting finite cover. We may do this in such a way that $\partial\Omega_c \cap \partial\Omega \supset \partial\Omega_* \cap \partial\Omega$. In particular, the integration-by-parts formulae

$$\int_{\Omega_*} - \Delta u_{k,i}^N (u_{k,i}^N - a_i)^- \, d\mathbf{x} - \int_{\Omega_*} |\nabla (u_{k,i}^N - a_i)^-|^2 \, d\mathbf{x}$$

$$= - \int_{\partial\Omega_*} \frac{\partial u_{k,i}^N}{\partial \nu} (u_{k,i}^N - a_i)^- \, d\sigma, \tag{2.5.17a}$$

$$\int_{\Omega_*} - \Delta u_{k,i}^N (u_{k,i}^N - b_i)^+ \, d\mathbf{x} - \int_{\Omega_*} |\nabla (u_{k,i}^N - b_i)^+|^2 \, d\mathbf{x}$$

$$= - \int_{\partial\Omega_*} \frac{\partial u_{k,i}^N}{\partial \nu} (u_{k,i}^N - b_i)^+ \, d\sigma \tag{2.5.17b}$$

hold on the approximation domains Ω_*. Note that, for Ω_c a component of Ω_- and for $d(\Omega_*, \Omega_c)$ sufficiently small, it follows, from the restriction of the closed zero set of f_i to the open region strictly between Q_{a_i} and Q_{b_i} (see (2.5.6)), that

$$(u_{k,i}^N - a_i)^- = 0 \qquad \text{on} \quad \partial\Omega_* \backslash \partial\Omega_* \cap \partial\Omega. \tag{2.5.18}$$

In particular,

$$\int_{\Omega_*} - \Delta u_{k,i}^N (u_{k,i}^N - a_i)^- \, d\mathbf{x} - \int_{\Omega_*} |\nabla (u_{k,i}^N - a_i)^-|^2 \, d\mathbf{x}$$
$$= \int_{\partial\Omega_* \cap \partial\Omega} \omega_i u_{k,i}^N (u_{k,i}^N - a_i)^- \, d\sigma \geqslant 0, \tag{2.5.19}$$

where we have used the Robin boundary condition and

$$\omega_i u_{k,i}^N (u_{k,i}^N - a_i)^- = \omega_i [a_i + (u_{k,i}^N - a_i)](u_{k,i}^N - a_i)^- \geqslant 0.$$

In a similar way, we deduce

$$\int_{\Omega_*} - \Delta u_{k,i} (u_{k,i} - b_i)^+ \, d\mathbf{x} - \int_{\Omega_*} |\nabla (u_{k,i}^N - b_i)^+|^2 \, d\mathbf{x}$$
$$= \int_{\partial\Omega_* \cap \partial\Omega} \omega_i u_{k,i}^N (u_{k,i}^N - b_i)^+ \, d\sigma \geqslant 0 \tag{2.5.20}$$

for Ω_c a component of Ω_+ and $d(\Omega_*, \Omega_c)$ sufficiently small. Thus, taking limits in (2.5.19) and (2.5.20) (as $d(\Omega_*, \Omega_c) \to 0$), we obtain (2.5.16). ■

Proposition 2.5.3. If $\{\mathbf{u}_k^N\}_{k=1}^{M(N)}$ satisfies (2.5.1)–(2.5.4), then the range of $\mathbf{u}_k^N$ is in Q.

Proof: We use the pointwise relations

$$(t_k^N - t_{k-1}^N)^{-1}(u_{k,i}^N - u_{k-1,i}^N) - D_i \Delta u_{k,i}^N + g_{k,i}^N = 0, \tag{2.5.21a}$$

which hold almost everywhere on Ω for $i \in \delta$ and $1 \leqslant i \leqslant i_0$, and

$$(t_k^N - t_{k-1}^N)^{-1}(u_{k,i}^N - u_{k-1,i}^N) + g_{k,i}^N = 0, \tag{2.5.21b}$$

for $i \notin \delta$ and $1 \leqslant i \leqslant i_0$. The proof now proceeds as follows. By (1.3.6c) and (2.5.6), we conclude that $u_{k,i}^N \geqslant a_i$ holds almost everywhere in Ω or else $\text{meas}(\Omega_-) > 0$, and these are not necessarily mutually exclusive alternatives. We shall show that the second of these implies the first. Multiplication of (2.5.21) by $(u_{k,i} - a_i)^-$, and integration over $\Omega_c \subset \Omega_-$ yields, after addition and subtraction of a_i,

$$0 \leqslant \int_{\Omega_c} (u_{k,i}^N - a_i)(u_{k,i}^N - a_i)^- \, d\mathbf{x} \leqslant \int_{\Omega_c} (u_{k-1,i}^N - a_i)(u_{k,i}^N - a_i)^- \, d\mathbf{x} \leqslant 0, \tag{2.5.22}$$

where we have used $g_{k,i}^N \leqslant 0$ in Ω_-, (2.5.16), and the inductive hypothesis. We conclude from (2.5.22) that $u_{k,i}^N \geqslant a_i$ in Ω_-, and, using (1.3.6d), that $u_{k,i}^N \geqslant a_i$ in Ω. A parallel argument shows that $u_{k,i}^N \leqslant b_i$ in Ω or else meas$(\Omega_+) > 0$, and that the second of these alternatives actually implies the first, upon multiplication of (2.5.21) by $(u_{k,i}^N - b_i)^+$, and integration over $\Omega_c \subset \Omega_+$. In particular, the range of $\mathbf{u}_k^N$ lies in Q. ■

2.6 BIBLIOGRAPHICAL REMARKS

The 'negative' norms introduced in Chapter 1, as equivalent dual space norms, are variations of those introduced by Lax [11] into the theory of partial differential equations (see also Leray [13]). Their use in numerical analysis dates from the work of Bramble and Osborn [3] on eigenvalue problems. The convergence results contained in Propositions 2.1.2, 2.2.5, and 2.3.2 depend upon the systematic use of these norms. They will be employed again, particularly in Chapter 4. The explicit smoothing of Section 2.1 was introduced in [8], whereas the smoothing of Section 2.3 is new to the writer.

The idea of using the lifted equations as primary pointwise relations, defined by the operators N_σ, D_0, and R_ω, appeared in the paper of the writer and Rose [9], where Proposition 2.1.2 was derived, as well as the uniqueness of solutions in the two-phase Stefan problem. The operators N_σ, $\sigma > 0$, make possible the derivation of uniqueness results for models involving nonlinear terms f with monotone components, not necessarily Lipschitz. This is due to the fact that N_σ is (pointwise) nonnegative for $\sigma > 0$. This remark also applies to D_0 and R_ω.

The writer is indebted to Veron (Introduction [9]) for the statement and proof of Proposition 2.2.3. The discrete Gronwall inequalities contained in Proposition 2.2.1 and Corollary 2.2.2 are familiar to numerical analysts, although the formulations and proofs are our own. The derivation of the semidiscrete stability relations in $\mathbf{L}^\infty$, for the case of the porous-medium equation, is closely tied to the existence of u_t as an $\mathbf{L}^2$ function in this problem (see the introductory part of Section 5.1 for details). We could, of course, have derived comparable $\mathbf{L}^p$ estimates, but their limit-case validity for the solution of the porous-medium equation would necessitate a relaxation in the regularity prescribed in (1.2.16a), with associated changes elsewhere.

The character of the results on invariant regions is due to the earlier investigations of Weinberger [15], and Chueh *et al.* [4], although the approach we have developed here is independent of these investigations. Also, the writer has not observed a mixing of the concentration and potential variables in any analysis of reaction–diffusion systems. The explicit determination of invariant regions for a specific reaction–diffusion system with given vector field is a major step in the application of these results. Some examples, including nonrectangular invariant regions, are discussed in Chueh *et al.* [4]. The FitzHugh–Nagumo system exclusively is considered by Rauch and Smoller in [14], where critical lower bounds and upper bounds on size are obtained for contracting rectangles.

A study of some specific models, such as predator–prey models, reveals the inadequacy of attempts at global existence proofs based upon *a priori* estimates insensitive to the location and/or size of the initial datum. Some type of comparison or weak maximum principle appears necessary even to obtain existence in such cases. Sections 2.2, 2.4, and 2.5 derive various formulations of weak maximum principles at discrete times within space-time cylinders. (see also some alternative discrete time investigations of Weinberger [16]).

A qualitative study of asymptotic and stationary behavior of solutions of reaction–diffusion systems and associated properties, such as bifurcation of solutions, is outside the scope of this analysis. For this, we refer the reader to the associated literature, for example, Aronson and Weinberger [1], Auchmuty and Nicolis [2], Fife [5], Fife and McLeod [6], Greenberg *et al.* [7], as well as to the bibliographies of these references. A treatment alternative to invariant rectangles is given by Lopes [*J. Differential Equations* **44**, 400–413 (1982)].

REFERENCES

[1] D. G. Aronson and H. F. Weinberger, Nonlinear diffusion in population genetics, combustion and nerve propagation, *in* "Partial Differential Equations and Related Topics" (J. A. Goldstein, ed.), *Lect. Notes Math.* **446**, pp. 5–49. Springer-Verlag, Berlin and New York, 1975; Multidimensional nonlinear diffusions arising in population genetics, *Adv. Math.* **30**, 33–76 (1978).

[2] J. F. G. Auchmuty and G. Nicolis, Bifurcation analysis of nonlinear reaction–diffusion equations: I. Evolution equations and the steady state solutions, *Bull. Math. Biol.* **37**, 323–365 (1975); III. Chemical oscillations, *ibid.* **38**, 325–349 (1975).

[3] J. H. Bramble and J. E. Osborn, Approximation of Steklov eigenvalues of non-selfadjoint second order elliptic operators, *in* "The Mathematical Foundations of the Finite Element Method" (I. Babuska and A. K. Aziz, eds.), pp. 387–408. Academic Press, New York, 1972; Rate of convergence estimates for non self-adjoint eigenvalue approximations, *Math. Comp.* **27**, 525–549 (1973).

[4] K. Chueh, C. Conley, and J. Smoller, Positively invariant regions for systems of nonlinear parabolic equations, *Indiana Univ. Math. J.* **26**, 373–392 (1977).

[5] P. C. Fife, Stationary patterns for reaction–diffusion equations, *in* "Nonlinear Diffusion" (W. E. Fitzgibbon and H. F. Walker, eds.), *Res. Notes Math.* **14**, pp. 81–121. Pitman, London, 1977.

[6] P. C. Fife and J. B. McLeod, The approach of solutions of nonlinear diffusion equations to traveling front solutions, *Arch. Rational Mech. Anal.* **65**, 335–361 (1977).

[7] J. M. Greenberg, B. D. Hassard, and S. P. Hastings, Pattern formation and periodic structures in systems modeled by reaction–diffusion equations, *Bull. Amer. Math. Soc.* **84**, 1296–1327 (1978).

[8] J. Jerome, Existence and approximation of weak solutions of nonlinear Dirichlet problems with discontinuous coefficients, *SIAM J. Math. Anal.* **9**, 730–742 (1978).

[9] J. Jerome and M. Rose, Error estimates for the multidimensional two-phase Stefan problem, *Math. Comp.* **39** (1982), in press.

[10] O. A. Ladyženskaja, V. A. Solonnikov, and N. N. Ural'ceva, "Linear and Quasilinear Equations of Parabolic Type," *Trans. Math. Monographs* **23**. American Mathematical Society, Providence, Rhode Island, 1968.

[11] P. Lax, On Cauchy's problem for hyperbolic equations and the differentiability of solutions of elliptic equations, *Comm. Pure Appl. Math.* **8**, 615–633 (1955).

[12] E. B. Leach and M. Sholander, Extended mean values, *Amer. Math. Monthly* **85**, 84–90 (1978).

[13] J. Leray, "Hyperbolic Differential Equations." Princeton Univ. Press, Princeton, New Jersey, 1952.

[14] J. Rauch and J. Smoller, Qualitative theory of the FitzHugh–Nagumo Equations, *Adv. Math.* **27**, 12–44 (1978).

[15] H. F. Weinberger, Invariant sets for weakly coupled parabolic and elliptic systems, *Rend. Mat. Univ. Roma* **8**, 295–310 (1975).

[16] H. F. Weinberger, Asymptotic behavior of a class of discrete–time models in population genetics, *in* "Applied Nonlinear Analysis" (V. Lakshmikantham, ed.), pp. 407–422. Academic Press, New York, 1979.

3 NONLINEAR ELLIPTIC EQUATIONS AND INEQUALITIES

3.0 INTRODUCTION

The method of semidiscretization in time, when applied to nonlinear parabolic evolutions, gives rise to elliptic equations and/or inequalities. The same is true of various steady-state theories. In this chapter, we provide the required theoretical foundations for studying such elliptic problems. The principal results of Section 3.1 are Tarski's fixed-point theorem (see Proposition 3.1.10) and an existence theorem for "equations" of the form $f \in \partial\phi(u) + \mathrm{A}(u)$, where ϕ is proper, convex, and lower-semicontinuous, and A is pseudomonotone (see Corollary 3.1.7). Here, $\partial\phi$ is the multivalued subdifferential mapping with range in an appropriate dual space.

In Section 3.2, four applications are presented. Perhaps the second, which specializes $\partial\phi(u) + \mathrm{A}(u)$ to those applications met in Chapter 1, and the third, related to the Navier–Stokes equations, are the most pertinent. An application of the Tarski fixed-point theorem to a quasi-variational inequality is also presented.

In Section 3.3, standard interpolation and extrapolation methods are discussed for the purpose of defining semidiscretizations for nonlinear evolution equations. Although these methods are quite routine in the theory of initial-value problems for ordinary differential equations, the novelty here is that they arise naturally from quadrature formulae for the *lifted* versions of parabolic evolutions. This demonstrates the theoretical interest of the lifting operators, over and above their technical use as demonstrated in the uniqueness and error analysis of the previous two chapters.

3.1 GENERAL OPERATOR RESULTS IN BANACH SPACES AND ORDERED SPACES

The development of this section will proceed to general Banach space results, after a fairly thorough treatment of the finite-dimensional case, which proves pivotal. The initial concept is that of the proximity mapping, which generalizes the notion of a projection onto a closed, convex set in Hilbert space.

Definition 3.1.1. Let $\mathbf{V}$ be a Hilbert space and let $\phi:\mathbf{V}\to(-\infty,\infty]$ satisfy $\phi\not\equiv\infty$, that is ϕ is proper. We suppose ϕ is convex and lower-semicontinuous on $\mathbf{V}$. By the proximity mapping, $\mathrm{prox}_\phi:\mathbf{V}\to\mathbf{V}$, is meant the mapping

$$x\to\mathrm{prox}_\phi(x)=u, \tag{3.1.1a}$$

where u (uniquely) satisfies

$$F(u,x)=\min_{v\in\mathbf{V}}F(v,x), \tag{3.1.1b}$$

and $F(\cdot,\cdot)$ is given by

$$F(v,x)=\tfrac{1}{2}\|v-x\|^2+\phi(v). \tag{3.1.1c}$$

Remark 3.1.1. The existence of a minimum of F is now a standard result (see Ekeland and Temam [13] p. 35). Use is made of the weak lower-semicontinuity and coerciveness of F. Note that the assumed lower-semicontinuity of ϕ is preserved in passing from the strong to the weak topology. The coerciveness property, $F(v)\to\infty$ as $\|v\|\to\infty$, follows from the fact that ϕ is bounded from below by a continuous affine function. The uniqueness follows from the strict convexity of $\|\cdot\|^2$, hence of F.

Just as the standard projection in Hilbert space has a familiar variational inequality characterization, the same is true of the generalized proximity mapping. For convenience, we quote the result (see [13] Sections 2.2 and 2.3).

Lemma 3.1.1. The element $u = \text{prox}_\phi x$ is characterized by either of the following equivalent conditions:

$$(u - x, v - u) + \phi(v) - \phi(u) \geqslant 0, \qquad \text{for all} \quad v \in \mathbf{V}, \tag{3.1.2a}$$

$$(v - x, v - u) + \phi(v) - \phi(u) \geqslant 0, \qquad \text{for all} \quad v \in \mathbf{V}. \tag{3.1.2b}$$

Moreover, prox_ϕ is nonexpansive, that is

$$\|\text{prox}_\phi x - \text{prox}_\phi y\| \leqslant \|x - y\|, \qquad \text{for all} \quad x, y \in \mathbf{V}. \tag{3.1.3}$$

Remark 3.1.2. Let ϕ be a proper, convex, lower-semicontinuous function on $\mathbf{V}$ as above. Let

$$\textbf{Dom}\ \phi = \{v \in \mathbf{V}: \phi(v) < \infty\}. \tag{3.1.4}$$

Then **Dom** ϕ is a convex subset of $\mathbf{V}$ and prox_ϕ has range in **Dom** ϕ.

Example 3.1.1. Let $\mathbf{K}$ be a nonempty, closed convex subset of $\mathbf{V}$. If $\phi_\mathbf{K}: \mathbf{K} \to (-\infty, \infty]$ is a given proper, convex, lower-semicontinuous function, then so is the extension,

$$\phi(x) = \begin{cases} \phi_\mathbf{K}(x), & x \in \mathbf{K}, \\ \infty, & x \notin \mathbf{K}, \end{cases} \tag{3.1.5}$$

to $\mathbf{V}$. An application of the previous lemma and remark gives the following characterization of

$$u = \text{prox}_\phi x = \text{prox}_{\phi_\mathbf{K}} x.$$

$$(u - x, v - u) + \phi_\mathbf{K}(v) - \phi_\mathbf{K}(u) \geqslant 0, \qquad \text{for all} \quad v \in \mathbf{K}, \tag{3.1.6a}$$

$$(v - x, v - u) + \phi_\mathbf{K}(v) - \phi_\mathbf{K}(u) \geqslant 0, \qquad \text{for all} \quad v \in \mathbf{K}. \tag{3.1.6b}$$

Definition 3.1.2. Let $\mathbf{S}$ be a finite-dimensional normed linear space with dual $\mathbf{S}^*$, and let $\phi: \mathbf{S} \to (-\infty, \infty]$ be a proper, convex, lower-semicontinuous function. Let $\mathrm{A}: \mathbf{S} \to \mathbf{S}^*$ satisfy

$$\mathrm{A} \text{ is continuous,} \tag{3.1.7a}$$

and, if **Dom** ϕ is unbounded,

$$\frac{\langle \mathrm{A}(v), v - v_0 \rangle + \phi(v)}{\|v\|} \geqslant \omega(\|v\|), \qquad \|v\| \geqslant \rho, \tag{3.1.7b}$$

where $\omega: [\rho, \infty) \to [0, \infty)$ is a monotone increasing function with $\omega^{-1}(t)$ bounded for each t. Here, v_0 is assumed to be a fixed element of $\mathbf{S}$ with $0 \leqslant \|v_0\| \leqslant \rho$ and ρ is a fixed positive number.

Proposition 3.1.2.† Let S, A, ϕ, ω, v_0, and ρ be given as in the previous definition. If **Dom** ϕ is bounded in **S**, then the inequality

$$\langle A(u) - f, v - u \rangle + \phi(v) - \phi(u) \geqslant 0, \qquad \text{for all} \quad v \in \mathbf{S}, \tag{3.1.8}$$

has a solution $u \in$ **Dom** ϕ for $f \in \mathbf{S}^*$. If **Dom** ϕ is unbounded, let ϕ_{r_0} denote the function, with **Dom** ϕ_{r_0} bounded, given by

$$\phi_{r_0}(v) = \begin{cases} \phi(v), & \|v\| \leqslant r_0, \\ \infty, & \|v\| > r_0, \end{cases} \tag{3.1.9a}$$

for $r_0 > \rho$ fixed, and let u_{r_0} be a solution of the inequality

$$\langle A(u_{r_0}) - f, v - u_{r_0} \rangle + \phi_{r_0}(v) - \phi_{r_0}(u_{r_0}) \geqslant 0, \qquad \text{for all} \quad v \in \mathbf{S}. \tag{3.1.9b}$$

If $u_{r_0} \neq 0$, and

$$\sigma_0 = \|f\| \left[1 + \frac{\|v_0\|}{\|u_{r_0}\|} \right] + |\phi(v_0)|/\|u_{r_0}\| \leqslant \sigma \in \text{Range}\, \omega, \tag{3.1.10}$$

then the inequality (3.1.8) has a solution $u \in \overline{\mathbf{B}(0, c)}$, where $c = \sup\{t : \omega(t) = \sigma\}$. If (3.1.9b) has only the solution $u_{r_0} = 0$ for all $r_0 > \rho$, then $u = 0$ is a solution of (3.1.8).

In order to indicate the usefulness of Proposition 3.1.2, we delay its proof to present two corollaries.

Corollary 3.1.3. If $\omega(\|v\|) \to \infty$ as $\|v\| \to \infty$, then (3.1.10) is necessarily satisfied for every solution $u_{r_0} \neq 0$ of (3.1.9b), and (3.1.8) has a solution for each $f \in \mathbf{S}^*$ under the hypotheses of Definition 3.1.2.

Corollary 3.1.4. Let **K** be a nonempty, closed, convex subset of **S** containing 0. Then the inequality

$$\langle A(u) - f, v - u \rangle \geqslant 0, \qquad \text{for all} \quad v \in \mathbf{K}, \tag{3.1.11}$$

has a solution $u \in \text{Cl}\, \mathbf{B}(0, \sup \omega^{-1}(\|f\|))$, if $\|f\| \in \text{Range}\, \omega$. In this case, the equation

$$A(u) = f \tag{3.1.12}$$

holds if u is an interior point (in $\mathbf{K}^0$), or if 0 is an interior point and

$$\langle A(w) - f, w \rangle \geqslant 0, \qquad \text{for all} \quad w \in \partial \mathbf{K}. \tag{3.1.13}$$

† This is a slight generalization of Theorem 3.1 (p. 41) of Ref. [13].

Proof of Corollary 3.1.4: We choose $v_0 = 0$ and ϕ to be the indicator function of **K**, that is,

$$\phi(x) = \begin{cases} 0, & x \in \mathbf{K}, \\ \infty, & x \notin \mathbf{K}. \end{cases} \tag{3.1.14}$$

The choice $\sigma_0 = \sigma = \|f\|$ in (3.1.10) is now valid by hypothesis and the definition of v_0 and ϕ. Hence, the proposition applies. If u is an interior point, (3.1.12) is immediate upon choosing $v = u \pm \varepsilon w$, $w \in \mathbf{S}$. If $u \in \partial\mathbf{K}$, then $\langle \mathrm{A}(u) - f, v \rangle \geqslant 0$ for all $v \in \mathbf{K}$, which implies (3.1.12), since 0 is an assumed interior point in this case. ■

Proof of Proposition 3.1.2: If **Dom** ϕ is bounded, then the mapping

$$v \mapsto \mathrm{prox}_\phi(v + f - \mathrm{A}(v)), \tag{3.1.15}$$

obtained by identifying f and $\mathrm{A}(v)$ with elements of **S**, is a continuous mapping of the closed, convex, bounded set $\overline{\mathbf{Dom}\ \phi}$ into itself by Lemma 3.1.1 and Remark 3.1.2. Hence, by the Brouwer fixed-point theorem (see Chapter 4 [26] pp. 18 and 19) this mapping has a fixed point $u \in \overline{\mathbf{Dom}\ \phi}$ which satisfies (3.1.8). In particular, $u \in \mathbf{Dom}\ \phi$.

Now suppose that **Dom** ϕ is unbounded. By the first part of the proposition, (3.1.9b) has a solution for each positive r_0, in particular for $r_0 > \rho$. If (3.1.9b) has only the zero solution for all $r_0 > \rho$, then taking limits in (3.1.9b) as $r_0 \to \infty$ shows that $u = 0$ is a solution of (3.1.8). Otherwise, let u_{r_0} satisfy (3.1.9b), with $0 \neq u_{r_0}$ and $r_0 > \rho$, and select a sequence $\{r_i\}_{i=1}^{\infty}$, satisfying $r_0 < r_i \to \infty$. If the u_{r_i} solve

$$\langle \mathrm{A}(u_{r_i}) - f, v - u_{r_i} \rangle + \phi_{r_i}(v) - \phi_{r_i}(u_{r_i}) \geqslant 0, \tag{3.1.16a}$$

where ϕ_{r_i} is defined, in analogy with (3.1.9a), by

$$\phi_{r_i}(v) = \begin{cases} \phi(v), & \|v\| \leqslant r_i, \\ \infty, & \|v\| > r_i, \end{cases} \tag{3.1.16b}$$

then we suppose $\|u_{r_0}\| \leqslant \|u_{r_i}\|$, $i = 1, 2, \ldots$. Then we obtain, from (3.1.7b) and (3.1.16a),

$$\begin{aligned} \omega(\|u_{r_i}\|) &\leqslant \frac{1}{\|u_{r_i}\|} \{\langle \mathrm{A}(u_{r_i}), u_{r_i} - v_0 \rangle + \phi(u_{r_i})\} \\ &\leqslant \frac{1}{\|u_{r_i}\|} \{-\langle f, v_0 - u_{r_i} \rangle + \phi(v_0)\} \\ &\leqslant \sigma_0 \leqslant \sigma, \end{aligned} \tag{3.1.17}$$

where σ_0 and σ are defined by (3.1.10). Since $\sigma \in$ Range ω, we have

$$\|u_{r_i}\| \leqslant \sup \omega^{-1}(\sigma) = c, \qquad i \geqslant 0, \tag{3.1.18}$$

so that $\{u_{r_i}\} \subset \overline{\mathbf{B}(0,c)} \subset \mathbf{S}$, and is, thus, a relatively compact sequence, with accumulation point u. Upon taking the limit supremum in (3.1.16a), we obtain (3.1.8) for $u \in \overline{\mathbf{B}(0,c)}$. If the sequences $\{r_i\}$ and $\{u_{r_i}\}$ cannot be chosen so that $\|u_{r_i}\| \geqslant \|u_{r_0}\|$, all i, then a sequence $\{u_{r_i}\}$ necessarily is contained in $\mathbf{B}(0, \|u_{r_0}\|)$. An identical argument, as above, in conjunction with (3.1.18), yields (3.1.8). ■

Remark 3.1.3. The theory developed thus far gives sufficient conditions for the duality inequality (3.1.8) to possess a solution in a finite-dimensional space **S**. The formulation permits various choices of ϕ to obtain, for example, variational inequalities and equations. Also covered is the important case of multivalued subdifferentials of convex functions. We briefly introduce some terminology to discuss this case.

Definition 3.1.3. Let **X** be a normed linear space with dual $\mathbf{X}^*$, and let $F:\mathbf{X} \to (-\infty, \infty]$ be a proper, convex, lower-semicontinuous function. By the subdifferential ∂F, defined for u, such that $F(u)$ is finite, is meant the set-valued function

$$\partial F(u) = \{f \in \mathbf{X}^*: F(v) - F(u) \geqslant \langle f, v - u\rangle \qquad \text{for all} \quad v \in \mathbf{X}\}. \tag{3.1.19}$$

If **X** is a Hilbert space, it is common to identify the functionals of $\partial F(u)$ with the set of representers or subgradients.

Remark 3.1.4. $\partial F(u)$ is a weak-* closed, convex, but possibly empty, set for a given $u \in \mathbf{X}$. Suppose now that u is a solution of (3.1.8). Then $f - \mathrm{A}(u) \in \partial\phi(u)$ and, of course, the converse holds. Equivalently,

$$f \in \partial\phi(u) + \mathrm{A}(u) \tag{3.1.20}$$

characterizes solutions u of (3.1.8). If ϕ is Gâteaux differentiable at u, with derivative $\phi'(u;\cdot)$, then (3.1.20) is, in fact, the equation

$$f = \phi'(u;\cdot) + \mathrm{A}(u). \tag{3.1.21}$$

Remark 3.1.5. We shall introduce additional properties required of the mapping A, which will permit us to pass from the finite-dimensional formulation to the infinite-dimensional formulation. A natural class of such operators is the class of pseudomonotone operators, which we now describe. Note that, whereas for the previous results, we required $A(\cdot)$ to be defined on the entire space, we now relax this requirement.

Definition 3.1.4. Let $\mathbf{K}$ be a convex subset of a normed linear space $\mathbf{X}$. Suppose $A:\mathbf{K} \to \mathbf{X}^*$ satisfies, for all $v \in \mathbf{X}$,

$$\langle A(u), u - v\rangle \leqslant \liminf_{i\to\infty} \langle A(u_i), u_i - v\rangle, \tag{3.1.22a}$$

whenever

$$u_i \rightharpoonup u \text{ (weakly)} \qquad \text{and} \qquad \limsup\langle A(u_i), u_i - u\rangle \leqslant 0, \tag{3.1.22b}$$

for $u_i, u \in \mathbf{K}$. Suppose also that

$$A(\mathbf{K}_b) \qquad \text{is bounded in} \quad \mathbf{X}^*, \tag{3.1.22c}$$

for every bounded subset $\mathbf{K}_b$ of $\mathbf{K}$. Then A is called a pseudomonotone operator. Note that the weak convergence $u_i \rightharpoonup u$ is specified as sequential convergence, rather than the more general convergence of nets; (3.1.22c) is separate in [6] and [12].

Remark 3.1.6. Pseudomonotone operators, with domains consisting of open subsets of finite-dimensional spaces, are characterized, in an elementary way, by continuity (see Brézis [6] p. 132). In particular, Proposition 3.1.2 is valid for a pseudomonotone operator A, defined on $\mathbf{K} = \mathbf{S}$, since A is, thus, continuous on $\mathbf{S}$. A more subtle problem arises when $\mathbf{K}$, say, is simply a compact, convex subset of $\mathbf{S}$ and $\mathbf{Dom}\ \phi \subset \mathbf{K}$. In this case, a standard reduction to $\mathbf{K}^0 = (\mathbf{Dom}\ \phi)^0 \neq \varnothing$ is possible, and, by expressing $\mathbf{K}^0$ as a union of compact, convex sets and employing the continuity of A on these sets, Brézis obtains a solution on $\mathbf{K}^0$, by passing to a limit, via pseudomonotone methods. The transition to $\mathbf{K}$ is standard. We refer the reader to Brézis [6], pp. 134–136, for details. This passage to the limit, however, is quite similar to that in passing from the finite- to the infinite-dimensional case, which we discuss in detail. Note that, if $\mathbf{K}$ is not bounded, some coerciveness condition, such as (3.1.7b) and (3.1.10), is required. Rather than state the finite-dimensional result, we avoid repetition and proceed directly to the Banach space result.

Proposition 3.1.5. Let $\mathbf{K}$ be a closed, convex subset of a reflexive Banach space $\mathbf{X}$, with $0 \in \mathbf{K}$, and suppose that $A:\mathbf{K} \to \mathbf{X}^*$ is pseudomonotone, and $\phi:\mathbf{K} \to (-\infty, \infty]$ is a proper, convex, lower-semicontinuous function satisfying $\phi(0) = 0$. Consider the inequality, for the unknown $u \in \mathbf{K}$,

$$\langle A(u) - f, v - u\rangle + \phi(v) - \phi(u) \geqslant 0, \qquad \text{for all} \quad v \in \mathbf{K}, \tag{3.1.23}$$

for f specified in $\mathbf{X}^*$. Then (3.1.23) has a solution $u \in \mathbf{K}$, if $\mathbf{K}$ is bounded. On the contrary, suppose $\mathbf{K}$ is unbounded. If

$$\frac{\langle \mathrm{A}(v), v\rangle + \phi(v)}{\|v\|} \geqslant \omega(\|v\|), \qquad v \in \mathbf{K}, \quad \|v\| \geqslant \rho, \tag{3.1.24}$$

with ω and ρ specified in Definition 3.1.2, and

$$\|f\| \leqslant \sigma \in \operatorname{Range} \omega, \tag{3.1.25}$$

then (3.1.23) has a solution $u \in \mathrm{Cl}\, \mathbf{B}(0, \sup \omega^{-1}(\sigma))$.

Proof: Let $z \in \mathbf{K}$ satisfy $\phi(z) < \infty$, and let $\mathscr{F}$ denote the family of finite-dimensional subspaces $\mathbf{F}$ of $\mathbf{X}$, such that $z \in \mathbf{F}$. For $\mathbf{F} \in \mathscr{F}$, let $\mathbf{K_F} = \mathbf{K} \cap \mathbf{F}$, and let $\mathrm{A_F}$ and $\phi_\mathbf{F}$ denote the restrictions of A and ϕ to $\mathbf{K_F}$. Applying the finite-dimensional result of [6] (see Remark 3.1.6) to $\mathbf{K_F}$, we conclude that the set

$$\begin{aligned} \mathbf{S_F} = \{u_\mathbf{F} \in \mathbf{K_F} \cap \overline{\mathbf{B}(0,c)} : c = \sup \omega^{-1}(\sigma) \qquad &\text{and} \\ \langle \mathrm{A}(u_\mathbf{F}) - f, v - u_\mathbf{F}\rangle + \phi(v) - \phi(u_\mathbf{F}) \geqslant 0, \qquad &\text{for all} \quad v \in \mathbf{K_F}\} \end{aligned} \tag{3.1.26}$$

is nonempty. In the event $\mathbf{K}$ is bounded, we replace $\mathbf{K_F} \cap \overline{\mathbf{B}(0,c)}$ by $\mathbf{K_F}$.

For each $\mathbf{F} \in \mathscr{F}$, denote by $\mathbf{V_F} \subset \mathbf{K}$ the weak closure of $\bigcup \{\mathbf{S_{F'}} : \mathbf{F} \subset \mathbf{F}'\}$. The family

$$\mathscr{V} = \{\mathbf{V_F} : \mathbf{F} \in \mathscr{F}\} \tag{3.1.27}$$

of weakly closed subsets of the weakly compact set $\mathbf{K} \cap \overline{\mathbf{B}(0,c)}$ has the finite-intersection property, and, hence, has a nonvoid intersection, say, $u \in \bigcap \{\mathbf{V_F} : \mathbf{F} \in \mathscr{F}\}$. We shall demonstrate that u satisfies (3.1.23).

Let $v \in \mathbf{K}$ and let $\mathbf{F} \in \mathscr{F}$ be chosen so that u and v are in $\mathbf{F}$; in particular, u and v are in $\mathbf{K_F}$. Since u is an accumulation or adherance point, in the weak topology, of the relatively weakly compact subset $\bigcup \{\mathbf{S_{F'}} : \mathbf{F} \subset \mathbf{F}'\}$ of the reflexive Banach space $\mathbf{X}$, it follows that u is a weak sequential limit point (see Browder [12] p. 81) of this set. Thus, there is a sequence $u_j = u_{\mathbf{F}'_j} \in \mathbf{S}_{\mathbf{F}'_j}$, such that $u_j \rightharpoonup u$, where $\mathbf{F} \subset \mathbf{F}'_j$. We, thus, have, for each j,

$$\langle \mathrm{A}(u_j) - f, u_j - v\rangle \leqslant \phi(v) - \phi(u_j), \tag{3.1.28}$$

so that, taking the limit infimum and limit supremum of left- and right-hand sides gives

$$\begin{aligned} \liminf_{j\to\infty} \langle \mathrm{A}(u_j), u_j - v\rangle &\leqslant \limsup_{j\to\infty} \; [\phi(v) - \phi(u_j)] + \langle f, u - v\rangle \\ &= \phi(v) - \liminf_{j\to\infty} \phi(u_j) + \langle f, u - v\rangle \\ &\leqslant \phi(v) - \phi(u) + \langle f, u - v\rangle. \end{aligned} \tag{3.1.29}$$

If we can show that

$$\limsup_{j\to\infty} \langle A(u_j), u_j - u\rangle \leqslant 0, \tag{3.1.30}$$

then the pseudomonotone property applied to (3.1.29) yields

$$\langle A(u), u - v\rangle \leqslant \phi(v) - \phi(u) + \langle f, u - v\rangle,$$

which is (3.1.24). To establish (3.1.30), we choose $v = u$ in (3.1.28), and take the limit supremum of both sides. This gives

$$\begin{aligned}\limsup_{j\to\infty} \langle A(u_j), u_j - u\rangle &\leqslant \limsup_{j\to\infty} [\phi(u) - \phi(u_j)] \\ &= \phi(u) - \liminf_{j\to\infty} \phi(u_j) \\ &\leqslant \phi(u) - \phi(u) = 0,\end{aligned}$$

which is (3.1.30). ■

Corollary 3.1.6. Let $\mathbf{X}$ be a reflexive Banach space, and let A be a pseudomonotone mapping defined on $\overline{\mathbf{B}(0,\rho)}$. Suppose that $g(x) \geqslant 0$ for $v \in \partial\mathbf{B}(0,\rho)$, where

$$g(v) = \langle A(v), v\rangle, \qquad v \neq 0; \tag{3.1.31}$$

then the equation $A(u) = 0$ has a solution $\|u\| \leqslant \rho$.

Proof: Since $\overline{\mathbf{B}(0,\rho)}$ is bounded, the existence of a solution u of the variational inequality

$$\langle A(u), v - u\rangle \geqslant 0, \qquad \|v\| \leqslant \rho$$

is guaranteed in $\overline{\mathbf{B}(0,\rho)}$ by Proposition 3.1.5. By the hypotheses, we conclude $\langle A(u), v\rangle \geqslant 0$ for all v, $\|v\| \leqslant \rho$. This yields $A(u) = 0$. ■

Remark 3.1.7. The previous result admits of generalization to closed convex sets $\mathbf{K}$. In particular, if $g \geqslant 0$ outside the relative interior $\mathbf{K}_1^0$ of a compact, convex subset $\mathbf{K}_1$ of $\mathbf{K}$, with $0 \in \mathbf{K}_1^0$, then $A(u) = 0$ has a solution in $\mathbf{K}$ (see [6] p. 137 for details). We now record another corollary for use in the sequel.

Corollary 3.1.7. Let $\mathbf{X}$ be a reflexive Banach space, let $\mathbf{X}_0$ be a closed linear subspace of $\mathbf{X}$, and suppose that $A: \mathbf{X}_0 \to \mathbf{X}^*$ is a pseudomonotone mapping and that $\phi: \mathbf{X} \to (-\infty, \infty]$ is a proper, convex, lower-semicontinuous function with $\phi(0) = 0$. Suppose that $\omega: [\rho, \infty) \to [0, \infty)$ is a monotone increasing function with $\omega^{-1}(t)$ bounded for each t and $\rho > 0$, such that

$$\frac{\langle \mathrm{A}(v), v\rangle + \phi(v)}{\|v\|} \geqslant \omega(\|v\|), \qquad \|v\| \geqslant \rho. \tag{3.1.32}$$

If $\|f\| \leqslant \sigma \in$ Range ω, then there exists $u \in \mathbf{X}_0$ such that

$$\|u\| \leqslant \sup \omega^{-1}(\sigma), \tag{3.1.33a}$$

$$f \in \partial\phi(u) + \mathrm{A}(u). \tag{3.1.33b}$$

Remark 3.1.8. The obvious extension of Corollary 3.1.4 carries over if $\mathrm{A}:\mathbf{K} \subset \mathbf{X} \to \mathbf{X}^*$ is pseudomonotone, and $\mathbf{X}$ is a reflexive Banach space. An important subclass of the class of pseudomonotone operators is the class of monotone. hemicontinuous mappings. These are defined later in this section (see Definition 3.1.6). Aside from the important surjectivity property described by (3.1.33b), the pseudomonotone operators enjoy the following property when M is bounded:

$$\mathrm{B} = \mathrm{A} + \mathrm{M} \quad \text{is pseudomonotone,} \tag{3.1.34}$$

if A is pseudomonotone and M is hemicontinuous and monotone. Property (3.1.34) is stated precisely later (see Proposition 3.1.9). The class of pseudomonotone operators defined on $\mathbf{X}$ is, in fact, a subclass of a still larger class introduced by Brézis [6] for the purpose of describing an existence theory for the equation $\mathrm{A}(u) = f$. The operators of this class are termed operators of type M. It is not our intent to discuss these operators at length. However, we shall present their definition and the basic surjectivity result (see also (3.1.33b)) quoted with some modifications from Brézis [6], pp. 124–128.

Definition 3.1.5. Let $\mathbf{X}$ be a normed linear space. Then $\mathrm{A}:\mathbf{X} \to \mathbf{X}^*$ is said to be of type M if

The restrictions of A to finite-dimensional subspaces of $\mathbf{X}$ are continuous; and (3.1.35a)

$\mathrm{A}(u) = f$ whenever $u_i \rightharpoonup u$, $\mathrm{A}u_i \rightharpoonup f$,† and $\limsup_{j\to\infty} \langle \mathrm{A}(u_i), u_i\rangle \leqslant \langle f, u\rangle$. (3.1.35b)

These operators enjoy the following surjectivity property.

Proposition 3.1.8. Let $\mathbf{X}$ be a reflexive Banach space, and suppose $\mathrm{A}:\mathbf{X}\to\mathbf{X}^*$ is of type M. Suppose the coerciveness hypothesis $\|f\| \leqslant \sigma \in$ Range ω holds

† The weak star convergence in the dual space, implied by the weak convergence in the dual space, usually suffices.

for ω, described in Definition 3.1.2 and satisfying

$$\frac{\langle \mathrm{A}(v), v\rangle}{\|v\|} \geqslant \omega(\|v\|), \qquad \|v\| \geqslant \rho > 0. \tag{3.1.36}$$

Then the equation

$$\mathrm{A}(u) = f \tag{3.1.37}$$

has a solution u, $\|u\| \leqslant \sup \omega^{-1}(\sigma)$.

The definition of a monotone, hemicontinuous operator is now given.

Definition 3.1.6. Let $\mathbf{K}$ be a convex subset of a normed linear space $\mathbf{X}$. A mapping $\mathrm{A}:\mathbf{K} \to \mathbf{X}^*$ is said to be hemicontinuous if, for any $x \in \mathbf{K}$, $y \in \mathbf{X}$, and any sequence $\{t_i\}$ of positive real numbers, such that $x + t_i y \in \mathbf{K}$, $i = 1, 2, \ldots,$ then

$$\mathrm{A}(x + t_i y) \rightharpoonup \mathrm{A}(x), \qquad i \to \infty. \tag{3.1.38}$$

The mapping A is said to be monotone if

$$\langle \mathrm{A}(x) - \mathrm{A}(y), x - y\rangle \geqslant 0, \tag{3.1.39}$$

for all $x, y \in \mathbf{K}$. Further, A is strictly monotone if A is monotone and equality in (3.1.39) implies $x = y$.

Remark 3.1.9. It is an easy exercise to show that solutions of (3.1.23) are uniquely determined if A is strictly monotone. The following proposition is proved by straightforward arguments in Brézis [6], p. 133.

Proposition 3.1.9. Let $\mathbf{K}$ be a convex subset of a normed linear space $\mathbf{X}$. Let $\mathrm{A}:\mathbf{K} \to \mathbf{X}^*$ and $\mathrm{M}:\mathbf{K} \to \mathbf{X}^*$ be, respectively, a pseudomonotone and a monotone, hemicontinuous, bounded mapping. Then the mapping $\mathrm{B} = \mathrm{A} + \mathrm{M}$ is pseudomonotone. In particular, $\mathrm{B} = \mathrm{M}$ is pseudomonotone.

We briefly state now the connection of the previously developed ideas with the topic of multivalued maximal monotone operators.

Definition 3.1.7. Let $\mathbf{X}$ be a normed linear space, and let $\mathbf{K}$ be a subset of $\mathbf{X}$. A (multivalued) mapping $\mathrm{A}: \mathbf{K} \to 2^{\mathbf{X}^*}$ is said to be monotone if the graph of A,†

$$\mathbf{gr}(\mathrm{A}) = \{(x, y): x \in \mathbf{K}, y \in \mathrm{A}(x)\}, \tag{3.1.40a}$$

† The conventional ordering is preserved. It is standard to reverse this order in the duality pairing.

is monotone, that is,

$$\langle y_1 - y_2, x_1 - x_2 \rangle \geqslant 0, \tag{3.1.40b}$$

for $y_1 \in \mathrm{A}(x_1)$, $y_2 \in \mathrm{A}(x_2)$. The mapping A is maximal monotone if the graph of A has no proper monotone extension to $\mathbf{X} \times \mathbf{X}^*$.

Remark 3.1.10. An example of a maximal monotone mapping is any hemicontinuous, monotone mapping defined on all of a Banach space $\mathbf{X}$ (see Opial [25], p. 77). Another example is the subdifferential

$$\mathrm{A} = \partial\phi : \mathbf{Dom}\,\mathrm{A} \to 2^{\mathbf{X}^*} \tag{3.1.41}$$

of a proper, convex, lower-semicontinuous function (see Barbu [3] Chapter 2, Section 1]). Here $\mathbf{Dom}\,\mathrm{A} = \{v \in \mathbf{Dom}\,\phi : \partial\phi(v) \neq \varnothing\}$. Equation (3.1.33b) may be paraphrased to say that the coercive sum of the pseudomonotone operator A and the maximal monotone operator $\mathrm{B} = \partial\phi$ is surjective. In fact, this result remains valid (see Brézis [7] Theorème 1) if B is any maximal monotone operator and A is pseudomonotone on $\overline{\mathbf{Dom}\,\mathrm{B}}$. Additional examples will be discussed in the next section. The reader, interested in what subclass of maximal monotone mappings is characterized by (3.1.41), is referred to the cyclically monotone mappings of [8]. If $\mathbf{X} = \mathbb{R}^1$, the classes coincide. Related ideas are discussed by Rockafellar [26].

Remark 3.1.11. It is sometimes useful to have fixed-point theorems for monotone (increasing) mappings, which do not depend on topological properties, such as continuity. We shall present this result, sometimes called Tarski's fixed-point theorem, after an initial definition, which describes the ordering with respect to which the monotonicity is defined. The term isotone is often applied in the literature to such monotone mappings.

Definition 3.1.8. A set $\mathscr{A}$ is said to be (partially) ordered if there exists a reflexive, antisymmetric, transitive relation defined by a subset of $\mathscr{A} \times \mathscr{A}$. A chain in $\mathscr{A}$ is a subset $\mathscr{C}$, such that $x, y \in \mathscr{C}$ imply $x \leqslant y$ or $y \leqslant x$. If $\mathscr{B}$ is a subset of $\mathscr{A}$, then $z \in \mathscr{A}$ is an upper bound for $\mathscr{B}$, if $x \leqslant z$ for all $x \in \mathscr{B}$, and z is a least upper bound for $\mathscr{B}$ if, in addition, $z \leqslant y$ for all upper bounds y of $\mathscr{B}$. Lower bounds and greatest lower bounds are defined similarly. The set $\mathscr{A}$ is said to be a lattice if each pair of elements possesses a greatest lower bound and a least upper bound. A lattice $\mathscr{A}$ is inductive if every chain $\mathscr{C}$ in $\mathscr{A}$ has a greatest lower bound a in $\mathscr{A}$ and a least upper bound b in $\mathscr{A}$. A lattice is complete and, in particular, inductive, if every subset $\mathscr{B}$ of $\mathscr{A}$ has a greatest lower bound a in $\mathscr{A}$ and a least upper bound b in $\mathscr{A}$. Finally, an element x in a lattice $\mathscr{A}$ is a minimal element (respectively maximal element) if $y \leqslant x$ implies $y = x$ (respectively $x \leqslant y$ implies $y = x$) and $A:\mathscr{A} \to \mathscr{A}$ is isotone (increasing) if $x \leqslant y \Rightarrow A(x) \leqslant A(y)$.

Proposition 3.1.10. Let $\mathbf{E}$ be a partially ordered set, and suppose $u_0 \leqslant v_0$ are elements of $\mathbf{E}$, with the property that the interval $\mathbf{I} = \{v \in \mathbf{E}: u_0 \leqslant v \leqslant v_0\}$ is an inductive lattice. Suppose that A is an isotone mapping of $\mathbf{I}$ into $\mathbf{E}$, such that

$$u_0 \leqslant A(u_0), \qquad A(v_0) \leqslant v_0. \tag{3.1.42}$$

Then the set of fixed points u of A satisfying $u_0 \leqslant u \leqslant v_0$ is nonempty and possesses a minimal element $\underline{u}$ and a maximal element $\bar{u}$.

Proof: We make the initial observation that A maps $\mathbf{I}$ into itself; this follows from the increasing property of A in conjunction with (3.1.42). Now set $x \geqslant y$ if $y \leqslant x$, and set

$$\mathscr{U} = \{u \in \mathbf{E}: u_0 \leqslant u \leqslant v_0, u \leqslant A(u)\}, \tag{3.1.43a}$$

$$\mathscr{V} = \{v \in \mathbf{E}: u_0 \leqslant v \leqslant v_0, A(v) \leqslant v\}. \tag{3.1.43b}$$

Then $u_0 \in \mathscr{U}$ and $v_0 \in \mathscr{V}$. Set

$$\mathscr{W} = \{w \in \mathscr{V}: w \geqslant u \quad \text{for all} \quad u \in \mathscr{U}\}, \tag{3.1.44a}$$

$$\mathscr{Y} = \{y \in \mathscr{U}: y \leqslant v \quad \text{for all} \quad v \in \mathscr{V}\}. \tag{3.1.44b}$$

Then $v_0 \in \mathscr{W}$ and $u_0 \in \mathscr{Y}$. We claim that $\mathscr{W}$ and $\mathscr{Y}$ are inductive. Let $\mathscr{C}$ be a chain in $\mathscr{W}$. Then, since the interval $\mathbf{I} = \{w \in \mathbf{E}: u_0 \leqslant w \leqslant v_0\}$ is inductive, we may select a greatest lower bound $a \in \mathbf{I}$ for $\mathscr{C}$. We shall show that $a \in \mathscr{W}$. Indeed, $a \leqslant c$ for all $c \in \mathscr{C}$ so that, since A is increasing and $c \in \mathscr{V}$, $A(a) \leqslant A(c) \leqslant c$. This shows that $A(a)$ is a lower bound in $\mathbf{I}$ for $\mathscr{C}$ and, since the greatest lower bound is comparable with all lower bounds in $\mathbf{I}$, we have $A(a) \leqslant a$, that is, $a \in \mathscr{V}$. Now let $u \in \mathscr{U}$. Since $c \geqslant u$ for all $c \in \mathscr{C}$, we deduce that u is a lower bound for $\mathscr{C}$ and, hence, $a \geqslant u$. Thus, $a \in \mathscr{W}$. The proofs for the other three cases are similar, and we conclude that $\mathscr{W}$ and $\mathscr{Y}$ are inductive. It now follows from Zorn's lemma (see Kelley [17] p. 33) that $\mathscr{W}$ (respectively $\mathscr{Y}$) possesses a *minimal* element $\bar{u}$ (respectively *maximal* element $\underline{u}$). These may be shown to satisfy the statement of the proposition. For example, $\bar{u} \geqslant u$ for $u \in \mathscr{U}$, so that, since A is increasing,

$$A(\bar{u}) \geqslant A(u) \geqslant u$$

and, thus, $A(\bar{u}) \in \mathscr{W}$; the inequality

$$A(\bar{u}) \leqslant \bar{u} \tag{3.1.45}$$

is immediate from $\bar{u} \in \mathscr{V}$. Thus, $\bar{u}$ is a fixed point, since $\bar{u}$ is of minimal type. Similar arguments show that $\underline{u}$ is a fixed point. Maximality and minimality are immediate. ■

3.2 APPLICATIONS AND EXAMPLES

Example 3.2.1. A Dirichlet Problem in Semiconductor Modeling

The boundary-value problem

$$-\Delta u + e^u - e^{\rho - u} = q \qquad \text{in} \quad \Omega, \tag{3.2.1a}$$

$$u = 0 \qquad \text{on} \quad \partial\Omega, \tag{3.2.1b}$$

where ρ is a nonnegative constant, arises in the study of semiconductors. We shall identify the weak form of (3.2.1) with the equation

$$\phi'(u; g) = (q, g)_{\mathbf{L}^2(\Omega)}, \qquad \text{for all} \quad g \in \mathbf{H}_0^1(\Omega), \tag{3.2.2}$$

for an appropriate proper, convex, and lower-semicontinuous function $\phi: \mathbf{L}^2(\Omega) \to (-\infty, \infty]$, where $\phi'(u; \cdot) := \partial\phi(u)$. Here, $\phi'(u; g)$ will be computed by the usual Gâteaux differentiation formula for $g \in \mathbf{L}^\infty(\Omega) \cap \mathbf{H}_0^1(\Omega)$ to have the value given by

$$\phi'(u; g) = \int_\Omega \nabla u \cdot \nabla g + \int_\Omega (e^u - e^{\rho - u}) g. \tag{3.2.3}$$

By a density argument, it can then be shown that $\phi'(u; \cdot)$ has the form given by (3.2.3) on all of $\mathbf{H}_0^1(\Omega)$, so that $\partial\phi(u)$ is indeed a singleton. Thus, set

$$F(s) = \int_0^s e^\sigma \, d\sigma = e^s - 1, \qquad -\infty < s < \infty, \tag{3.2.4a}$$

$$G(t) = e^\rho \int_0^t (-e^{-\tau}) \, d\tau = e^\rho (e^{-t} - 1), \qquad -\infty < t < \infty, \tag{3.2.4b}$$

and note that F and G are convex functions, satisfying $F(0) = G(0) = 0$. We define ϕ by

$$\phi(f) = \infty, \qquad f \in \mathbf{L}^2(\Omega) \backslash \mathbf{H}_0^1(\Omega), \tag{3.2.5a}$$

$$\phi(f) = \infty, \qquad f \in \mathbf{H}_0^1(\Omega), \quad \text{if} \quad \int_\Omega F(f(\mathbf{x})) \, d\mathbf{x} = \infty \tag{3.2.5b}$$

$$\text{or if} \quad \int_\Omega G(f(\mathbf{x})) \, d\mathbf{x} = \infty,$$

$$\phi(f) = \tfrac{1}{2} \int_\Omega |\nabla f|^2 \, d\mathbf{x} + \int_\Omega [F(f(\mathbf{x})) + G(f(\mathbf{x}))] \, d\mathbf{x}, \qquad \text{otherwise.} \tag{3.2.5c}$$

The function ϕ is clearly proper and convex, $\phi(0) = 0$, and satisfies the coerciveness condition

$$\frac{\phi(u)}{\|u\|_{\mathbf{L}^2(\Omega)}} \to \infty \qquad \text{as} \quad \|u\|_{\mathbf{L}^2(\Omega)} \to \infty, \tag{3.2.6}$$

since for some $C > 0$

$$\left\{\int_\Omega |\nabla u|^2\right\}^{1/2} \geqslant C\|u\|_{\mathbf{L}^2(\Omega)} \qquad \text{for} \quad u \in \mathbf{H}_0^1(\Omega), \tag{3.2.7a}$$

$$F(s) \geqslant -1, \qquad G(t) \geqslant -e^\rho. \tag{3.2.7b}$$

We shall show that ϕ is lower semicontinuous. In particular, we must show that

$$\forall a \in \mathbb{R}^1, \qquad \mathbf{B} = \{f \in \mathbf{L}^2(\Omega) : \phi(f) \leqslant a\} \quad \text{is closed.} \tag{3.2.8}$$

Thus, if $f_k \to f$, $\phi(f_k) \leqslant a$, we have by (3.2.5c) and (3.2.7b) that $\{f_k\}$ is bounded in $\mathbf{H}_0^1(\Omega)$ and, hence, is weakly convergent (the entire sequence) in $\mathbf{H}_0^1(\Omega)$ to f. We have

$$\liminf_{k\to\infty} \int_\Omega |\nabla f_k|^2 \geqslant \int_\Omega |\nabla f|^2, \tag{3.2.9}$$

and, by Fatou's lemma (see (3.2.7)),

$$\begin{aligned} \liminf_{k\to\infty} \int_\Omega F(f_k(\mathbf{x}))\,d\mathbf{x} &\geqslant \int_\Omega \liminf_{k\to\infty} F(f_k(\mathbf{x}))\,d\mathbf{x} \\ &\geqslant \int_\Omega F\left(\liminf_{k\to\infty} f_k(\mathbf{x})\right)d\mathbf{x} \\ &= \int_\Omega F(f(\mathbf{x}))\,d\mathbf{x}, \end{aligned} \tag{3.2.10}$$

via the lower semicontinuity of F, and the fact that $\liminf_{k\to\infty} f_k(\mathbf{x}) = f(\mathbf{x})$. Similarly,

$$\liminf_{k\to\infty} \int_\Omega G(f_k(\mathbf{x}))\,d\mathbf{x} \geqslant \int_\Omega G(f(\mathbf{x}))\,d\mathbf{x}. \tag{3.2.11}$$

Thus, by (3.2.9), (3.2.10), and (3.2.11),

$$\begin{aligned} \phi(f) &\leqslant \liminf_{k\to\infty} \int_\Omega \tfrac{1}{2}|\nabla f_k|^2\,d\mathbf{x} + \liminf_{k\to\infty} \int_\Omega F(f_k(\mathbf{x}))\,d\mathbf{x} + \liminf_{k\to\infty} \int_\Omega G(f_k(\mathbf{x}))\,d\mathbf{x} \\ &\leqslant \liminf_{k\to\infty} \left[\int_\Omega [\tfrac{1}{2}|\nabla f_k|^2 + F(f_k(\mathbf{x})) + G(f_k(\mathbf{x}))]\,d\mathbf{x}\right] \\ &= \liminf_{k\to\infty} \phi(f_k) \\ &\leqslant a, \end{aligned}$$

which proves that **B** is closed. Corollary 3.1.7 is now applicable with $\mathbf{X} = \mathbf{X}_0 = \mathbf{L}^2(\Omega)$, and $A \equiv 0$. Note that (3.1.32) and the condition on $\|q\|_{\mathbf{L}^2(\Omega)}$ follow from (3.2.6) and (3.2.7), with ω a suitable affine function. We, thus, obtain a function u, such that $q \in \partial\phi(u)$, where q has been identified with its functional. Now select $g \in \mathbf{L}^\infty(\Omega) \cap \mathbf{H}_0^1(\Omega)$. By computing the standard difference quotients, for $t > 0$ and $t < 0$, defined by $t^{-1}[\phi(u + tg) - \phi(u)]$, and letting $t \to 0$, we see that

$$\chi(g) = \int_\Omega \nabla u \cdot \nabla g + \int_\Omega (e^u + e^{\rho - u})g, \tag{3.2.12}$$

for every $\chi \in \partial\phi(u)$. Standard arguments show that e^u and e^{-u} are in $\mathbf{L}^2(\Omega)$ (cf. Example 3.2.2), so that (3.2.10) holds for all $g \in H_0^1(\Omega)$ by a standard density argument.

This shows that $\partial\phi(u)$ is a singleton, so that the definition $\phi'(u;\cdot) := \partial\phi(u)$ and (3.2.3) are valid. As noted above, the existence theorem contained in Corollary 3.1.7 guarantees (3.2.2). In fact, this example is a special case of the following.

Example 3.2.2. Maximal Monotone Mappings and Perturbations

Let $\mathrm{gr}(\beta)$ be a maximal monotone graph in $\mathbb{R}^1 \times (-\infty, \infty)$,† induced by a multivalued function β. We suppose that 0 is an interior point of the domain of β. If we define $\bar{\beta}: \mathbf{L}^2(\Omega) \to 2^{\mathbf{L}^2(\Omega)}$ by

$$\begin{aligned} &f \in \bar{\beta}(u) \quad \text{if} \quad f, u \in \mathbf{L}^2(\Omega), \quad \text{and} \\ &f(x) \in \beta(u(x)) \quad \text{almost everywhere in} \quad \Omega, \end{aligned} \tag{3.2.13}$$

then the operator $\bar{\beta}$ is maximal monotone. We note that $\bar{\beta}$ may be realized as the subdifferential of a proper, convex, lower-semicontinuous functional. Indeed, if ℓ is the unique convex function on $(-\infty, \infty]$, such that $\ell(0) = 0$ and $\partial\ell = \beta$ (more precisely, the representers of $\partial\ell(t)$ coincide with $\beta(t)$), set

$$\gamma(u) = \begin{cases} \int_\Omega \ell(u(\mathbf{x}))\,d\mathbf{x}, & \text{if} \quad \ell \circ u \in \mathbf{L}^1(\Omega), \\ \infty, & \text{otherwise.} \end{cases} \tag{3.2.14}$$

Since ℓ is convex and lower semicontinuous, ℓ possesses an affine minorant, permitting the application of Fatou's lemma to the functional γ. In particular,

† Note that we do not permit β to assume infinite values.

γ is lower semicontinuous. Moreover, $\mathbf{gr}(\bar{\beta}) \subset \mathbf{gr}(\partial\gamma)$ (More precisely, $\mathbf{gr}(\bar{\beta})$ is contained in the graph of the representers of $\partial\gamma$). Since $\bar{\beta}$ is maximal monotone and $\partial\gamma$ is (maximal) monotone, $\bar{\beta} = \partial\gamma$, where we have identified functionals and subgradients. In summary, every maximal monotone graph induces an associated (multivalued) mapping on $\mathbf{L}^2(\Omega)$, which is the subdifferential of a proper, convex, lower-semicontinuous function.

As an illustration, consider the function H, arising in the Stefan problem (cf. Section 1.1). To obtain a maximal monotone graph, set

$$\beta(t) = \begin{cases} H(t), & t \neq 0, \\ [0, A], & t = 0. \end{cases} \tag{3.2.15}$$

In this case, ℓ is the primitive of H vanishing at 0. If $f: \mathbb{R}^1 \to \mathbb{R}^1$ is a continuous monotone function with primitive F vanishing at 0, then β may be taken equal to f, and ℓ may be taken equal to F. Similar remarks apply to sums $f + H$, etc.

To continue the development of the example, let L be a linear elliptic operator, defined in a weak sense on a closed linear subspace $\mathbf{G}$ of $\mathbf{H}^1(\Omega)$, with range in $\mathbf{G}^*$. We shall suppose that L is an isomorphism, defined by a continuous, coercive bilinear form $B(\cdot,\cdot)$ that is,

$$\langle \mathrm{L}u, v\rangle = B(u, v), \qquad \text{for all} \quad u, v \in \mathbf{G}, \tag{3.2.16}$$

where B is a bilinear form satisfying for appropriate constants

$$|B(u, v)| \leqslant c_1 \|u\|_{\mathbf{G}} \|v\|_{\mathbf{G}}, \tag{3.2.17a}$$

$$B(u, u) \geqslant c\|u\|_{\mathbf{G}}^2 - C\|u\|_{\mathbf{L}^2(\Omega)}^2. \tag{3.2.17b}$$

Special choices of L include the inverses of the operators N_σ, D_0, and R_ω of Sections 1.1, 1.2 and 1.3, where $B(\cdot,\cdot)$ is symmetric and $C = 0$.

Now L is pseudomonotone. In fact L is continuous, and hence, bounded, and also weakly continuous. Now suppose

$$u_i \rightharpoonup u \quad (\text{weakly in } \mathbf{G}), \qquad \limsup_{i\to\infty} \langle \mathrm{L}u_i, u_i - u\rangle \leqslant 0. \tag{3.2.18}$$

We shall show that

$$u_i \to u \quad (\text{in } \mathbf{G}), \tag{3.2.19}$$

which clearly implies

$$\langle \mathrm{L}u, u - v\rangle = \lim_{i\to\infty} \langle \mathrm{L}u_i, u_i - v\rangle. \tag{3.2.20}$$

Now,

$$u_i \rightharpoonup u \quad (\text{weakly in } \mathbf{G}) \Rightarrow \mathrm{L}u_i \rightharpoonup^* \mathrm{L}u \quad (\text{weak-* in } \mathbf{G}^*). \tag{3.2.21}$$

Moreover, by the compactness of the embedding $\mathbf{G} \to \mathbf{L}^2(\Omega)$, we conclude that

$$u_i \to u \quad (\text{in } \mathbf{L}^2(\Omega)). \tag{3.2.22}$$

Thus, using this convergence relation, and (3.2.17b) and (3.2.21), we obtain

$$\begin{aligned} 0 &\leqslant c \liminf_{i\to\infty} \|u_i - u\|_{\mathbf{G}}^2 \leqslant \liminf_{i\to\infty} B(u_i - u, u_i - u) \\ &= \liminf_{i\to\infty} \langle \mathrm{L}u_i - \mathrm{L}u, u_i - u\rangle = \liminf_{i\to\infty} \langle \mathrm{L}u_i, u_i - u\rangle, \end{aligned}$$

so that

$$\liminf_{i\to\infty} \langle \mathrm{L}u_i, u_i - u\rangle \geqslant 0. \tag{3.2.23}$$

Coupled with (3.2.18), this implies that

$$\lim_{i\to\infty} \langle \mathrm{L}u_i, u_i - u\rangle = 0, \tag{3.2.24}$$

which implies (3.2.19), via (3.2.17b), (3.2.21) and (3.2.22).

We have shown that L is pseudomonotone. In fact, similar arguments show that the operator $\mathrm{L} + \bar{\mathrm{r}}$ is pseudomonotone, if $\bar{\mathrm{r}}$ is any continuous, bounded operator from $\mathbf{L}^2(\Omega)$ into $\mathbf{L}^2(\Omega)$. In particular, this includes the operator $\bar{\mathrm{r}}$ induced by a Lipschitz continuous function r or an appropriate integral transform appearing in certain integrodifferential operator expressions. Suppose that $\beta = \partial\ell$ is a maximal monotone function, $\ell(0) = 0$, that γ is defined by (3.2.14) and that

$$\frac{(\bar{\mathrm{r}}(v), v)_{\mathbf{L}^2(\Omega)} + \gamma(v) - C(v, v)_{\mathbf{L}^2(\Omega)}}{\|v\|_{\mathbf{G}}} \geqslant -C_1 \qquad \text{as} \quad \|v\|_{\mathbf{G}} \to \infty, \tag{3.2.25}$$

where C is given by (3.2.17b) and $C_1 \geqslant 0$.

We now draw a final conclusion for this case in a summarizing proposition.

Proposition 3.2.1. Let β be a multivalued function defining a maximal monotone operator $\bar{\beta}: \mathbf{L}^2(\Omega) \to 2^{\mathbf{L}^2(\Omega)}$, via (3.2.13), such that $0 \in \beta(0)$. Let $\mathbf{G}$ be a closed subspace of $\mathbf{H}^1(\Omega)$, and suppose $\mathrm{L}: \mathbf{G} \to \mathbf{G}^*$ is an isomorphism, satisfying (3.2.16) and (3.2.17). Let $\bar{\mathrm{r}}$ be a bounded continuous transformation on $\mathbf{L}^2(\Omega)$, satisfying (3.2.25), where $\bar{\beta} = \partial\gamma$ and $\gamma(0) = 0$. Then, for any $f \in \mathbf{G}^*$, there exists a function $u \in \mathbf{G}$, satisfying $\bar{\mathrm{r}}(u) \in \mathbf{L}^2(\Omega)$, $\bar{\beta}(u) \subset \mathbf{L}^2(\Omega)$, and the relation

$$f \in \mathrm{L}u + \bar{\mathrm{r}}(u) + \bar{\beta}(u), \tag{3.2.26}$$

where we have identified $\overline{\Gamma}(u) + \overline{\beta}(u)$ with the induced linear functionals. Moreover, there is an element $\tilde{\beta} \in \overline{\beta}(u)$, such that the relation

$$\langle \mathrm{L}u, v \rangle + (\overline{\Gamma}(u) + \tilde{\beta}, v)_{\mathbf{L}^2(\Omega)} = \langle f, v \rangle, \qquad \text{for all} \quad v \in \mathbf{G}, \tag{3.2.27}$$

is valid. The solution is unique if

$$\begin{aligned} &\langle \mathrm{L}(v - w), v - w \rangle + (\overline{\Gamma}(v) - \overline{\Gamma}(w), v - w)_{\mathbf{L}^2(\Omega)} \\ &\quad + (\overline{\beta}(v) - \overline{\beta}(w), v - w)_{\mathbf{L}^2(\Omega)} \geqslant \delta \|w - v\|^2_{\mathbf{L}^2(\Omega)}, \end{aligned} \tag{3.2.28}$$

for some $\delta > 0$ and all $v, w \in \mathbf{G}$.

Proof: The coercive relation

$$\frac{\langle \mathrm{L}v, v \rangle + \gamma(v) + (\overline{\Gamma}(v), v)_{\mathbf{L}^2(\Omega)}}{\|v\|_{\mathbf{G}}} \to \infty \qquad \text{as} \quad \|v\|_{\mathbf{G}} \to \infty \tag{3.2.29}$$

is immediate from (3.2.25) and (3.2.17b). We have seen that $\mathrm{L} + \overline{\Gamma}$ is pseudo-monotone, and that γ is a proper, convex, lower-semicontinuous function on $\mathbf{G}$. The existence of a solution u satisfying (3.2.26) now follows from Corollary 3.1.7, if we set $\mathbf{X} = \mathbf{X}_0 = \mathbf{G}$, $\mathrm{A}(v) = \mathrm{L}v + \overline{\Gamma}(v)$, and $\phi(v) = \gamma(v)$. Note that $\overline{\beta}(u) \subset \mathbf{L}^2(\Omega)$ by the definition of $\overline{\beta}$. Equation (3.2.27) is simply a restatement of (3.2.26). The uniqueness follows directly from (3.2.28). ■

Example 3.2.3. Abstract Stationary Navier–Stokes Theory

This example is related to the stationary Navier–Stokes theory or, more properly, to the generalization introduced in Section 1.4. The general situation considered here involves a separable, reflexive Banach space $\mathbf{X}$, a dense linear, separable, subspace $\mathbf{Y}$ continuously embedded in $\mathbf{X}$ and a mapping $\mathrm{A}:\mathbf{X} \to \mathbf{Y}^*$. The mapping A is defined explicitly by

$$\mathrm{A}(u) = \mathrm{L}u + a(u, u, \cdot), \tag{3.2.30}$$

where $\mathrm{L}:\mathbf{X} \to \mathbf{X}^*$ is a linear isomorphism, satisfying (3.2.16) and (3.2.17), with $C = 0$; note that in (3.2.17) we have identified $\mathbf{G}$ with $\mathbf{X}$ and $\mathbf{L}^2$ with a space $\mathbf{W}$ to be described shortly. Also $a(\cdot, \cdot, \cdot)$ is a continuous trilinear form on $\mathbf{X} \times \mathbf{X} \times \mathbf{Y}$, satisfying

$$a(v, v, v) \geqslant 0, \qquad \text{for all} \quad v \in \mathbf{Y}, \tag{3.2.31a}$$

$$a(u_m, u_m, v) \to a(u, u, v), \qquad \text{for all} \quad v \in \mathbf{Y}, \tag{3.2.31b}$$

if $u_m \rightharpoonup u$ (weakly in $\mathbf{X}$) and $u_m \to u$ (strongly in $\mathbf{W}$). Here, $\mathbf{X}$ is densely and compactly embedded in the Banach space $\mathbf{W}$. The reader correlating this with Section 1.4 should set $\mathbf{X} = \mathbf{V}$, $\mathbf{Y} = \mathbf{V}_s$, $\mathbf{W} = \mathbf{H}$, and $\mathrm{L} = -\Delta$.

The mapping A is clearly continuous, yet, because $\mathbf{Y}^*$ is, in general, strictly larger than $\mathbf{X}^*$, the infinite-dimensional theories of the previous section do not apply directly for the equation $\mathrm{A}(u) = f$, even if $f \in \mathbf{X}^*$, which will be required for this example. More precisely, given $f \in \mathbf{X}^*$, we shall solve $\mathrm{A}(u) = f_{|\mathbf{Y}}$. Suppose then that $\{\mathbf{Y}_k\}$ is a sequence of increasing finite-dimensional subspaces of $\mathbf{Y}$, such that $_k \cup \mathbf{Y}_k$ is dense in $\mathbf{Y}$. Consider the finite-dimensional problem of determining $u_k \in \mathbf{Y}_k$, such that

$$\mathrm{A}(u_k) = f_{|\mathbf{Y}_k} \tag{3.2.32a}$$

or, equivalently, such that

$$\langle \mathrm{L}u_k, v\rangle + a(u_k, u_k, v) = \langle f, v\rangle, \qquad \text{for all} \quad v \in \mathbf{Y}_k. \tag{3.2.32b}$$

To verify the existence of a solution u_k of (3.2.32), with the property that $\{u_k\}$ lies in a bounded subset of $\mathbf{X}$, we first impose upon $\mathbf{Y}_k$ the norm induced by $\underline{\mathbf{X}}$. Then, by (3.2.17b), with $C = 0$, and (3.2.31a), we deduce the coerciveness property

$$\frac{\langle \mathrm{A}(v), v\rangle}{\|v\|_{\mathbf{X}}} \to \infty \qquad \text{as} \quad \|v\|_{\mathbf{X}} \to \infty. \tag{3.2.33}$$

In particular, there is an r, namely, $r = \|f\|_{\mathbf{X}^*}/c$, in the notation of (3.2.17b), such that[†]

$$\frac{\langle \mathrm{A}(v), v\rangle}{\|v\|_{\mathbf{Y}_k}} \geqslant \|f_{|\mathbf{Y}_k}\|_{\mathbf{Y}_k^*}, \tag{3.2.34}$$

for $\|v\|_{\mathbf{Y}_k} = r$. Since $\mathrm{A}: \mathbf{Y}_k \to \mathbf{Y}_k^*$ is continuous, and hence, pseudomonotone, on the finite-dimensional space $\mathbf{Y}_k$, Corollary 3.1.6 applies to yield a solution $u_k \in \mathbf{Y}_k$ of (3.2.32), with the property that

$$\|u_k\|_{\mathbf{X}} \leqslant r. \tag{3.2.35}$$

Since $\mathbf{X}$ is assumed reflexive, and $\mathbf{X}$ is compactly embedded into $\mathbf{W}$, we may extract a subsequence u_{k_j}, and an element $u \in \mathbf{X}$, such that

$$u_{k_j} \rightharpoonup u \quad (\text{weakly in } \mathbf{X}), \qquad u_{k_j} \to u \quad (\text{strongly in } \mathbf{W}). \tag{3.2.36}$$

Now, let $w \in \bigcup \mathbf{Y}_k$. Then $w \in \mathbf{Y}_{k_0}$, for some k_0, and taking limits in (3.2.32b) yields

$$\langle \mathrm{L}u, w\rangle + a(u, u, w) = \langle f, w\rangle, \tag{3.2.37}$$

[†] Recall that the induced $\mathbf{X}$ norm is used in (3.2.34).

where we have used (3.2.31b). Finally, if $v \in \mathbf{Y}$, then letting $w_j \to v$ in $\mathbf{Y}$ yields

$$\langle \mathrm{L}u, v\rangle + a(u, u, v) = \langle f, v\rangle, \tag{3.2.38}$$

where we have used the continuous injection of $\mathbf{Y} \subset \mathbf{X}$. Note that (3.2.38) is just the statement that

$$\mathrm{A}(u) = f_{|\mathbf{Y}}, \tag{3.2.39a}$$

and (3.2.35) directly implies

$$\|u\|_{\mathbf{X}} \leqslant r. \tag{3.2.39b}$$

We shall summarize these results insofar as they apply to the case of Section 1.4.

Proposition 3.2.2. Let $\mathbf{V}$ and $\mathbf{V}_s$ be defined as in (1.4.20) and (1.4.21), and let $a(\cdot,\cdot,\cdot)$ be a continuous trilinear form on $\mathbf{V} \times \mathbf{V} \times \mathbf{V}_s$, satisfying (1.4.22) and (1.4.24a). Let $\mathbf{f} = (f_1, \ldots, f_n) \in \mathbf{V}^*$. Suppose the norm in $\mathbf{V}$ is defined by (equivalently)

$$\|\mathbf{v}\|_{\mathbf{V}}^2 = \eta \int_\Omega \nabla\mathbf{v} \cdot \nabla\mathbf{v} \qquad (\eta > 0) \tag{3.2.40a}$$

(see (1.4.26) for dot notation). Then there exists a $\mathbf{u} \in \mathbf{V}$, satisfying

$$\|\mathbf{u}\|_{\mathbf{V}} \leqslant \|\mathbf{f}\|_{\mathbf{V}^*}, \tag{3.2.40b}$$

$$\eta(\nabla u_i, \nabla v_i)_{\mathbf{L}^2(\Omega)} + a(u_i, u_i, v_i) = \langle f_i, v_i\rangle, \tag{3.2.40c}$$

for $\mathbf{v} = (v_1, \ldots, v_n) \in \mathbf{V}_s$, and $i = 1, \ldots, n$.

Proof: Set $\mathrm{L} = -\eta\Delta$, $\mathbf{X} = \mathbf{V}$, $\mathbf{Y} = \mathbf{V}_s$, and $\mathbf{W} = \mathbf{H}$. Note that $c = 1$ and, hence, $r = \|f\|_{\mathbf{V}^*}$. ■

Example 3.2.4. A Quasi-Variational Inequality

Let $\mathbf{X}$ be a reflexive Banach space with a strictly convex dual, and suppose that $\mathbf{X}$ possesses a lattice structure. In particular, $v = v^+ + v^-$, with $v^+ = \max(v, 0)$ and $v^- = \min(v, 0)$. Set $\mathbf{X}^+ = \{v \in \mathbf{X} : v \geqslant 0\}$, and suppose $\mathrm{J}:\mathbf{X} \to \mathbf{X}^*$ is the uniquely defined duality map, satisfying

$$\langle \mathrm{J}(v), v\rangle = \|v\|_{\mathbf{X}}^2, \qquad \|\mathrm{J}(v)\|_{\mathbf{X}^*} = \|v\|_{\mathbf{X}}, \qquad \text{for all} \quad v \in \mathbf{X}, \tag{3.2.41}$$

and the lattice compatibility condition

$$\langle \mathrm{J}(w_2) - \mathrm{J}(w_1), v^-\rangle \leqslant 0, \qquad \text{for all} \quad v \in \mathbf{X}, \tag{3.2.42}$$

and $w_2 \geqslant w_1$. Suppose that $\mathrm{M}:\mathbf{X}^+ \to \mathbf{X}^+$ satisfies $\mathrm{M}(0) \geqslant 0$, and that $\mathrm{A}:\mathbf{X} \to \mathbf{X}^*$ is a pseudomonotone operator, satisfying

$$\langle \mathrm{A}(u), u\rangle + \lambda_1\|u\|_{\mathbf{X}}^2 \geqslant \lambda_2\|u\|_{\mathbf{X}}^2, \tag{3.2.43}$$

for all $u \in \mathbf{X}$, where $\lambda_1 \geqslant 0$, $\lambda_2 > 0$. Suppose also that

$$\langle \mathrm{A}(u) - \mathrm{A}(v), (u - v)^{\pm}\rangle + \lambda_1\langle \mathrm{J}(u) - \mathrm{J}(v), (u - v)^{\pm}\rangle \geqslant 0, \tag{3.2.44a}$$

and

$$\langle \mathrm{A}(u) - \mathrm{A}(v), (u - v)^{\pm}\rangle + \lambda_1 \langle \mathrm{J}(u) - \mathrm{J}(v), (u - v)^{\pm}\rangle = 0$$
$$\Leftrightarrow (u - v)^+ = 0 \text{ (respectively} (u - v)^- = 0) \quad \text{if the positive/negative parts are chosen.} \tag{3.2.44b}$$

Suppose, moreover, that $f \in \mathbf{X}^*$ is given, with the property that a solution $v_0 \in \mathbf{X}^+$ exists for the equation

$$\mathrm{A}(v_0) = f. \tag{3.2.45}$$

Suppose that M is increasing on $\{v \in \mathbf{X}: 0 \leqslant v \leqslant v_0\}$, that is, $\mathrm{M}(u) - \mathrm{M}(v) \geqslant 0$ if $u - v \geqslant 0$. Finally, we set

$$\mathrm{A}_1 = \mathrm{A} + \lambda_1\mathrm{J}, \tag{3.2.46}$$

and observe that A_1 is strictly monotone by (3.2.44). Moreover, A_1 is pseudomonotone as the sum of a pseudomonotone and a bounded, continuous, monotone mapping (see Proposition 3.1.9). Also, by (3.2.43),

$$\frac{\langle \mathrm{A}_1(v), v\rangle}{\|v\|_{\mathbf{X}}} \geqslant \lambda_2\|v\|_{\mathbf{X}} \to \infty \qquad \text{as} \quad \|v\|_{\mathbf{X}} \to \infty. \tag{3.2.47}$$

In particular, we conclude the existence of a unique solution $u = \mathrm{S}(w)$ of the variational inequality (see Proposition 3.1.5),

$$\langle \mathrm{A}_1(u) - \lambda_1\mathrm{J}(w) - f, v - u\rangle + \phi(v) - \phi(u) \geqslant 0, \tag{3.2.48}$$

for all $v \in \mathbf{X}$, where ϕ is the indicator function of the closed (see Birkhoff [4]), convex, nonempty set

$$\mathbf{K}(w) = \{v \in X^+: v \leqslant \mathrm{M}(w)\}, \tag{3.2.49}$$

and $0 \leqslant w \leqslant v_0$. The inequality (3.2.48) may be rewritten as

$$u \leqslant \mathrm{M}(w), \tag{3.2.50a}$$

$$\langle \mathrm{A}_1(u), v - u\rangle \geqslant \langle \lambda_1\mathrm{J}(w) + f, v - u\rangle, \qquad v \leqslant \mathrm{M}(w). \tag{3.2.50b}$$

The implication $S(0) \geqslant 0$ is immediate from $S(0) \in \mathbf{X}^+$. The implication $S(v_0) \leqslant v_0$ follows from setting $w = v_0$, $u = S(v_0)$, and $v = S(v_0) - (S(v_0) - v_0)^+$ in (3.2.50b), and adding the resulting inequality to

$$\langle -A_1(v_0), (S(v_0) - v_0)^+ \rangle = \langle -\lambda_1 J(v_0) - f, (S(v_0) - v_0)^+ \rangle$$

to obtain

$$\langle A_1(S(v_0)) - A_1(v_0), (S(v_0) - v_0)^+ \rangle \leqslant 0. \tag{3.2.51}$$

Combining (3.2.51) with (3.2.44) gives $(S(v_0) - v_0)^+ = 0$, that is $S(v_0) \leqslant v_0$. If we can prove that S is increasing on the interval $0 \leqslant v \leqslant v_0$, then it will follow from Proposition 3.1.10 that S has a fixed point on this interval and, hence, that $u \leqslant M(u)$ and

$$\langle A_1(u), v - u \rangle \geqslant \langle \lambda_1 J(u) + f, v - u \rangle, \qquad \text{for all} \quad v \leqslant M(u), \tag{3.2.52}$$

provided the interval is inductive in the sense of Definition 3.1.10. In light of (3.2.46), we see that (3.2.52) is equivalent to solving on $\mathbf{X}^+$

$$\langle A(u), v - u \rangle \geqslant \langle f, v - u \rangle, \qquad \text{for all} \quad v \leqslant M(u), \tag{3.2.53}$$

for $u \leqslant M(u)$.

To prove that S is increasing, let $0 \leqslant w_1 \leqslant w_2 \leqslant v_0$. Then $u_1 = S(w_1)$ and $u_2 = S(w_2)$ satisfy $u_1 \leqslant M(w_1)$ and $u_2 \leqslant M(w_2)$, and

$$\langle A_1(u_1), v - u_1 \rangle \geqslant \langle \lambda_1 J(w_1) + f, v - u_1 \rangle, \qquad \text{for all} \quad v \leqslant M(w_1), \tag{3.2.54a}$$

$$\langle A_1(u_2), v - u_2 \rangle \geqslant \langle \lambda_1 J(w_2) + f, v - u_2 \rangle, \qquad \text{for all} \quad v \leqslant M(w_2). \tag{3.2.54b}$$

Since

$$u = u_2 - (u_2 - u_1)^- \leqslant \sup(u_1, u_2) \leqslant M(w_2),$$

we may set $v = u$ in (3.2.54b). With the choice

$$v = u_1 + (u_2 - u_1)^-$$

in (3.2.54a), we have, after addition of the two inequalities

$$0 \leqslant \langle A_1(u_2) - A_1(u_1), (u_2 - u_1)^- \rangle \leqslant \lambda_1 \langle J(w_2) - J(w_1), (u_2 - u_1)^- \rangle \\ \leqslant 0, \tag{3.2.55}$$

where we have used (3.2.42) and (3.2.44a). By (3.2.44b), we conclude $(u_2 - u_1)^- = 0$, so that S is increasing. We summarize this result in the following proposition.

Proposition 3.2.3.† Suppose that **X**, M, and A satisfy the hypotheses stated above. Then the quasi-variational inequality (3.2.53) has a solution u, satisfying $u \leqslant \mathrm{M}(u) \subset \mathbf{X}^+$.

3.3 SEMIDISCRETIZATIONS DEFINED BY QUADRATURE

Suppose we are considering a partial differential equation‡

$$\frac{d}{dt}[H(u)] + \mathrm{L}u + f(u) = 0, \qquad 0 < t < T_0, \tag{3.3.1}$$

for an appropriate elliptic operator L, and specified functions H and f, with imposed boundary conditions of homogeneous Dirichlet, Neumann, or Robin type. Although more general frameworks are possible, and sometimes desirable, we conceive of L as being defined on a linear subspace of $\mathbf{H}^1(\Omega)$. More precisely, let **G** be a closed subspace of $\mathbf{H}^1(\Omega)$, and let an inversion operator

$$\mathrm{T}: \mathbf{G}^* \to \mathbf{G} \subset \mathbf{H}^1(\Omega) \tag{3.3.2}$$

be specified, where T is a continuous linear mapping onto a closed linear subspace **G** of $\mathbf{H}^1(\Omega)$, satisfying

$$\mathrm{L} \circ \mathrm{T} = \text{identity}. \tag{3.3.3}$$

The operator L, thus, has domain **G** and range $\mathbf{G}^*$. Hence, as observed in Example 3.2.2, the homogeneous boundary conditions are embedded in **G**, via $\mathbf{G} = \mathrm{T}(\mathbf{G}^*)$. The special cases of interest were discussed in detail in Chapter 1. Formally, we transform (3.3.1) into

$$\frac{d}{dt}[\mathrm{T}H(u)] + u + \mathrm{T}f(u) = 0, \tag{3.3.4}$$

and we find, integrating from $t = t_{k-1}$ to $t = t_k$, within a given partition $\mathscr{P} = \{0 = t_0 < t_1 < \cdots < t_M = T_0\}$,

$$\mathrm{T}H(u(t_k)) + \int_{t_{k-1}}^{t_k} u(t)\,dt + \int_{t_{k-1}}^{t_k} \mathrm{T}f(u(t))\,dt = \mathrm{T}H(u(t_{k-1})). \tag{3.3.5}$$

† This generalizes Theorem 4.3 of [19], and its proof, contained in pp. 170–173.

‡ We anticipate uniqueness and require single-valuedness for $H(u)$; the **x**-dependence is implicit.

Suppose a quadrature rule for $\int_{t_{k-1}}^{t_k} v(t)\,dt$ is specified for functions $v:[0, T_0] \to \mathbf{G}$ as follows:

$$\int_{t_{k-1}}^{t_k} v(t)\,dt \approx \left[\sum_{j=0}^{k} \gamma_{j,k} v(t_j)\right](t_k - t_{k-1}) = Q_k(v), \qquad k = 1, \ldots, M, \quad (3.3.6)$$

where $\gamma_{j,k}$ are specified constants. Note that all values $v(t_0), \ldots, v(t_k)$ may be used in (3.3.6). Such a quadrature rule may be used to define semidiscretizations of (3.3.4) by direct application to (3.3.5).

In the literature on initial-value problems for ordinary differential equations, the application of quadrature formulae, such as (3.3.6), to the integrated form of a differential equation induces what is known as a multistep method. If $\gamma_{j,k} = 0$ for $j < k - 1$, the method is called, not surprisingly, a single-step method. It is advantageous to permit the application of different quadrature rules to the integrals appearing in (3.3.5). With the applications in mind, let us set $f = g + h$, in the notation of Chapters 1 and 2, and rewrite (3.3.5) as

$$\mathrm{T}H(u(t_k)) + \int_{t_{k-1}}^{t_k} [u(t) + \mathrm{T}g(u(t))]\,dt + \int_{t_{k-1}}^{t_k} \mathrm{T}h(u(t))\,dt = \mathrm{T}H(u(t_{k-1})). \quad (3.3.7)$$

If (3.3.6) is applied, with $\gamma_{j,k} = \alpha_{j,k}$ and $\gamma_{j,k} = \beta_{j,k}$, respectively, to (3.3.7), we obtain a corresponding formal approximation equation. We shall display this equation in the following definition.

Definition 3.3.1. Given a partition $\mathscr{P}$ of $[0, T_0]$, and a pair of quadrature rules $Q_{k,\boldsymbol{\alpha}}(\cdot)$ and $Q_{k,\boldsymbol{\beta}}(\cdot)$, defined by (3.3.6), for $k = 1, \ldots, M$, with $\boldsymbol{\alpha} = (\alpha_{j,k})$, $\boldsymbol{\beta} = (\beta_{j,k})$, we define the semidiscretization of the equation

$$\frac{d}{dt}[H(u)] + \mathrm{L}u + g(u) + h(u) = 0, \quad (3.3.8)$$

induced by the quadrature rules, to consist of the recursive sequence

$$(t_k - t_{k-1})^{-1} H(u_k) + \alpha_{k,k}[\mathrm{L}u_k + g(u_k)] + \beta_{k,k} h(u_k)$$
$$= -\sum_{j=0}^{k-1} \{\alpha_{j,k}[\mathrm{L}u_j + g(u_j)] + \beta_{j,k} h(u_j)\} + (t_k - t_{k-1})^{-1} H(u_{k-1}), \quad (3.3.9)$$

for $k = 1, \ldots, M$. For each k, it is required that the unknown functions $u_0, \ldots, u_{k-1}$ have already been computed, and that u_k is determined by (3.3.9). The functions $\{u_k\}$ are approximations to $u(t_k)$. We refer to the lifted form of (3.3.9) as the recursive sequence

$$(t_k - t_{k-1})^{-1}TH(u_k) + \alpha_{k,k}[u_k + Tg(u_k)] + \beta_{k,k}Th(u_k)$$
$$= -\sum_{j=0}^{k-1} \{\alpha_{j,k}[u_j + Tg(u_j)] + \beta_{j,k}Th(u_j)\} + (t_k - t_{k-1})^{-1}TH(u_{k-1})), \tag{3.3.10}$$

for $k = 1, \ldots, M$. The method is said to be implicit if either $\alpha_{k,k} \neq 0$ or $\beta_{k,k} \neq 0$, and explicit otherwise.

Example 3.3.1.

In this example, we shall define

$$Q_k(v) := \int_{t_{k-1}}^{t_k} \bar{v}(t)\,dt, \tag{3.3.11}$$

for appropriate combinations $\bar{v}$, of $v(t_{k-1})$ and $v(t_k)$, which are constant in t. If

$$\bar{v}(t) = v(t_{k-1}), \qquad t_{k-1} \leqslant t < t_k, \tag{3.3.12a}$$

then

$$Q_k(v) = v(t_{k-1})(t_k - t_{k-1}). \tag{3.3.12b}$$

In this case, $\gamma_{k-1,k} = 1$ and $\gamma_{j,k} = 0, j \neq k-1$. If $\alpha_{j,k}$ and $\beta_{j,k}$ are defined by these choices of $\gamma_{j,k}$, there results the well-known explicit or forward Euler semidiscretization. On the other hand, if

$$\bar{v}(t) = v(t_k), \qquad t_{k-1} \leqslant t < t_k, \tag{3.3.13a}$$

then

$$Q_k(v) = v(t_k)(t_k - t_{k-1}), \tag{3.3.13b}$$

and the fully implicit or backward Euler method results. In this case, $\gamma_{j,k} = \delta_{j,k}$. Of course, a mixing of the methods may be desirable (see Chapter 2), that is, $\alpha_{j,k}$ defined by (3.3.13) and $\beta_{j,k}$ defined by (3.3.12). The step functions defined by (3.3.12a) and (3.3.13a) are the special cases $\theta = 0$ and $\theta = 1$ of

$$\bar{v}(t) = \theta v(t_k) + (1-\theta)v(t_{k-1}), \qquad t_{k-1} \leqslant t < t_k \quad (0 \leqslant \theta \leqslant 1). \tag{3.3.14a}$$

If

$$Q_k(v) = [\theta v(t_k) + (1-\theta)v(t_{k-1})](t_k - t_{k-1}), \tag{3.3.14b}$$

then

$$\gamma_{k-1,k} = 1 - \theta, \qquad \gamma_{k,k} = \theta, \qquad \gamma_{j,k} = 0, \quad j \neq k-1, j \neq k. \quad (3.3.14c)$$

The choice $\theta = \frac{1}{2}$ leads to the Crank–Nicolson method when applied to $\alpha_{j,k}$, if the latter are defined by (3.3.14c).

Example 3.3.2.

The semidiscretizations discussed in the previous example were induced by quadrature rules defined by the exact integration of step function approximations. If, instead, piecewise linear interpolation is employed to define the approximation $\bar{v}$ of v, given by

$$\bar{v}(t) = (t_k - t_{k-1})^{-1}[(t - t_{k-1})v(t_k) + (t_k - t)v(t_{k-1})], \qquad t_{k-1} \leqslant t < t_k, \quad (3.3.15a)$$

then (3.3.11) yields

$$Q_k(v) = \tfrac{1}{2}(t_k - t_{k-1})[v(t_{k-1}) + v(t_k)], \qquad t_{k-1} \leqslant t < t_k. \quad (3.3.15b)$$

This is precisely (3.3.14b) with $\theta = \frac{1}{2}$, leading once again to the Crank–Nicolson method. This suggests higher-order approximation properties of this method than might at first be apparent. The explicit Adams extrapolation method is obtained by employing a variant of (3.3.15b), in which $Q_{k+1}(v)$ is defined by exact integration of the extension of the linear interpolant of $v(t_{k-1})$ and $v(t_k)$, that is,

$$\begin{aligned} Q_{k+1}(v) &:= \int_{t_k}^{t_{k+1}} \{(t_k - t_{k-1})^{-1}[(t - t_{k-1})v(t_k) + (t_k - t)v(t_{k-1})]\}\, dt \\ &= \tfrac{1}{2}(t_k - t_{k-1})^{-1}\{(t_{k+1} - t_k)(t_{k+1} + t_k - 2t_{k-1})v(t_k) \\ &\quad - (t_k - t_{k+1})^2 v(t_{k-1})\}. \end{aligned} \quad (3.3.16)$$

If (3.3.16) is employed for $k = 1, \ldots, M-1$, then it must be used in conjunction with a quadrature rule Q_1 on $[t_0, t_1]$.

Example 3.3.3.

Let $y_0, \ldots, y_k$ be arbitrary real numbers. Then there is a unique natural cubic spline function s, defined by the properties:

$$s^{(4)}(t) = 0, \qquad t_{i-1} < t < t_i, \quad i = 1, \ldots, k, \quad (3.3.17a)$$

$$s(t_i) = y_i, \qquad i = 0, \ldots, k, \quad (3.3.17b)$$

$$s \in \mathbf{C}^2([t_0, t_k]; \mathbb{R}^1), \tag{3.3.17c}$$

$$s''(t_0+) = s''(t_k-) = 0. \tag{3.3.17d}$$

If $y_i = \delta_{ij}$, then the corresponding spline s_j is called a cardinal spline. We may uniquely express the general spline s as

$$s = \sum_{j=0}^{k} y_j s_j. \tag{3.3.18a}$$

Now given v, we may define

$$\bar{v}(t) = \sum_{j=0}^{k} v(t_j) s_j(t), \qquad t_{k-1} \leqslant t < t_k, \tag{3.3.18b}$$

and

$$Q_k(v) := \int_{t_{k-1}}^{t_k} \bar{v}(t)\, dt = \sum_{j=0}^{k} \left\{ v(t_j) \int_{t_{k-1}}^{t_k} s_j(t)\, dt \right\}. \tag{3.3.18c}$$

Although formula (3.3.18) would suggest that $Q_k(v)$ depends upon $v(t_0), \ldots,$ $v(t_k)$, in reality only $v(t_{k-3})$, $v(t_{k-2})$, $v(t_{k-1})$, and $v(t_k)$ appear explicitly. This is due to local basis representations in terms of B-splines.

Remark 3.3.1. Various other rules could obviously be constructed.† The formulae of Newton–Cotes type are those which utilize polynomial interpolation. Although spline interpolation appears attractive because of associated variational principles, there are, in fact, questions of stability for higher-order smooth spline interpolation, even in the case of initial-value problems for ordinary differential equations (see Böhmer [5] for discussion and references). Splines of Hermite type are generally free of such instabilities, though they are unsuitable in pristine form for partial differential equations, since (time) derivatives would be required at nodal points t_i. The replacement of the derivatives by difference quotients is, of course, a possibility. We shall not pursue any of this, however.

Remark 3.3.2. The interval of integration in (3.3.5) was chosen as $[t_{k-1}, t_k]$. One could easily choose an interval of the form $[t_{k-i}, t_k]$, which would lead to the replacement of $(t_k - t_{k-1})^{-1}$ by $(t_k - t_{k-i})^{-1}$, and $H(u_{k-1})$ by $H(u_{k-i})$ in (3.3.9). In the event that the partition is uniform, one can achieve the same result by successive elimination of $H(u_{k-1})$, $H(u_{k-2})$, etc. This is sometimes

† This would include the implicit Runge-Kutta methods, not discussed here. The author is indebted to Kenneth Jackson for this observation.

desirable in the use of spline quadrature rules, such as that for rules of Milne–Simpson type. For example, if $i = 2$, and one defines $\overline{v}$ on $[t_{k-2}, t_k]$ by the unique quadratic interpolant of $v(t_{k-2})$, $v(t_{k-1})$, and $v(t_k)$, then, for $(t_k - t_{k-1}) = (t_{k-1} - t_{k-2}) = \Delta t$, one obtains the standard formula

$$Q_k(v) := \int_{t_{k-2}}^{t_k} \overline{v}(t)\,dt = \frac{\Delta t}{3}\left[v(t_{k-2}) + 4v(t_{k-1}) + v(t_k)\right]. \tag{3.3.19}$$

Remark 3.3.3. We shall note the application of Propositions 3.2.1 and 3.2.2 to the semidiscrete equations (3.3.9) for the models introduced in Chapter 1. For the applications of Sections 1.1, 1.2, and 1.3, we set $\mathrm{L} = \alpha_{k,k}\mathrm{N}_\sigma^{-1}$, $\mathrm{L} = \alpha_{k,k}\mathrm{D}_0^{-1}$, and $\mathrm{L} = \alpha_{k,k}\mathrm{R}_\omega^{-1}$, respectively, in (3.2.26), where $\sigma > 0$ is required to satisfy only $\lambda T_0^{-1} > \alpha_{k,k}\sigma$. Each of these operators satisfies (3.2.17), with $C = 0$, provided $\alpha_{k,k} > 0$. In the first case of the Stefan problem, we have $-\alpha_{k,k}\Delta = \mathrm{L} - \alpha_{k,k}\sigma\mathrm{I}$, so that the function choices

$$\beta = (t_k - t_{k-1})^{-1}H + \alpha_{k,k}g - \alpha_{k,k}\sigma id + \beta_{k,k}h, \qquad r = 0,$$

lead to the operator identity

$$-\alpha_{k,k}\Delta + (t_k - t_k)^{-1}\mathrm{H} + \alpha_{k,k}\mathrm{g} + \beta_{k,k}\mathrm{h} = \mathrm{L} + \overline{\beta},$$

and (3.2.25) is satisfied, with

$$\mathbf{G} = [\mathbf{H}^1(\Omega)]_\sigma, \qquad C = 0, \qquad \text{and} \qquad C_1 = |\beta_{k,k}|\,|h(0)|\,|\Omega|^{1/2}/\sigma^{1/2},$$

provided

$$[\lambda(t_k - t_{k-1})^{-1} - \alpha_{k,k}\sigma] \geqslant |\beta_{k,k}|\,\|h\|_{\mathbf{Lip}}. \tag{3.3.20}$$

We, thus, have, as an immediate consequence of Proposition 3.2.1, the following.

Proposition 3.3.1. Let H, g, h, and λ be given as in Section 1.1, with the initial datum $H(u_0)$, and consider the formal semidiscretization (3.3.9) of (1.1.9), rewritten here as

$$\begin{aligned}(t_k - t_{k-1})^{-1}H(u_k) &- \alpha_{k,k}\Delta u_k + \alpha_{k,k}g(u_k) + \beta_{k,k}h(u_k) \\ &= -\sum_{j=0}^{k-1}\{\alpha_{j,k}[-\Delta u_j + g(u_j)] + \beta_{j,k}h(u_j)\} + (t_k - t_{k-1})^{-1}H(u_{k-1}),\end{aligned} \tag{3.3.21a}$$

with Neumann boundary condition

$$\frac{\partial u_k}{\partial v} = 0 \qquad \text{on} \quad \partial\Omega, \tag{3.3.21b}$$

$k = 1, \ldots, M$, where $\alpha_{k,k} > 0$ is assumed. Suppose that

$$\lambda(t_k - t_{k-1})^{-1} > |\beta_{k,k}| \, \|h\|_{\mathbf{Lip}}. \tag{3.3.22}$$

Then (3.3.21) has a unique weak solution pair $(u_k, H(u_k))$, such that

$$H(u_k(\mathbf{x})) \in [0, A], \qquad \text{if } u_k(\mathbf{x}) = 0, \tag{3.3.23a}$$

$$H(u_k) \in \mathbf{L}^2(\Omega), \tag{3.3.23b}$$

$$u_k \in \mathbf{H}^1(\Omega), \tag{3.3.23c}$$

and (3.3.21) holds, when interpreted as an identity in $[\mathbf{H}^1(\Omega)]^*$. Equivalently, if $\sigma > 0$ is chosen to satisfy (3.3.20), then (3.3.21) may be understood in the lifted sense, when N_σ is applied.

Proof: For a given k, select σ, such that (3.3.20) holds. With the identifications of Remark 3.3.3, the result follows from Proposition 3.2.1, in conjunction with an argument employing $J = H^{-1}$, which shows that $H(u_k)$ is also uniquely determined, via the norm in $[\mathbf{H}^1(\Omega)]^*$, and standard lifting by means of N_σ. ■

The application to the case of the porous-medium equation is similar. We state this explicitly as follows. Note that no condition on the partition is required.

Proposition 3.3.2. Let H be given as in Section 1.2, with the initial datum u_0, and consider the formal semidiscretization (3.3.9) of (1.2.13), rewritten as

$$(t_k - t_{k-1})^{-1} H(u_k) - \alpha_{k,k} \Delta u_k = \sum_{j=0}^{k-1} \alpha_{j,k} \Delta u_j + (t_k - t_{k-1})^{-1} H(u_{k-1}), \tag{3.3.24a}$$

with Dirichlet boundary condition

$$u_k = 0 \qquad \text{on} \quad \partial\Omega, \tag{3.3.24b}$$

$k = 1, \ldots, M$, where $\alpha_{k,k} > 0$ is assumed. Then (3.3.24) has a unique weak solution $u_k \in \mathbf{H}_0^1(\Omega)$, and (3.3.24) holds, when interpreted as an identity in $\mathbf{H}^{-1}(\Omega)$. Equivalently, (3.3.24) may be understood in the lifted sense, when $\mathbf{D}_0$ is applied.

The reader will recall from Section 2.5 that the semidiscretization defined there had the property of maintaining the range of the solution in the specified invariant region Q, under the corresponding supposition on $\mathbf{u}_0$ and certain hypotheses on the vector field $\mathbf{f}$. Parallel theorems for other semidiscretizations would be extremely useful, but we shall not attempt them here. We analyze now the existence and uniqueness theory for (2.5.1)–(2.5.4). For those equations for which $D_i \neq 0$, Proposition 3.2.1 is applicable. However, if $D_i = 0$, direct arguments are required. We present these now. If i is the index of a concentration variable, then it is required to determine an $\mathbf{H}^1(\Omega)$ solution, satisfying

$$u^N_{k,i} = G_i(u^N_{k-1,1}, \ldots, u^N_{k-1,i-1}, u^N_{k,i}, u^N_{k-1,i+1}, \ldots, u^N_{k-1,m}, u^N_{k-1,i}), \tag{3.3.25a}$$

where

$$G_i(x_1, \ldots, x_{m+1}) = f_i(x_1, \ldots, x_m)(t^N_k - t^N_{k-1}) + x_{m+1}. \tag{3.3.25b}$$

If i is the index of a potential variable, then G_i is written as

$$G_i(x_1, \ldots, x_{m+1}) \\ = \{g_i(x_i) + h_i(x_1, \ldots, x_{i-1}, x_{m+1}, x_{i+1}, \ldots, x_m)\}(t^N_k - t^N_{k-1}) + x_{m+1}. \tag{3.3.25c}$$

We make here the obvious induction hypothesis that $\mathbf{u}^N_{k-1}$ is determined. If it can be shown that, for fixed $\mathbf{x}' = (x_1, \ldots, x_{i-1}, x_{i+1}, \ldots, x_{m+1})$, the equation

$$R(x_1, \ldots, x_{i-1}, x_{i+1}, \ldots, x_{m+1}) = G_i(x_1, \ldots, x_{i-1}, R(\cdots), x_{i+1}, \ldots, x_{m+1}) \tag{3.3.26}$$

has a solution which is Lipschitz in the variables denoted by $\mathbf{x}'$, then we may proceed as follows. Given $\mathbf{u}^N_{k-1} \in \mathbf{H}^1(\Omega; \mathbb{R}^m)$, define $u^N_{k,i}$ by (3.3.25). Since

$$R: \prod_1^m \mathbf{H}^1(\Omega) \to \mathbf{H}^1(\Omega),$$

we conclude that $u^N_{k,i} \in \mathbf{H}^1(\Omega)$ for each fixed $i = 1, \ldots, m$, such that $D_i = 0$. We shall now determine a fixed compact interval K, such that the mapping

$$\xi \mapsto G_i(x_1, \ldots, x_{i-1}, \xi, x_{i+1}, \ldots, x_{m+1}) \tag{3.3.27}$$

has domain and range in K for $\mathbf{x}'$ suitably restricted, and with restrictions on $(t^N_k - t^N_{k-1})$; these will guarantee (3.3.27) is a strict contraction. Let us now use the *a priori* bounds derived in Proposition 2.5.1. Thus, by the latter, let σ denote half the length of a side of a closed hypercube K_σ, centered at $\mathbf{0}$ in $\mathbb{R}^m$, within which the range of $\mathbf{u}^N_{k-1}$ is constrained to lie by the bounds (2.5.11). The interval K will now be determined independently of $\mathbf{x}'$, $|x'_j| \leqslant \sigma$. Let upper bounds be defined by

$$M_i = \sup\left\{\left|\frac{\partial f_i}{\partial y_i}\right| : \mathbf{y} \in K_{2\sigma}\right\}, \qquad 1 \leqslant i \leqslant i_0, \tag{3.3.28a}$$

$$M_i = \sup\left\{\frac{\partial g_i}{\partial \xi} : -2\sigma \leqslant \xi \leqslant 2\sigma\right\}, \qquad i_0 + 1 \leqslant i \leqslant m. \tag{3.3.28b}$$

Then $K = [-2\sigma, 2\sigma]$ is invariant under (3.3.27), provided

$$(|f_i(0)| + M_i 2\sigma)(t_k^N - t_{k-1}^N) < \sigma, \qquad 1 \leqslant i \leqslant i_0, \tag{3.3.29a}$$

$$(|h_i(0)| + \|h_i\|_{\mathbf{Lip}}\sigma + M_i 2\sigma)(t_k^N - t_{k-1}^N) < \sigma, \qquad i_0 + 1 \leqslant i \leqslant m, \tag{3.3.29b}$$

and moreover the mapping is a contraction. We emphasize that $\mathbf{x}'$ is constrained to lie in K_σ. The assumptions (3.3.29) not only guarantee that (3.3.26) has a solution, but that R is Lipschitz in $\mathbf{x}' \in K_\sigma$. This completes the verification of existence if $D_i = 0$. The uniqueness follows from the *a priori* estimates of Proposition 2.5.1 and the contractive property of (3.3.26).

Proposition 3.3.3. Let the system (1.3.4) be given, with the initial datum $\mathbf{u}_0$, and let σ denote the maximum of the upper bounds in (2.5.11). Then unique solutions exist satisfying (2.5.1)–(2.5.4), provided (3.3.29) is satisfied when $D_i = 0$, and provided

$$(t_k^N - t_{k-1}^N)\sup\left\{\left|\frac{\partial f_i}{\partial u_j}\right| : |u_j| \leqslant \sigma, 1 \leqslant j \leqslant m\right\} < 1, \tag{3.3.30}$$

where $D_i > 0$ and $1 \leqslant i \leqslant i_0$.

Proof: The existence of unique solutions of (2.5.2b) corresponding to the potential variables follows directly from Proposition 3.2.1, with $\mathrm{L} = \mathrm{R}_\omega^{-1}$, $\bar{\beta} = (t_k^N - t_{k-1}^N)^{-1}\mathrm{I} + \mathrm{g}_i$, and $\bar{\mathrm{r}} = 0$, provided $D_i \neq 0$. A similar statement holds, under hypothesis (3.3.30), for the system (2.5.2a), with $\mathrm{L} = \mathrm{R}_\omega^{-1}$, $\bar{\beta} = (t_k^N - t_{k-1}^N)^{-1}\mathrm{I} + \mathrm{f}_i(u_{k-1,1}^N, \ldots, u_{k-1,\,i-1}^N, \cdot, \ldots, u_{k-1,m}^N)$, and $\bar{\mathrm{r}} = 0$. For those equations, such that $D_i = 0$, it has already been demonstrated that unique solutions exist under hypothesis (3.3.29). ■

Our final statement of this section is an application of Proposition 3.2.2 to the semidiscretization of the generalized Navier–Stokes system (1.4.25).

Proposition 3.3.4. Let $a(\cdot,\cdot,\cdot)$ be a continuous trilinear form on $\mathbf{V} \times \mathbf{V} \times \mathbf{V}_s$, satisfying (1.4.22b) and (1.4.24a), where $\mathbf{V}$ and $\mathbf{V}_s$ are defined by (1.4.20) and (1.4.21) in terms of an orthogonal projection P. Consider a formal semidiscretization of (1.4.25) given by

$$(t_k - t_{k-1})^{-1}\mathbf{u}_k - \alpha_{k,k}\eta\,\Delta\mathbf{u}_k + \beta_{k,k}a(\mathbf{u}_k, \mathbf{u}_k, \cdot)$$
$$= \sum_{j=0}^{k-1}\{(\alpha_{j,k}\eta\,\Delta\mathbf{u}_j) - \beta_{j,k}a(\mathbf{u}_j, \mathbf{u}_j, \cdot)\} + (t_k - t_{k-1})^{-1}\mathbf{u}_{k-1}, \tag{3.3.31a}$$

and

$$\mathbf{P}\mathbf{u}_k = \mathbf{0}; \qquad \mathbf{u}_k = \mathbf{0} \qquad \text{on} \quad \partial\Omega, \tag{3.3.31b}$$

$k = 1, \ldots, M$, for a given initial datum $\mathbf{u}_0$, where $\alpha_{k,k} > 0$ and $\beta_{k,k} \geqslant 0$ are assumed. Then a solution $\mathbf{u}_k \in \mathbf{V}$ exists for (3.3.31), when the latter is interpreted as an identity in $\mathbf{V}_s^*$.

Proof: The continuous trilinear form $a(\cdot,\cdot,\cdot)$ of Proposition 3.2.2 is realized here as

$$((t_k - t_{k-1})^{-1}\mathbf{u}, \cdot)_{\mathbf{L}^2(\Omega)} + \beta_{k,k} a(\mathbf{u}, \mathbf{u}, \cdot),$$

when applied to $\mathbf{u} \in \mathbf{V}$. The properties (1.4.22b) and (1.4.24) clearly remain intact. ■

Remark 3.3.4. An important special case of (3.3.31) is the semidiscretization

$$(t_k - t_{k-1})^{-1}(\mathbf{u}_k - \mathbf{u}_{k-1}) - \theta\eta\,\Delta\mathbf{u}_k - (1-\theta)\eta\,\Delta\mathbf{u}_{k-1} + a(\mathbf{u}_k, \mathbf{u}_k, \cdot) = 0, \tag{3.3.32a}$$

$$\mathbf{P}\mathbf{u}_k = \mathbf{0}; \qquad \mathbf{u}_k = \mathbf{0} \qquad \text{on} \quad \partial\Omega, \tag{3.3.32b}$$

where $\frac{1}{2} < \theta \leqslant 1$. An observation necessary for the existence analysis of the evolution equation is the fact that†

$$\sum_{k=1}^{M} \|\mathbf{u}_k - \mathbf{u}_{k-1}\|^2_{\mathbf{L}^2(\Omega;\,\mathbb{R}^n)} + \sum_{k=1}^{M} \|\mathbf{u}_k\|^2_{\mathbf{V}}(t_k - t_{k-1}) \leqslant C, \tag{3.3.33}$$

for some constant C, not depending on the partition selected for $[0, T_0]$. For the most part, we have relegated such estimates to Section 5.1. However, we shall discuss the derivation of (3.3.33) here, since the technique requires a preliminary finite-dimensional estimate and subsequent passage to the limit, as discussed in Example 3.2.3. Thus, with an obvious change of notation, we obtain, from Proposition 3.2.2 and its proof, a sequence $\{\mathbf{u}_k^N\}$ of solutions of the equations

$$\mathrm{A}_{k,N}(\mathbf{u}_k^N) = f_{k|\mathbf{Y}^N}, \qquad \mathbf{u}_k^N \in \mathbf{Y}^N, \tag{3.3.34a}$$

where $\{\mathbf{Y}^N\}$ is a sequence of increasing finite-dimensional subspaces of $\mathbf{V}_s$, such that $\bigcup_N \mathbf{Y}^N$ is dense in $\mathbf{V}_s$. Here $\mathrm{A}_{k,N}$ and f_k are defined through

† A similar estimate is obtained by Temam; for example, see Chapter 1 [23] (Lemma 4.4, p. 325).

the relation

$$(t_k - t_{k-1})^{-1}(\mathbf{u}_k^N - \mathbf{u}_{k-1}^N, \boldsymbol{\phi})_{\mathbf{L}^2(\Omega;\,\mathbb{R}^n)} + \theta\eta(\nabla\mathbf{u}_k^N, \nabla\boldsymbol{\phi})_{\mathbf{L}^2(\Omega;\,\mathbb{R}^n)} + (1-\theta)\eta(\nabla\mathbf{u}_{k-1}^N, \nabla\boldsymbol{\phi})_{\mathbf{L}^2(\Omega;\,\mathbb{R}^n)} + a(\mathbf{u}_k^N, \mathbf{u}_k^N, \boldsymbol{\phi}) = 0, \qquad \text{for all} \quad \boldsymbol{\phi} \in \mathbf{Y}^N \tag{3.3.34b}$$

in an evident recursive manner for $k = 1, \ldots, M$. For $\mathbf{u}_0^N$, we may conveniently choose the orthogonal projection in $\mathbf{V}$ of $\mathbf{u}_0$ onto $\mathbf{Y}^N$. The proof of Proposition 3.2.2 permits a passage to the limit in (3.3.34), from which (3.3.32) results. Moreover, standard lower semicontinuity guarantees (3.3.33) if the corresponding *a priori* estimate can be obtained, independent of N, for the solutions of (3.3.34). By selecting $\boldsymbol{\phi} = \mathbf{u}_k^N$ in (3.3.34b), and using (1.4.22b) and (2.2.38) ($p = q = 2$), we obtain, upon summing from $k = 1$ to $k = M$,

$$\tfrac{1}{2}\|\mathbf{u}_M^N\|^2_{\mathbf{L}^2(\Omega;\,\mathbb{R}^n)} + \tfrac{1}{2}\sum_{k=1}^{M} \|\mathbf{u}_k^N - \mathbf{u}_{k-1}^N\|^2_{\mathbf{L}^2(\Omega;\,\mathbb{R}^n)} + (2\theta - 1)\sum_{k=1}^{M} \|\mathbf{u}_k^N\|^2_{\mathbf{V}}(t_k - t_{k-1}) \leqslant \tfrac{1}{2}\|\mathbf{u}_0^N\|^2_{\mathbf{L}^2(\Omega;\,\mathbb{R}^n)} + \frac{T_0}{2}(1-\theta)\|\mathbf{u}_0^N\|^2_{\mathbf{V}}, \tag{3.3.35}$$

where the norm of (3.2.40a) has been used, together with the identity

$$(\mathbf{u}, \mathbf{v}) = \tfrac{1}{2}[\|\mathbf{u}\|^2 - \|\mathbf{v}\|^2 + \|\mathbf{u} - \mathbf{v}\|^2] \tag{3.3.36}$$

in $\mathbf{L}^2(\Omega; \mathbb{R}^n)$. Now the right-hand side of (3.3.35) is bounded by a constant times $\|\mathbf{u}_0\|^2_{\mathbf{V}}$. We summarize this result in Lemma 3.3.5.

Lemma 3.3.5. If $\frac{1}{2} < \theta \leqslant 1$, then the estimate (3.3.33) holds for the semidiscrete solutions of (3.3.32). Here, C depends upon $\|\mathbf{u}_0\|_{\mathbf{V}}$ and T_0, but not upon the partition selected for $[0, T_0]$.

3.4 BIBLIOGRAPHICAL REMARKS

The proximity mapping corresponding to a given proper, convex, lower-semicontinuous function was introduced by Moreau [24], and is a natural device with which to construct a theory of variational inequalities. The origin of this idea is attributed to Brézis [6] by Ekeland and Temam [13], who describe a slightly less general version of our finite-dimensional result contained in Proposition 3.1.2. The latter result attempts to match the coerciveness hypothesis directly to the given $f \in \mathbf{X}^*$. We have essentially followed Brézis [6] in the development of Proposition 3.1.5 and Corollaries 3.1.6 and

3.1.7, via pseudomonotone operators. For simplicity, however, we have developed the theory in reflexive Banach spaces in terms of sequential weak convergence. Nonetheless, it is not possible to avoid transfinite induction in the proof of Proposition 3.1.5, and we have here adapted the corresponding proof of Opial [25], pp. 98 and 99. We have not presented here an alternative constructive approach, via contraction mappings, which is possible when A is a monotone, hemicontinuous operator. This idea is probably due to Brézis [6].

Pseudomonotone operators represent a precise generalization of continuous mappings defined on open subsets of finite-dimensional spaces $\mathbf{X}_n$, with range in $\mathbf{X}_n^*$. Continuity properties for more restricted classes were discovered by Browder [9] and Kato [16]. The origin of monotonicity methods appears to be found in the study of integral equations. Minty [23], pp. 67 and 68 credits Golomb [14] with perhaps the earliest ideas. Perhaps the first systematic identification and development are due to Zarantonello [31] and Vaĭnberg [30]. The reader is referred to the lecture notes of Opial [25] for elaboration and historical development through 1966. Early work of Minty [21], Browder [10], and Hartman and Stampacchia [15] was decisive. In addition, Minty [22] drew the important connection between convex functions and their subgradients. The early applications to partial differential equations were recognized by Browder [11]. The books of Brézis [8] and Barbu [3] give further elaboration, as does the book-length article of Browder [12]. These three references are concerned with evolution equations in Hilbert and/or Banach spaces. In an attempt to obtain regular solutions via nonexpansive semigroup methods, the mappings are defined from $\mathbf{X}$ to $\mathbf{X}$, rather than from $\mathbf{X}$ to $\mathbf{X}^*$. In this context, it is the accretive mappings, with monotone realizations in Hilbert space, which prove decisive. The reader is referred to these sources for details.

The Tarski fixed-point theorem, described as Proposition 3.1.10, is set forth by Birkhoff [4], p. 115 (see also Tarski [28]). Our proof proceeds under the apparently weaker hypothesis that $\mathbf{I}$ is inductive rather than complete. However, we do not consider commuting families of increasing functions as did Tarski originally. Interesting applications are given by Amann [1]. The application given in Example 3.2.4 was motivated by a similar application in Lions [19], pp. 169–173 (see also Tartar [29]). It is now known that quasi-variational inequalities model many physical and probabilistic phenomena (see Kikuchi and Oden [18] for applications to seepage problems). Moreover, mathematical analogies exist which identify the solutions of certain optimal stopping-time problems with Stefan problems. In these cases, the primary formulation is a quasi-variational inequality.

Some surprising facts concerning stability emerged when spline functions were employed to obtain quadratures for initial-value problems. One of the earliest studies was carried out by Loscalzo and Talbot [20] (see also Böhmer [5]). It is now known that a number of traditional methods, such as the Milne–Simpson method, may be represented by such quadratures. A comprehensive analysis of discrete methods for initial-value problems has been given by Stetter [27]. We note, finally, that the numerical solution of the elliptic boundary-value problem in Example 3.2.1 has been achieved by descent methods by Bank and Rose [2].

REFERENCES

[1] H. Amann, Fixed point equations and nonlinear eigenvalue problems in ordered Banach spaces, *SIAM Rev.* **18**, 620–709 (1976).

[2] R. Bank and D. Rose, Parameter selection for Newton-like methods applicable to nonlinear partial differential equations, *SIAM J. Numer. Anal.* **17**, 806–822 (1980).

[3] V. Barbu, "Nonlinear Semigroups and Differential Equations in Banach Spaces." Noordhoff, Leyden, 1976.

[4] G. Birkhoff, "Lattice Theory," 3rd ed. American Mathematical Society Colloq. Publ. 25, Providence, Rhode Island, 1973.

[5] K. Böhmer, "Spline-Funktionen." Teubner, Stuttgart, 1974.

[6] H. Brézis, Équations et inéquations non linéaires dans les espaces vectoriels en dualité, *Ann. Inst. Fourier (Grenoble)* **18**, 115–175 (1968).

[7] H. Brézis, Perturbation non linéaire d'opérateurs maximaux monotones, *C. R. Acad. Sci. Paris Sér. A-B* **269**, 566–569 (1969).

[8] H. Brézis, "Opérateurs Maximaux Monotones et Semi-groupes de Contractions dans les Espaces de Hilbert." North–Holland Publ., Amsterdam and American Elsevier, New York, 1973.

[9] F. E. Browder, Continuity properties of monotone nonlinear operators in Banach spaces, *Bull. Amer. Math. Soc.* **70**, 551–553 (1964).

[10] F. E. Browder, Nonlinear monotone operators and convex sets in Banach space, *Bull. Amer. Math. Soc.* **71**, 176–183 (1965).

[11] F. E. Browder, Existence and uniqueness theorems for solutions of nonlinear boundary value problems, *Amer. Math. Soc. Proc. Symp. Appl. Math.* **17**, pp. 24–49 (1965).

[12] F. E. Browder, "Nonlinear Operators and Nonlinear Equations of Evolution in Banach Spaces," *Amer. Math. Soc. Proc. Symp. Pure Math.* **18** (Part 2). Providence, Rhode Island, 1976.

[13] I. Ekeland and R. Temam, "Convex Analysis and Variational Problems." North–Holland Publ., Amsterdam and American Elsevier, New York, 1976.

[14] M. Golomb, Zur Theorie der nichtlinearen Integralgleichungen, Integralgleichungsysteme und allgemeinen Funktionalgleichungen, *Math. Z.* **39**, 45–75 (1934).

[15] P. Hartman and G. Stampacchia, On some nonlinear elliptic differential functional equations, *Acta Math.* **115**, 271–310 (1966).

[16] T. Kato, Demicontinuity, hemicontinuity and monotonicity, *Bull. Amer. Math. Soc.* **70**, 548–550 (1964).

[17] J. L. Kelley, "General Topology." Van Nostrand-Reinhold, New York, 1955.

[18] N. Kikuchi and J. T. Oden, Theory of variational inequalities with applications to problems of flow through porous media, *Internat. J. Eng. Sci.* **18**, 1173–1284 (1980).

[19] J. L. Lions, "Sur Quelques Questions d'Analyse de Méchanique et de Contrôle Optimal." Univ. of Montreal Press, Montreal, 1976.

[20] F. Loscalzo and T. D. Talbot, Spline function approximation for solutions of ordinary differential equations, *SIAM J. Numer. Anal.* **4**, 433–445 (1967).

[21] G. Minty, Monotone (nonlinear) operators in Hilbert space, *Duke Math. J.* **29**, 341–346 (1962).

[22] G. Minty, On the monotonicity of the gradient of a convex function, *Pacific J. Math.* **14**, 243–247 (1964).

[23] G. Minty, On some aspects of the theory of monotone operators, *in Proc. NATO Adv. Study Inst. Theory and Appl. Monotone Operators.* Edizioni Oderisi, Gubbio, Italy, 1969.

[24] J. J. Moreau, Proximité et dualité dans un espace hilbertian, *Bull. Soc. Math. France* **93**, 273–299 (1965).

[25] Z. Opial, "Nonexpansive and Monotone Mappings in Banach Spaces," Lecture Notes, Division of Applied Mathematics, Brown Univ., 1967.

[26] R. T. Rockafellar, On the maximality of sums of nonlinear monotone operators, *Trans. Amer. Math. Soc.* **149**, 75–88 (1970).

[27] J. J. Stetter, "Analysis of Discretization Methods for Ordinary Differential Equations." Springer–Verlag, Berlin, and New York, 1973.

[28] A. Tarski, A lattice-theoretical fixpoint theorem and its applications, *Pacific J. Math.* **5**, 285–309 (1955).

[29] L. Tartar, Inéquations quasi-variationelles abstraites, *C. R. Acad. Sci. Paris* **278**, 1193–1196 (1974).

[30] M. M. Vaĭnberg, On the convergence of the method of steepest descent, *Sibirsk. Mat. Z.* **2**, 201–220 (1961)

[31] E. H. Zarantonello, "Solving Functional Equations by Contractive Averaging, MRC Technical Summary Rep. 160. Univ. of Wisconsin, Madison, Wisconsin (1960).

4 NUMERICAL OPTIMALITY AND THE APPROXIMATE SOLUTION OF DEGENERATE PARABOLIC EQUATIONS

4.0 INTRODUCTION

If a square-integrable function u, on a bounded space–time domain $\mathscr{D}$ of dimension $n + 1$, possesses a square-integrable gradient and time derivative, then the theory of approximation asserts that the optimal asymptotic approximation order in $\mathbf{L}^2(\mathscr{D})$, by subspaces of dimension N, is $N^{-1/(n+1)}$, provided the estimate is to be uniform over the unit ball of $\mathbf{H}^1(\mathscr{D})$. The approximation concept, which expresses this property, is known as the (Kolmogorov) N-width or N-dimensional diameter, which is developed in Section 4.2. It follows that any numerical procedure for approximating the solution of a degenerate parabolic equation, for example, the two-phase Stefan problem or porous-medium equation, for which the solution u is known to belong only to $\mathbf{H}^1(\mathscr{D})$, should be constructed with this optimal order in mind. In Proposition 4.3.4, we prove that the continuous-time Galerkin (Faedo–Galerkin) procedure, based on piecewise linear finite elements of diameter h, is convergent in $\mathbf{L}^2(\mathscr{D})$, with order $h|\ln(1/h)|^{d(n)/2}$. Here, $d(n) = 0$ if $n = 1$, and $d(n) = 2$ if $n > 1$. This includes both the case of the two-phase Stefan problem and the porous-medium equation. One sees, intuitively, that this order is essentially optimal if the continuous-time Galerkin approximation is replaced by its piecewise-linear interpolant based on $t_k = kh$. The approximation order is unchanged, and the resultant finite-dimensional spaces are of dimension $N \approx ch^{-(n+1)}$. The presence of the logarithmic factor is accounted for by the occurrence in the equation of the term $\partial H(u)/\partial t$, which is not, in general, square integrable. It is, however, essentially bounded in time as a regular Baire measure, which permits the application of the recent $\mathbf{L}^\infty$ finite-element estimates, for linear elliptic problems, via duality.

The plan of this chapter is actually more ambitious than described in the previous paragraph. For example, the exposition on N-widths gives a fairly complete discussion of the general abstract problem, and describes, as a major application, the determination of the widths in $\mathbf{L}^q(\Omega)$ of the unit ball of $\mathbf{W}^{\ell,p}(\Omega)$, when the embedding is compact. The general result is stated in Remark 4.2.5. A complete proof for all of the subcases would simply be too lengthy here. We present our own proofs for the subcases $p = q$ and $1 < p \leqslant q \leqslant 2$. These are variationally based, and lead naturally into the case $p = q = 2$, where very precise results are obtainable (see Theorem 4.2.14). The upper N-width bounds for $p = q$ and $1 \leqslant p \leqslant q \leqslant 2$ are obtained conveniently in Section 4.1 by means of the Sobolev integral-representation formula, so that the first two sections mesh nicely. However, the standard linear finite-element estimates depend upon the results of Section 4.1, and we use them in Section 4.3, where a very general version of the Aubin–Nitsche result is obtained in Proposition 4.3.1. Only minor modifications are required to extend this Hilbert space result to a Banach space duality result.

We believe the reader is entitled to an explanation of why the case $p \neq q$ is considered in such detail in this chapter. It is precisely because mixed norms enter in such a fundamental way in nonlinear partial differential equations. For example, in Proposition 4.4.1, where the convergence of the horizontal-line method is analyzed, a natural choice is $q = 2$ and $p = 1 + (1/\gamma)$ for the expression of the estimates (see (4.4.5)).

We mention, finally, that a really complete mixed-norm analysis, extending the first two sections, would provide for different orders of differentiation in the various independent variables, together with different $\mathbf{L}^p$ norms of the respective variables. This is the situation which actually occurs in nonlinear partial differential equations. The symmetric subcase of a uniform order ℓ and uniform $\mathbf{L}^p$ norm, hence a standard $\mathbf{W}^{\ell,p}$ space, actually impinges on the relatively innocuous linear equation $-\Delta u = f$, $f \in \mathbf{L}^2$. By the Sobolev embedding theorem, $u \in \mathbf{L}^q$, $q > 2$. Thus, even this simple linear problem has a natural mixed-norm interpretation with $p = 2$ and $q > 2$.

4.1 REPRESENTATIONS OF SOBOLEV-TYPE AND UPPER-BOUND ESTIMATES

We suppose that Ω is a bounded open domain in $\mathbb{R}^n$, with diameter d, which is star-shaped, with respect to every point in an open ball $B \subset \Omega$, and that $\chi \in \mathbf{C}_0^\infty(B)$ has integral value unity and is fixed, $\chi \geqslant 0$.

Proposition 4.1.1.[†] Suppose that $\ell > 0$ is a specified integer, that

$$k(\mathbf{x}, \mathbf{y}) = \int_0^1 s^{-n-1}\chi(\mathbf{x} + s^{-1}(\mathbf{y} - \mathbf{x}))\,ds, \qquad \mathbf{x}, \mathbf{y} \in \mathbb{R}^n, \quad \mathbf{x} \neq \mathbf{y}, \tag{4.1.1a}$$

and for $\boldsymbol{\alpha}$ a multi-integer, $|\boldsymbol{\alpha}| = \ell$, that the kernels $k_{\boldsymbol{\alpha}}$ are given by

$$k_{\boldsymbol{\alpha}}(\mathbf{x}, \mathbf{y}) = (\ell/\boldsymbol{\alpha}!)(\mathbf{x} - \mathbf{y})^{\boldsymbol{\alpha}} k(\mathbf{x}, \mathbf{y}), \qquad \mathbf{x}, \mathbf{y} \in \mathbb{R}^n. \tag{4.1.1b}$$

Then $k(\mathbf{x}, \cdot)$ is a continuous function on $\mathbb{R}^n \backslash \{\mathbf{x}\}$, is supported in the convex hull $K_{\mathbf{x}}$ of $\{\mathbf{x}\} \cup \text{support}(\chi)$, and satisfies the estimate, for $\mathbf{y} \in \mathbb{R}^n$,

$$|k(\mathbf{x}, \mathbf{y})| \leqslant \|\chi\|_{\mathbf{L}^\infty(B)} d^n |\mathbf{x} - \mathbf{y}|^{-n}/n, \qquad \mathbf{x} \in \Omega, \quad \mathbf{x} \neq \mathbf{y}. \tag{4.1.2}$$

In particular, $k_{\boldsymbol{\alpha}}(\mathbf{x}, \cdot)$ is supported in $K_{\mathbf{x}}$, and

$$|k_{\boldsymbol{\alpha}}(\mathbf{x}, \mathbf{y})| \leqslant C\ell \|\chi\|_{\mathbf{L}^\infty(B)} d^n |\mathbf{x} - \mathbf{y}|^{-n+|\boldsymbol{\alpha}|}/n, \qquad \mathbf{x} \in \Omega, \quad \mathbf{x} \neq \mathbf{y}, \tag{4.1.3}$$

for a constant C depending only on n. When $f \in \mathbf{C}^\infty(\Omega)$, then a natural projection onto the polynomials of total degree less than ℓ on Ω is induced by the representation of Sobolev type

$$f(\mathbf{x}) = \mathrm{P}_\ell f(\mathbf{x}) + \mathrm{R}_\ell f(\mathbf{x}), \tag{4.1.4a}$$

where

$$\mathrm{P}_\ell f(\mathbf{x}) = \sum_{|\boldsymbol{\alpha}| < \ell} \int_B \chi(\mathbf{y}) \frac{(\mathbf{x} - \mathbf{y})^{\boldsymbol{\alpha}}}{\boldsymbol{\alpha}!} f^{(\boldsymbol{\alpha})}(\mathbf{y})\,d\mathbf{y} \tag{4.1.4b}$$

is a polynomial of degree less than ℓ, and

$$\mathrm{R}_\ell f(\mathbf{x}) = \sum_{|\boldsymbol{\alpha}| = \ell} \int_\Omega k_{\boldsymbol{\alpha}}(\mathbf{x}, \mathbf{y}) f^{(\boldsymbol{\alpha})}(\mathbf{y})\,d\mathbf{y}. \tag{4.1.4c}$$

In particular, the definition of $\mathrm{P}_\ell f$ depends only upon $f_{|B}$.

[†] Propositions (4.1.1–4.1.3) and Corollary (4.1.4) reproduced with modifications from [13].

Proof: We note that the semi-infinite ray $\{\mathbf{x} + s^{-1}(\mathbf{z} - \mathbf{x}):0 < s \leqslant 1\}$, for $\mathbf{z} \neq \mathbf{x}$, intersects the support of χ if and only if $\mathbf{z}$ is in $K_{\mathbf{x}}$. It follows that $\{\mathbf{z}:k(\mathbf{x},\mathbf{z}) > 0,\ \mathbf{x} \neq \mathbf{z}\} \subset K_{\mathbf{x}}$ and, hence, by a simple closure argument, that $\operatorname{support}(k(\mathbf{x},\cdot)) \subset K_{\mathbf{x}}$. To verify (4.1.2) we have, for $|\mathbf{y} - \mathbf{x}| \leqslant d$, $\mathbf{y} \neq \mathbf{x}$,

$$
\begin{aligned}
|k(\mathbf{x},\mathbf{y})| &= \left| \int_0^1 \chi(\mathbf{x} + s^{-1}(\mathbf{y} - \mathbf{x}))s^{-n-1}\,ds \right| \\
&= \left| \int_{|\mathbf{y}-\mathbf{x}|/d}^1 \chi(\mathbf{x} + s^{-1}(\mathbf{y} - \mathbf{x}))s^{-n-1}\,ds \right| \\
&\leqslant \|\chi\|_{\mathbf{L}^\infty(B)} \int_{|\mathbf{y}-\mathbf{x}|/d}^1 s^{-n-1}\,ds \\
&= \|\chi\|_{\mathbf{L}^\infty(B)} \left[\left(\frac{d}{|\mathbf{x} - \mathbf{y}|} \right)^n - 1 \right] \Big/ n,
\end{aligned}
$$

from which (4.1.2) follows for $|\mathbf{y} - \mathbf{x}| \leqslant d$. Of course, $k(\mathbf{x},\mathbf{y}) = 0$ for $|\mathbf{y} - \mathbf{x}| \geqslant d$. To verify (4.1.3), we let C denote an equivalence constant, for which

$$
|(\mathbf{x} - \mathbf{y})^{\boldsymbol{\alpha}}| \leqslant \|\mathbf{x} - \mathbf{y}\|_{\ell^\infty(\mathbb{R}^n)}^{|\boldsymbol{\alpha}|} \leqslant C|x - y|^{|\boldsymbol{\alpha}|} = C\|\mathbf{x} - \mathbf{y}\|_{\ell^2(\mathbb{R}^n)}^{|\boldsymbol{\alpha}|}, \tag{4.1.5}
$$

and then use (4.1.2). Note that $|\cdot|$ has been used as a Euclidean norm on $\mathbb{R}^1$ and $\mathbb{R}^n$, respectively, in (4.1.5).

Now suppose $\mathbf{x} \in \Omega$ and $\mathbf{y} \in B$. By Taylor's theorem,

$$
f(\mathbf{x}) = \sum_{|\boldsymbol{\alpha}| < \ell} \frac{(\mathbf{x} - \mathbf{y})^{\boldsymbol{\alpha}}}{\boldsymbol{\alpha}!} f^{(\boldsymbol{\alpha})}(\mathbf{y}) + \ell \sum_{|\boldsymbol{\alpha}| = \ell} \frac{(\mathbf{x} - \mathbf{y})^{\boldsymbol{\alpha}}}{\boldsymbol{\alpha}!} \int_0^1 s^{\ell-1} f^{(\boldsymbol{\alpha})}(\mathbf{x} + s(\mathbf{y} - \mathbf{x}))\,ds, \tag{4.1.6}
$$

and multiplication by $\chi(\mathbf{y})$, followed by integration with respect to $\mathbf{y}$, yield

$$
f(\mathbf{x}) = \mathrm{P}_\ell f(\mathbf{x}) + \ell \sum_{|\boldsymbol{\alpha}| = \ell} \frac{1}{\boldsymbol{\alpha}!} \int_B \chi(\mathbf{y})(\mathbf{x} - \mathbf{y})^{\boldsymbol{\alpha}} \int_0^1 s^{\ell-1} f^{(\boldsymbol{\alpha})}(\mathbf{x} + s(\mathbf{y} - \mathbf{x}))\,ds\,d\mathbf{y}. \tag{4.1.7}
$$

We shall analyze the latter integrals. By (4.1.3), the iterated integrals

$$
I(\boldsymbol{\alpha},\mathbf{x}) = \frac{\boldsymbol{\alpha}!}{\ell} \int_{K_{\mathbf{x}}} k_{\boldsymbol{\alpha}}(\mathbf{x},\mathbf{z}) f^{(\boldsymbol{\alpha})}(\mathbf{z})\,d\mathbf{z} \tag{4.1.8}
$$

are finite for $|\boldsymbol{\alpha}| = \ell$, and, by the Fubini theorem, we have equality with the product integral expressed by

$$
I(\boldsymbol{\alpha},\mathbf{x}) = \int_0^1 \int_{K_{\mathbf{x}}} \{\chi(\mathbf{x} + s^{-1}(\mathbf{z} - \mathbf{x}))(\mathbf{x} - \mathbf{z})^{\boldsymbol{\alpha}} s^{-1} f^{(\boldsymbol{\alpha})}(\mathbf{z}) s^{-n}\}\,ds\,d\mathbf{z}. \tag{4.1.9}
$$

For $s \in (0,1]$ and $z \in K_{\mathbf{x}}$, consider the mapping

$$
\phi(s,\mathbf{z}) = (s,\mathbf{y}) = (s,\mathbf{x} + s^{-1}(\mathbf{z} - \mathbf{x})) \in (0,1] \times \mathbb{R}^n. \tag{4.1.10}
$$

One easily sees that ϕ is injective onto its range, since $\phi(s_1, \mathbf{z}_1) = \phi(s_2, \mathbf{z}_2)$ implies $s_1 = s_2$ and, hence, $\mathbf{z}_1 = \mathbf{z}_2$. Now, by the Lebesgue–Radon–Nikodým theorem ([14] p. 230) the substitution of (4.1.10) is permitted in (4.1.9), yielding

$$I(\boldsymbol{\alpha}, \mathbf{x}) = \int_{\text{Range}(\phi)} \chi(\mathbf{y})(\mathbf{x} - \mathbf{y})^{\boldsymbol{\alpha}} s^{\ell - 1} f^{(\boldsymbol{\alpha})}(\mathbf{x} + s(\mathbf{y} - \mathbf{x}))\, ds\, d\mathbf{y}. \quad (4.1.11)$$

However,

$$\text{Range}(\phi) \supset (0, 1] \times \text{support}(\chi), \quad (4.1.12)$$

and χ vanishes for $\mathbf{y} \notin \text{support}(\chi)$. It follows that

$$I(\boldsymbol{\alpha}, \mathbf{x}) = \int_0^1 \int_B \chi(\mathbf{y})(\mathbf{x} - \mathbf{y})^{\boldsymbol{\alpha}} s^{\ell - 1} f^{(\boldsymbol{\alpha})}(\mathbf{x} + s(\mathbf{y} - \mathbf{x}))\, ds\, d\mathbf{y}. \quad (4.1.13)$$

If the Fubini theorem is applied to (4.1.13), and the resultant substituted into (4.1.7), we obtain (4.1.4). Note that here we have used the facts that $\text{support}(k_{\boldsymbol{\alpha}}(\mathbf{x}, \cdot)) \subset K_{\mathbf{x}}$ and that the latter set is contained in the star-shaped region Ω. ■

Remark 4.1.1. For $\mathbf{x} \neq \mathbf{y}$, we may generalize (4.1.2) and (4.1.3) to

$$\left| \left(\frac{\partial}{\partial \mathbf{x}}\right)^{\boldsymbol{\beta}} \left(\frac{\partial}{\partial \mathbf{y}}\right)^{\boldsymbol{\gamma}} k(\mathbf{x}, \mathbf{y}) \right| \leqslant C(n, d, \chi, |\boldsymbol{\beta}|, |\boldsymbol{\gamma}|) |\mathbf{x} - \mathbf{y}|^{-n - |\boldsymbol{\beta}| - |\boldsymbol{\gamma}|}, \quad (4.1.14a)$$

and

$$\left| \left(\frac{\partial}{\partial \mathbf{x}}\right)^{\boldsymbol{\beta}} \left(\frac{\partial}{\partial \mathbf{y}}\right)^{\boldsymbol{\gamma}} k_{\boldsymbol{\alpha}}(\mathbf{x}, \mathbf{y}) \right| \leqslant C(n, \ell, d, \chi, |\boldsymbol{\beta}|, |\boldsymbol{\gamma}|) |\mathbf{x} - \mathbf{y}|^{|\boldsymbol{\alpha}| - n - |\boldsymbol{\beta}| - |\boldsymbol{\gamma}|}. \quad (4.1.14b)$$

An elementary integration by parts shows that

$$(P_\ell f)(\mathbf{x}) = \sum_{|\boldsymbol{\alpha}| < \ell} (-1)^{|\boldsymbol{\alpha}|} \int_B f(\mathbf{y}) \left(\frac{\partial}{\partial \mathbf{y}}\right)^{\boldsymbol{\alpha}} \left[\chi(\mathbf{y}) \frac{(\mathbf{x} - \mathbf{y})^{\boldsymbol{\alpha}}}{\boldsymbol{\alpha}!} \right] d\mathbf{y}, \quad (4.1.15)$$

which shows that P_ℓ may be defined for arbitrary distributions f on Ω by

$$\mathrm{P}_\ell f(\mathbf{x}) = \sum_{|\boldsymbol{\alpha}| < \ell} f^{(\boldsymbol{\alpha})}(\chi(\cdot)(\mathbf{x} - \cdot)^{\boldsymbol{\alpha}} / \boldsymbol{\alpha}!). \quad (4.1.16)$$

Note that P_ℓ is a linear projection operator. In fact, the idempotent property of P_ℓ follows from the representation (4.1.4), since $\mathrm{R}_\ell = 0$ if f is a polynomial of degree less than ℓ.

Proposition 4.1.2. For $f \in \mathscr{D}'(\Omega)$, the commutation result

$$\left(\frac{\partial}{\partial \mathbf{x}}\right)^{\boldsymbol{\alpha}} \mathrm{P}_\ell f = \mathrm{P}_{\ell - |\boldsymbol{\alpha}|} \left(\frac{\partial}{\partial \mathbf{x}}\right)^{\boldsymbol{\alpha}} f \quad (4.1.17)$$

holds for $|\boldsymbol{\alpha}| < \ell$, for every positive integer ℓ.

Proof: It suffices to prove (4.1.17) for $f \in \mathbf{C}^\infty(\Omega)$, since this set is dense in $\mathcal{D}'(\Omega)$, and the composed operations in (4.1.17) are continuous on $\mathcal{D}'(\Omega)$. For $f \in \mathbf{C}^\infty(\Omega)$, $\mathbf{x} \in \Omega$, and $\mathbf{y} \in B$, set

$$\mathrm{T}_{\mathbf{y}}^{\ell} f(\mathbf{x}) = \sum_{|\alpha| < \ell} f^{(\alpha)}(\mathbf{y}) \frac{(\mathbf{x} - \mathbf{y})^{\alpha}}{\alpha!}, \tag{4.1.18}$$

the Taylor polynomial of f. A proof by induction on ℓ shows that

$$\left[\left(\frac{\partial}{\partial \mathbf{x}}\right)^{\beta} (\mathrm{T}_{\mathbf{y}}^{\ell} f)\right](\mathbf{x}) = \left[\mathrm{T}_{\mathbf{y}}^{\ell - |\beta|}\left(\frac{\partial}{\partial \mathbf{x}}\right)^{\beta} f\right](\mathbf{x}), \qquad |\beta| < \ell. \tag{4.1.19}$$

Thus, differentiating the expression

$$\mathrm{P}_{\ell} f(\mathbf{x}) = \int_B \chi(\mathbf{y}) \mathrm{T}_{\mathbf{y}}^{\ell} f(\mathbf{x})\, d\mathbf{y}$$

under the integral sign, and making use of (4.1.19), we are led to (4.1.17). ■

Definition 4.1.1. Let s be a real number, satisfying $0 < s < n$. By the Riesz potentials are meant the convolutions

$$\mathrm{I}_s f(\mathbf{x}) = \int_{\Omega} |\mathbf{x} - \mathbf{y}|^{-n+s} f(\mathbf{y})\, d\mathbf{y}, \qquad \mathbf{x} \in \Omega, \tag{4.1.20}$$

when the latter converge absolutely for almost every $\mathbf{x}$.

Remark 4.1.2. The convolution (4.1.20) represents an adaptation of the classical Riesz potential, which is typically defined on all of $\mathbb{R}^n$ with a constant multiplier to reflect the Fourier transform of a fractional power of the Laplacian. A fundamental question is whether the mapping $f \to \mathrm{I}_s f$ is bounded from $\mathbf{L}^p(\Omega)$ to $\mathbf{L}^q(\Omega)$. This is discussed in the following.

Proposition 4.1.3. For $0 < s < n$, $1 \leqslant p \leqslant \infty$, $1 \leqslant q \leqslant \infty$ and

$$\frac{1}{q} - \frac{1}{p} + \frac{s}{n} \geqslant 0, \tag{4.1.21}$$

the mapping $\mathrm{I}_s : \mathbf{L}^p(\Omega) \to \mathbf{L}^q(\Omega)$ is bounded if at least one of the two numbers $\eta_1 = (1/q) - (1/p) + (s/n)$ and $\eta_2 = (1 - 1/p)(1/q)$ is positive. For $s \geqslant n$, I_s is uniformly bounded for all p and q.

Proof: Set

$$K_s(\mathbf{x}) = \begin{cases} |\mathbf{x}|^{s-n}, & |\mathbf{x}| \leqslant d, \\ 0, & |\mathbf{x}| > d, \end{cases} \tag{4.1.22a}$$

and note the convolution identity

$$\mathrm{I}_s f(\mathbf{x}) = K_s * \tilde{f}(\mathbf{x}), \qquad \mathbf{x} \in \Omega, \tag{4.1.22b}$$

if $\tilde{f}$ is the extension of f which is zero outside Ω. We consider first the most delicate case,† namely,

$$\frac{1}{q} = \frac{1}{p} - \frac{s}{n}, \qquad 1 < p < q < \infty, \tag{4.1.23}$$

which corresponds to $\eta_1 = 0, \eta_2 > 0, 0 < s < n$, and show that I_s is bounded from $\mathbf{L}^p(\Omega)$ to $\mathbf{L}^q(\Omega)$. Construct the ray, in the $(1/p), (1/q)$ plane, from $(1, (1/q_*))$ to $((1/p), (1/q))$, where $1/q_* = 1 - (s/n)$. If this ray of slope one is extended to a first quadrant point $((1/p_1), (1/q_1))$, such that $((1/p), (1/q))$ is an interior point of the ray, then the Marcinkiewicz interpolation theorem (see Stein [38] p. 272) will yield the desired result, provided I_s is of weak type $(1, q_*)$, (p_1, q_1). We are thus led to prove the following statement. The mapping I_s is of weak type (p, q) for

$$\frac{1}{q} = \frac{1}{p} - \frac{s}{n}, \qquad 1 \leqslant p < q < \infty. \tag{4.1.24}$$

Thus, we show, in this case, that there exist positive constants $A_{p,q}$, such that

$$m\{\mathbf{x} \in \Omega : |K_s * \tilde{f}(\mathbf{x})| > \lambda\} \leqslant \left(A_{p,q} \frac{\|f\|_{\mathbf{L}^p(\Omega)}}{\lambda}\right)^q, \qquad \lambda > 0, \tag{4.1.25}$$

for all $f \in \mathbf{L}^p(\Omega)$. Clearly, if (4.1.25) holds for $\|f\|_{\mathbf{L}^p(\Omega)} = 1$, it holds for general f, upon division by $\|f\|_{\mathbf{L}^p(\Omega)}$, when $f \neq 0$. Similarly, an adjustment of $A_{p,q}$ permits the replacement of λ by 2λ on the left-hand side of (4.1.25). Given $\mu > 0$, we introduce the kernels

$$K_{s,\mu,1}(\mathbf{x}) = \begin{cases} K_s(\mathbf{x}), & |\mathbf{x}| \leqslant \mu, \\ 0, & |\mathbf{x}| > \mu, \end{cases} \tag{4.1.26a}$$

and

$$K_{s,\mu,\infty}(\mathbf{x}) = \begin{cases} K_s(\mathbf{x}), & |\mathbf{x}| > \mu, \\ 0, & |\mathbf{x}| \leqslant \mu, \end{cases} \tag{4.1.26b}$$

and note that $K_s = K_{s,\mu,1} + K_{s,\mu,\infty}$. For $\lambda > 0$, define

$$\mu = \mu(\lambda) = \left(\frac{\omega_{n-1} q}{p' n}\right)^{q/(np')} \lambda^{-q/n}, \tag{4.1.27}$$

where $(1/p') + (1/p) = 1$, and ω_{n-1} is the measure of the unit sphere in $\mathbb{R}^n$. As we shall see below, $K_{s,\mu,1} \in \mathbf{L}^1(\mathbb{R}^n)$, and $K_{s,\mu,\infty} \in \mathbf{L}^{p'}(\mathbb{R}^n)$. Now, by the

† Proof adapted from [38], p. 119–121.

triangle inequality,

$$m\{\mathbf{x} \in \Omega: |K_s * \tilde{f}(\mathbf{x})| > 2\lambda\}$$
$$\leqslant m\{\mathbf{x} \in \Omega: |K_{s,\mu,1} * \tilde{f}(\mathbf{x})| > \lambda\} + m\{\mathbf{x} \in \Omega: |K_{s,\mu,\infty} * \tilde{f}(\mathbf{x})| > \lambda\}. \quad (4.1.28)$$

We shall show that the second term on the right-hand side of (4.1.28) is zero. Indeed, by Young's inequality,

$$\begin{aligned}\|K_{s,\mu,\infty} * \tilde{f}\|_{\mathbf{L}^\infty(\Omega)} &\leqslant \|K_{s,\mu,\infty}\|_{\mathbf{L}^{p'}(\mathbb{R}^n)} \|\tilde{f}\|_{\mathbf{L}^p(\mathbb{R}^n)} \\ &= \left(\int_{|\mathbf{x}|>\mu} |\mathbf{x}|^{(-n+s)p'}\, dx\right)^{1/p'} \\ &= \left(\frac{\omega_{n-1} q}{p' n}\right)^{1/p'} \mu^{-n/q} = \lambda.\end{aligned}$$

Note that this computation has used the relation

$$(-n+s)p' + n = -np'/q < 0,$$

which follows from (4.1.24), and has used the definition (4.1.27). It remains to estimate the first term. Thus,

$$\begin{aligned}m\{\mathbf{x} \in \Omega: |K_{s,\mu,1} * \tilde{f}(\mathbf{x})| > \lambda\} &\leqslant \frac{\|K_{s,\mu,1} * \tilde{f}\|^p_{\mathbf{L}^p(\Omega)}}{\lambda^p} \\ &\leqslant \frac{\|K_{s,\mu,1}\|^p_{\mathbf{L}^1(\mathbb{R}^n)} \|\tilde{f}\|^p_{\mathbf{L}^p(\mathbb{R}^n)}}{\lambda^p} \\ &= \left(\frac{\int_{|\mathbf{x}|\leq\mu} |\mathbf{x}|^{-n+s}\, d\mathbf{x}}{\lambda}\right)^p \\ &= \frac{(\omega_{n-1}/s)^p \mu^{sp}}{\lambda} \\ &= A^q_{p,q} \lambda^{-q},\end{aligned}$$

where $A_{p,q}$ is given explicitly by

$$A_{p,q} = \left(\frac{\omega_{n-1} q}{p' n}\right)^{sp/(np')} \left(\frac{\omega_{n-1}}{s}\right)^{p/q}. \quad (4.1.29)$$

This completes the verification of (4.1.25) and, hence, the boundedness in the case (4.1.23).

The case $\eta_1 > 0$, $0 < s < n$, that is,

$$\frac{1}{q} - \frac{1}{p} + \frac{s}{n} > 0, \qquad 0 < s < n, \quad (4.1.30)$$

follows by a routine application of Young's inequality:

$$\|\mathrm{I}_s f\|_{\mathbf{L}^q(\Omega)} = \|K_s * \tilde{f}\|_{\mathbf{L}^q(\Omega)} \leqslant \|K_s\|_{\mathbf{L}^r(\mathbb{R}^n)} \|f\|_{\mathbf{L}^p(\Omega)}, \tag{4.1.31a}$$

for

$$\frac{1}{r} = 1 - \frac{1}{p} + \frac{1}{q} > 1 - \frac{s}{n}. \tag{4.1.31b}$$

In this case, $n + (s - n)r > 0$, and

$$\|K_s\|_{\mathbf{L}^r(\mathbb{R}^n)} = \left(\int_{|\mathbf{x}|<d} |\mathbf{x}|^{(-n+s)r}\, d\mathbf{x} \right)^{1/r}$$
$$= \left(\frac{\omega_{n-1} d^{(s-n)r+n}}{n + (s-n)r} \right)^{1/r}. \tag{4.1.32}$$

In the case $s \geqslant n$, K_s is bounded, so that

$$\|\mathrm{I}_s f\|_{\mathbf{L}^\infty(\Omega)} \leqslant d^{s-n} \|f\|_{\mathbf{L}^1(\Omega)},$$

and the boundedness of I_s for all p and q, follows from Hölder's inequality. ■

Corollary 4.1.4. Let m and ℓ be integers, such that $0 \leqslant m < \ell$, and suppose $1 \leqslant p \leqslant \infty$, $1 \leqslant q \leqslant \infty$. Suppose that

$$\frac{1}{q} - \frac{1}{p} + \frac{\ell - m}{n} \geqslant 0, \tag{4.1.33}$$

and, if equality holds in (4.1.33), that $(1 - (1/p))(1/q) \neq 0$. Then there are constants $C(n, \ell, d, \chi, m, p, q)$, such that

$$\|f - \mathrm{P}_\ell f\|_{\mathbf{W}^{m,q}(\Omega)} \leqslant C(n, \ell, d, \chi, m, p, q) |f|_{\mathbf{W}^{\ell,p}(\Omega)}, \tag{4.1.34}$$

for all $f \in \mathbf{W}^{\ell,p}(\Omega)$.

Proof: Fix $\boldsymbol{\beta}$, $|\boldsymbol{\beta}| \leqslant m$. By the commutation property of P_ℓ, we have

$$\left(\frac{\partial}{\partial \mathbf{x}}\right)^{\boldsymbol{\beta}} (f - \mathrm{P}_\ell f) = f^{(\boldsymbol{\beta})} - \mathrm{P}_{\ell-|\boldsymbol{\beta}|} f^{(\boldsymbol{\beta})} = \mathrm{R}_{\ell-|\boldsymbol{\beta}|} f^{(\boldsymbol{\beta})},$$

for $f \in \mathbf{W}^{\ell,p}(\Omega)$. Thus, by (4.1.3) and (4.1.4),

$$\left\| \left(\frac{\partial}{\partial \mathbf{x}}\right)^{\boldsymbol{\beta}} (f - \mathrm{P}_\ell f) \right\|_{\mathbf{L}^q(\Omega)} = \|\mathrm{R}_{\ell-|\boldsymbol{\beta}|} f^{(\boldsymbol{\beta})}\|_{\mathbf{L}^q(\Omega)}$$
$$\leqslant C(\ell - |\beta|, \chi, d, n) \sum_{|\boldsymbol{\alpha}| = \ell - |\boldsymbol{\beta}|} \|\mathrm{I}_{\ell-|\boldsymbol{\beta}|} f^{(\boldsymbol{\alpha}+\boldsymbol{\beta})}\|_{\mathbf{L}^q(\Omega)},$$

and Proposition 4.1.3 applies, with $s = \ell - |\boldsymbol{\beta}|$, to yield (4.1.34), upon summation over $|\boldsymbol{\beta}| \leqslant m$. Here, we use the inequality, for $1 \leqslant p < \infty$, and $C_{p,n} = n^{p-1}$ (see Beckenbach and Bellman [5]),

$$\left\{\sum_{i=1}^{n} |a_i|^p\right\}^{1/p} \leqslant \sum_{i=1}^{n} |a_i| \leqslant C_{p,n}\left\{\sum_{i=1}^{n} |a_i|^p\right\}^{1/p}. \quad \blacksquare \tag{4.1.35}$$

Remark 4.1.3. The conclusion of the previous corollary holds for more general domains than star-shaped domains, in particular for the connected, finite union of such domains Ω_j. Somewhat surprisingly, the polynomial $\mathrm{P}_\ell f$ may be defined with respect to any one of the balls B_j. The extended corollary may, thus, be applied to connected bounded domains satisfying the restricted-cone condition, but also to other domains, such as slit domains. The proof of this fact is very easily demonstrated in the case $\Omega = \Omega_1 \cup \Omega_2$, $\Omega_1 \cap \Omega_2 \neq \varnothing$. Thus, setting $p_i = \mathbf{P}_{\ell,i} f$, $i = 1, 2$, we have

$$\|f - p_1\|_{\mathbf{W}^{m,q}(\Omega)} \leqslant \|f - p_1\|_{\mathbf{W}^{m,q}(\Omega_1)} + \|f - p_2\|_{\mathbf{W}^{m,q}(\Omega_2)} + \|p_1 - p_2\|_{\mathbf{W}^{m,q}(\Omega_2)}. \tag{4.1.36}$$

If the equivalence of norms on the finite-dimensional space $\mathscr{P}_\ell$ of polynomials of degree less than ℓ is used in (4.1.36), with respect to $\|\cdot\|_{\mathbf{W}^{m,q}(\Omega_2)}$ and $\|\cdot\|_{\mathbf{W}^{m,q}(\Omega_1 \cap \Omega_2)}$, followed by the triangle inequality, we obtain

$$\begin{aligned}
\|f - p_1\|_{\mathbf{W}^{m,q}(\Omega)} &\leqslant C_1\|f - p_1\|_{\mathbf{W}^{m,q}(\Omega_1)} + C_2\|f - p_2\|_{\mathbf{W}^{m,q}(\Omega_2)} \\
&\leqslant C_1'|f|_{\mathbf{W}^{\ell,p}(\Omega_1)} + C_2'|f|_{\mathbf{W}^{\ell,p}(\Omega_2)} \\
&\leqslant (C_1' + C_2')|f|_{\mathbf{W}^{\ell,p}(\Omega)}.
\end{aligned}$$

A parallel argument holds for approximation by p_2.

Remark 4.1.4. An immediate consequence of Corollary 4.1.4 is the equivalence of the norm

$$\|f\|_{\ell,p} = |||\mathrm{P}_\ell f||| + |f|_{\mathbf{W}^{\ell,p}(\Omega)} \tag{4.1.37}$$

with the standard Sobolev norm for $1 \leqslant p \leqslant \infty$ and $\ell \geqslant 1$, where $|||\cdot|||$ is any norm on $\mathscr{P}_\ell$. Indeed, we have the one-sided estimate

$$\begin{aligned}
\|f\|_{\mathbf{W}^{\ell,p}(\Omega)} &\leqslant \|\mathrm{P}_\ell f\|_{\mathbf{W}^{\ell,p}(\Omega)} + \|f - \mathrm{P}_\ell f\|_{\mathbf{W}^{\ell,p}(\Omega)} \\
&\leqslant C_1|||\mathrm{P}_\ell f||| + \|f - \mathrm{P}_\ell f\|_{\mathbf{W}^{\ell-1,p}(\Omega)} + |f|_{\mathbf{W}^{\ell,p}(\Omega)} \\
&\leqslant C_1|||\mathrm{P}_\ell f||| + C_2|f|_{\mathbf{W}^{\ell,p}(\Omega)} + |f|_{\mathbf{W}^{\ell,p}(\Omega)} \\
&\leqslant C\|f\|_{\ell,p},
\end{aligned}$$

in conjunction with the trivial estimate

$$\begin{aligned}\|f\|_{\ell,p} &\leqslant C\|\mathrm{P}_\ell f\|_{\mathbf{W}^{\ell-1,p}(\Omega)} + |f|_{\mathbf{W}^{\ell,p}(\Omega)} \\ &\leqslant CC_1\|f\|_{\mathbf{W}^{\ell-1,p}(\Omega)} + |f|_{\mathbf{W}^{\ell,p}(\Omega)} \\ &\leqslant C_2\|f\|_{\mathbf{W}^{\ell,p}(\Omega)},\end{aligned}$$

where we have used the continuity of P_ℓ as a linear mapping of $\mathbf{W}^{\ell-1,p}(\Omega)$ onto $\mathscr{P}_\ell \subset \mathbf{W}^{\ell-1,p}(\Omega)$.

Remark 4.1.5. Another consequence of Corollary 4.1.4 is the inequality, for $1 \leqslant p \leqslant \infty$ and $\ell \geqslant 1$,

$$|f|_{\mathbf{W}^{\ell,p}(\Omega)} \leqslant \|[f]\|_{\mathbf{W}^{\ell,p}(\Omega)/\mathscr{P}_\ell} \leqslant C|f|_{\mathbf{W}^{\ell,p}(\Omega)}, \tag{4.1.38}$$

which shows the equivalence of $|\cdot|_{\mathbf{W}^{\ell,p}(\Omega)}$ and the standard norm on the quotient space $\mathbf{W}^{\ell,p}(\Omega)/\mathscr{P}_\ell$. The first inequality in (4.1.38) is trivial, whereas the second uses the estimate

$$\inf_{p \in \mathscr{P}_\ell} \|f - p\|_{\mathbf{W}^{\ell,p}(\Omega)} \leqslant \|f - \mathrm{P}_\ell f\|_{\mathbf{W}^{\ell,p}(\Omega)} \leqslant C|f|_{\mathbf{W}^{\ell,p}(\Omega)}. \tag{4.1.39}$$

Inequality (4.1.38) is the basis of the well-known Bramble–Hilbert lemma (see Bramble and Hilbert [7] p. 114), which derives the estimate

$$|F(u)| \leqslant C|u|_{\mathbf{W}^{\ell,p}(\Omega)}, \tag{4.1.40}$$

for continuous linear functionals F on $\mathbf{W}^{\ell,p}(\Omega)$ which annihilate functions in $\mathscr{P}_\ell$. Bramble and Hilbert did not claim (4.1.40) in the case $p = 1$, since their fundamental Lemma 2 ([7] p. 114) requires the weak compactness of the unit ball in $\mathbf{W}^{\ell,p}(\Omega)$. This point has been emphasized by Shapiro [37]. However, the approach sketched above yields (4.1.40) for $p = 1$.

Remark 4.1.6. It is of interest to inquire whether other polynomial projection operators $\tilde{\mathrm{P}}_\ell$ may be employed in (4.1.37). Interesting examples would include projections defined by relations, such as

$$\int_\Omega \mathrm{D}^\alpha(\tilde{\mathrm{P}}_\ell f - f) = 0, \qquad |\alpha| < \ell, \tag{4.1.41a}$$

and

$$\int_\Omega \mathbf{x}^\alpha(\tilde{\mathrm{P}}_\ell f - f) = 0, \qquad |\alpha| < \ell. \tag{4.1.41b}$$

The key estimate permitting an affirmative answer, provided $\tilde{\mathrm{P}}_\ell$ is continuous from $\mathbf{W}^{\ell-1,p}(\Omega)$ onto $\mathscr{P}_\ell$, is

$$\|f - \tilde{\mathrm{P}}_\ell f\|_{\mathbf{W}^{\ell,p}(\Omega)} \leqslant C|f|_{\mathbf{W}^{\ell,p}(\Omega)}. \tag{4.1.42}$$

However, (4.1.42) is verified only for $p > 1$ in Morrey ([25] p. 85), where the special operator defined in (4.1.41a) is employed, though the argument is completely general. In particular, if $s = \binom{n+\ell-1}{\ell-1}$ and $\{\lambda_i\}_1^s$ is a set of continuous linear functionals on $\mathbf{W}^{\ell-1,p}(\Omega)$, which are linearly independent over $\mathscr{P}_\ell$, then

$$\|f\|_{\ell,p} = \left\{\sum_{i=1}^{s} |\lambda_i(f)|^p\right\}^{1/p} + |f|_{\mathbf{W}^{\ell,p}(\Omega)} \tag{4.1.43}$$

is equivalent to $\|\cdot\|_{\mathbf{W}^{\ell,p}(\Omega)}$ for $1 < p \leqslant \infty$. Note that (4.1.43) has the structure of (4.1.37) if $\tilde{\mathrm{P}}_\ell$ is defined by

$$\lambda_i \tilde{\mathrm{P}}_\ell f = \lambda_i f, \qquad i = 1, \ldots, s. \tag{4.1.44}$$

We shall now present an application of Corollary 4.1.4 to obtain upper bounds in the estimation of piecewise polynomial approximation on the unit hypercube.

Definition 4.1.2. Let $Q = \{\mathbf{x} \in \mathbb{R}^n : 0 \leqslant |x_i| \leqslant 1, i = 1, \ldots, n\}$ denote the unit hypercube in $\mathbb{R}^n$, and, for $k \geqslant 1$ a fixed integer, let $Q_{\boldsymbol{\alpha}}$ denote one of the k^n subcubes of Q, defined by $h = 1/k$ and the relation

$$Q_{\boldsymbol{\alpha}} = \{\mathbf{x} \in Q : \alpha_i h \leqslant x_i \leqslant (\alpha_i + 1)h, \qquad i = 1, \ldots, n\}, \tag{4.1.45}$$

for $0 \leqslant \alpha_i \leqslant k - 1$, $i = 1, \ldots, n$. Set $\mathbf{x}_{\boldsymbol{\alpha}} = (x_{\alpha_1}, \ldots, x_{\alpha_n}) = (\alpha_1 h, \ldots, \alpha_n h)$, and write

$$\mathbf{x} = \tau_{\boldsymbol{\alpha}}(\mathbf{y}) = \mathbf{x}_{\boldsymbol{\alpha}} + h\mathbf{y}, \qquad \mathbf{y} \in Q, \tag{4.1.46a}$$

so that

$$Q_{\boldsymbol{\alpha}} = \mathbf{x}_{\boldsymbol{\alpha}} + hQ = \tau_{\boldsymbol{\alpha}}(Q). \tag{4.1.46b}$$

Finally, for $u \in \mathbf{W}^{\ell,p}(Q)$, define $\sigma_{\boldsymbol{\alpha}} u = v \in \mathbf{W}^{\ell,p}(Q_{\boldsymbol{\alpha}})$ by

$$v = \sigma_{\boldsymbol{\alpha}} \circ u := u \circ \tau_{\boldsymbol{\alpha}}^{-1}. \tag{4.1.47}$$

Now, for $\ell \geqslant 1$, $1 \leqslant p \leqslant \infty$, $1 \leqslant q \leqslant \infty$, such that

$$\frac{1}{q} - \frac{1}{p} + \frac{\ell}{n} \geqslant 0,$$

and

$$\left(1 - \frac{1}{p}\right)\left(\frac{1}{q}\right) > 0 \quad \text{if} \quad \frac{1}{q} - \frac{1}{p} + \frac{\ell}{n} = 0,$$

consider the idempotent operator

$$\mathrm{P} : \mathbf{W}^{\ell,p}(Q) \to \mathbf{L}^q(Q), \tag{4.1.48a}$$

defined by

$$\mathrm{P}v_{|Q_\alpha} = \sigma_\alpha[\mathrm{P}_\ell \circ \sigma_\alpha^{-1} \circ (v_{|Q_\alpha})]. \tag{4.1.48b}$$

Remark 4.1.7. The following change of scale result is a routine computation:

$$|v|_{\mathbf{W}^{m,p}(Q_\alpha)} = h^{(n/p)-m}|u|_{\mathbf{W}^{m,p}(Q)} \qquad (m \geqslant 0), \tag{4.1.49}$$

where $v = \sigma_\alpha u$.

Proposition 4.1.5. Let p, q, ℓ, and k be specified as in Definition 4.1.2. Then there is a constant $C = C(n, \ell, p, q)$, such that

$$\|v - \mathrm{P}v\|_{\mathbf{L}^q(Q)} \leqslant Ck^{-\ell - n((1/q)-(1/p))}|v|_{\mathbf{W}^{\ell,p}(Q)}, \qquad p \leqslant q. \tag{4.1.50}$$

Here P is defined by (4.1.48).

Proof: By the change of scale result (4.1.49), we have

$$\|v - \mathrm{P}v\|_{\mathbf{L}^q(Q_\alpha)} = h^{n/q}\|u - \mathrm{P}_\ell u\|_{\mathbf{L}^q(Q)}, \tag{4.1.51}$$

where v and u are related by (4.1.47). The latter quantity is estimated, via Corollary 4.1.4, by

$$\|u - \mathrm{P}_\ell u\|_{\mathbf{L}^q(Q)} \leqslant C|u|_{\mathbf{W}^{\ell,p}(Q)}. \tag{4.1.52}$$

A change of scale now yields

$$|u|_{\mathbf{W}^{\ell,p}(Q)} = h^{\ell-(n/p)}|v|_{\mathbf{W}^{\ell,p}(Q_\alpha)}. \tag{4.1.53}$$

Coalescing (4.1.51)–(4.1.53), with $h = 1/k$, gives

$$\|v - \mathrm{P}v\|_{\mathbf{L}^q(Q_\alpha)} \leqslant C\left(\frac{1}{k}\right)^{\ell + n((1/q)-(1/p))} |v|_{\mathbf{W}^{\ell,p}(Q_\alpha)}. \tag{4.1.54}$$

The argument to this point has not required the hypothesis $p \leqslant q$, delineating (4.1.50). For the completion of the argument, we take qth powers of both sides of (4.1.54), sum over α, and then extract qth roots. Inequality (4.1.50) follows upon application of

$$\left\{\sum_{Q_\alpha \subset Q} |v|^q_{\mathbf{W}^{\ell,p}(Q_\alpha)}\right\}^{1/q} \leqslant \left\{\sum_{Q_\alpha \subset Q} |v|^p_{\mathbf{W}^{\ell,p}(Q_\alpha)}\right\}^{1/p} = |v|_{\mathbf{W}^{\ell,p}(Q)}, \qquad q \geqslant p, \tag{4.1.55}$$

which is an elementary version of Jensen's inequality. ■

Remark 4.1.8. The range of the projection P is the subspace $\mathscr{M}$ of $\mathbf{L}^q(Q)$, with dimension $N = k^n s$, such that $\omega \in \mathscr{M}$ satisfies $\omega_{|Q_\alpha} \in \mathscr{P}_\ell(Q_\alpha)$, each Q_α.

Here $s = \binom{n+\ell-1}{\ell-1}$ is the dimension of the space of polynomials of degree not exceeding $\ell - 1$. Thus, in terms of N, the order in (4.1.50) is $N^{-(\ell/n)+(1/p)-(1/q)}$ for $p \leqslant q$. Note that Hölder's inequality in the form

$$\|fg\|_{\mathbf{L}^r(\Omega)} \leqslant \|f\|_{\mathbf{L}^p(\Omega)}\|g\|_{\mathbf{L}^q(\Omega)}, \qquad \frac{1}{r} = \frac{1}{p} + \frac{1}{q}, \quad 1 \leqslant r, p, q \leqslant \infty, \tag{4.1.56}$$

implies the estimate

$$\|v - \mathbf{P}v\|_{\mathbf{L}^r(Q)} \leqslant Ck^{-\ell}|v|_{\mathbf{W}^{\ell,p}(Q)}, \qquad r \leqslant p, \tag{4.1.57}$$

when taken in conjunction with (4.1.50) with $q = p$. It is reasonable to ask whether other operators onto possibly different finite-dimensional spaces yield superior asymptotic estimates of approximation with respect to N for given values of p and q. The answer is negative, except in the range $p < 2 < q$. This point is amplified in the bibliographical remarks.

4.2 LOWER-BOUND ESTIMATES AND N-WIDTHS

We shall introduce the notion of the dispersion of a set from a linear manifold, followed by the notion of N-width or N-dimensional diameter. This will be followed by some abstract lower-bound estimates obtained (necessarily) by the Borsuk antipodal theorem. Specific applications to Sobolev classes will follow.

Definition 4.2.1. For a set $\mathbf{B}$ in a normed linear space $\mathbf{X}$, the dispersion of $\mathbf{B}$ from a linear manifold $\mathscr{M}$ is given by the (extended) real number

$$E(\mathbf{B}, \mathscr{M}) = E_{\mathbf{X}}(\mathbf{B}, \mathscr{M}) = \sup_{u \in \mathbf{B}} \inf_{v \in \mathscr{M}} \|u - v\|, \tag{4.2.1a}$$

and the (Kolmogorov) N-width by the (extended) real number

$$d_N(\mathbf{B}, \mathbf{X}) = \inf\{E(\mathbf{B}, \mathscr{M}) : \mathscr{M} \subset \mathbf{X}, \dim \mathscr{M} = N\}, \tag{4.2.1b}$$

for $N = 0, 1, \ldots$.

The following result is a superficial generalization of the Borsuk antipodal theorem.

Proposition 4.2.1. Let $\mathbf{X}_n$ and $\mathbf{X}_N$ be subspaces of a normed linear space $\mathbf{X}$, with dimensions n and N, respectively, where $0 < N < n$. Let $\mathbf{\Omega}_n$ be a bounded open subset of $\mathbf{X}_n$, symmetric about the origin, such that $0 \in \mathbf{\Omega}_n$ and let $\psi : \partial\mathbf{\Omega}_n \to \mathbf{X}_N$ be a continuous, odd mapping. Then there exists $x \in \partial\mathbf{\Omega}_n$, such that $\psi(x) = 0$.

Proof: The usual Borsuk theorem is the special case $\mathbf{X} = \mathbb{R}^n = \mathbf{X}_n$ and $\mathbf{X}_N = \mathbb{R}^N$ (see Nirenberg [26] p. 25). A simple reduction to this case is afforded by the map of $\mathrm{I}_{\mathbf{X}_n} \partial\mathbf{\Omega}_n$ into $\mathbb{R}^N$, given by

$$\tilde{\psi} = \mathrm{I}_{\mathbf{X}_N} \circ \psi \circ \mathrm{I}_{\mathbf{X}_n}^{-1},$$

where $\mathrm{I}_{\mathbf{X}_n}:\mathbf{X}_n \to \mathbb{R}^n$ and $\mathrm{I}_{\mathbf{X}_N}:\mathbf{X}_N \to \mathbb{R}^N$ are (linear) isomorphisms. ■

Remark 4.2.1. The classical Borsuk theorem is itself a consequence of degree theory. For a continuous map $\psi:\partial\Omega_n \subset \mathbb{R}^n \to \mathbb{R}^n\backslash\{0\}$, the degree of ψ may be defined, for any (connected) component ω_n of $\mathbb{R}^n\backslash\{\psi(\partial\Omega_n)\}$, by

$$\deg(\psi, \Omega_n, \omega_n) = \int_{\Omega_n} \mu \circ \psi \bigg/ \int_{\mathbb{R}^n} \mu,$$

where μ is any smooth n-form in $\mathbb{R}^n$, with support in ω_n and $\int_{\mathbb{R}^n} \mu \neq 0$. Here, we may choose any continuous extension of ψ to $\bar{\Omega}_n$. It is then an elementary fact that $\deg(\psi, \Omega_n, \omega_n) = 0$ if $\omega_n \subset \mathbb{R}^n\backslash\{\psi(\bar{\Omega}_n)\}$. This may be used, in conjunction with a basic fact about odd mappings, to prove the classical Borsuk theorem. We state this basic fact for the reader's interest in the form of a lemma† (for the proofs see Nirenberg [26] pp. 21–25).

Lemma 4.2.2. Let Ω_n be a bounded, open subset of $\mathbb{R}^n$, symmetric about the origin, such that $0 \in \Omega_n$, and let $\psi:\partial\Omega_n \to \mathbb{R}^n\backslash\{0\}$ be a continuous odd mapping. Then, if ω_n is that component of $\mathbb{R}^n\backslash\{\psi(\partial\Omega_n)\}$ containing 0, $\deg(\psi, \Omega_n, \omega_n)$ is odd.

Proposition 4.2.1 has an important consequence in terms of lower bounds for N-widths.

Proposition 4.2.3. Let $\mathbf{X}$ be a normed linear space, and suppose $\mathbf{X}_r \subset \mathbf{X}$, $\dim \mathbf{X}_r = r > 0$. Let $\mathbf{B}$ be a subset of $\mathbf{X}$, such that the set $\mathbf{B}_r = \mathbf{B} \cap \mathbf{X}_r$ is a bounded open subset of $\mathbf{X}_r$, symmetric about 0, with $0 \in \mathbf{B}_r$. Then, for each $0 \leqslant N < r$, and each subspace $\mathscr{M}$ of $\mathbf{X}$ of dimension N, there is an element $x = x(\mathscr{M}) \in \partial\mathbf{B}_r$, satisfying

$$\|x\| = \inf\{\|x - u\| : u \in \mathscr{M}\}. \tag{4.2.2}$$

In particular,

$$d_N(\mathbf{B}, \mathbf{X}) = d_N(\bar{\mathbf{B}}, \mathbf{X}) \geqslant \inf\{\|u\| : u \in \partial\mathbf{B}_r\}, \qquad 0 \leqslant N < r. \tag{4.2.3}$$

† Lemma 4.2.2 is referred to by Nirenberg [26] as the Borsuk theorem. Following what is possibly more common usage, we have preferred the formulation of Proposition 4.2.1.

Proof: Select $\mathbf{Z}$, of dimension not exceeding $r + N$, such that $\mathbf{Z} =$ linear span$(\mathbf{X}_r, \mathcal{M})$. Assume first that $\mathbf{Z}$ is strictly convex (see Chapter 1 [14] p. 17). Define a mapping $\psi: \partial\mathbf{B}_r \to \mathcal{M}$ by

$$\psi(y) = z, \qquad \|y - z\| = \inf\{\|y - u\| : u \in \mathcal{M}\}. \tag{4.2.4}$$

Here, we have used the strict convexity of $\mathbf{Z}$, which guarantees that z is uniquely determined. Now the mapping $\zeta: \mathbf{Z} \to \mathbb{R}^1$,

$$\zeta(y) = \inf\{\|y - u\| : u \in \mathcal{M}\}, \tag{4.2.5}$$

is continuous from the inequalities, for y_1 and y_2 in $\mathbf{Z}$,

$$\zeta(y_i) \leqslant \zeta(y_j) + \|y_i - y_j\|, \qquad i, j = 1, 2.$$

These hold irrespective of the strict convexity of $\mathbf{Z}$. It follows that ψ is continuous. Since ψ is odd, Proposition 4.2.1 applies to give $x \in \partial\mathbf{B}_r$, for which $\psi(x) = 0$. Thus, (4.2.4) implies (4.2.2) for $y = x$. If $\mathbf{Z}$ is not strictly convex, its norm can be approximated by strictly convex norms $\|\cdot\|_\varepsilon$, $\varepsilon > 0$, satisfying (see Chapter 1 [14] p. 138)

$$\|x\| \leqslant \|x\|_\varepsilon \leqslant (1 + \varepsilon)\|x\|, \qquad x \in \mathbf{Z}. \tag{4.2.6}$$

If $\varepsilon_m \to 0$ is chosen, and the preceding argument is applied for each m, a sequence $\{x_m\} \subset \partial\mathbf{B}_r$ is obtained, with accumulation point $x \in \partial\mathbf{B}_r$, such that

$$\|x_m\| \leqslant (1 + \varepsilon_m)\zeta(x_m), \tag{4.2.7}$$

where $\zeta(\cdot)$ is defined in (4.2.5) with respect to the fixed norm $\|\cdot\|$. Letting $\varepsilon_m \to 0$ in (4.2.7), we find that $\|x\| \leqslant \zeta(x)$, hence, $\|x\| = \zeta(x)$.

To verify (4.2.3), let $\mathcal{M}$ be a subspace of $\mathbf{X}$ of dimension $0 \leqslant N < r$, and let x satisfy (4.2.2). We have

$$E(\bar{\mathbf{B}}, \mathcal{M}) \geqslant E(\bar{\mathbf{B}}_r, \mathcal{M}) \geqslant \zeta(x) = \|x\| \geqslant \inf\{\|u\| : u \in \partial\mathbf{B}_r\}. \tag{4.2.8}$$

Taking the infimum over $\mathcal{M} \subset \mathbf{X}$, $\dim \mathcal{M} = N$, gives (4.2.3). ■

The relation (4.2.2) also suggests the crude upper bound for $d_N(\bar{\mathbf{B}}_r, \mathbf{X})$ of $\sup\{\|u\| : u \in \partial\mathbf{B}_r\}$. A much more precise estimate is possible, however, for certain sets when $N = r - 1$. We discuss this now.

Definition 4.2.2. A set $\mathbf{K}$ in a linear topological space $\mathbf{X}$ is said to be absorbing if, for each $x \in \mathbf{X}$, there exists $0 < \gamma = \gamma(x)$, such that $x \in \lambda\mathbf{K}$ for $|\lambda| \geqslant \gamma$. $\mathbf{K}$ is balanced if $\lambda\mathbf{K} \subset \mathbf{K}$ for $|\lambda| \leqslant 1$.

Remark 4.2.2. The class of closed, convex, balanced, absorbing sets **K**, for which 0 is an interior point, is in one-to-one correspondence with the class of continuous seminorms p on **X**, via the correspondence

$$\mathbf{K} = \{x \in \mathbf{X} : p(x) \leqslant 1\}. \tag{4.2.9}$$

Given **K**, p is termed the Minkowski functional of **K**. Moreover,

$$\operatorname{int} \mathbf{K} = \{x \in \mathbf{X} : p(x) < 1\}, \qquad \partial\mathbf{K} = \{x \in \mathbf{X} : p(x) = 1\} \tag{4.2.10}$$

(see Taylor [42] Chapter 3). Moreover, to each point $x_0 \in \partial\mathbf{K}$, there can be associated a continuous linear functional F, defined on **X**, such that

$$\|F\| = 1, \qquad F(x_0) = \sup\{|F(x)| : x \in \mathbf{K}\} \tag{4.2.11}$$

(see [42] p. 145).

Proposition 4.2.4. Let $\mathbf{X}_{N+1}$ be a normed linear space of finite dimension $N+1$. Let $\mathbf{K} \subset \mathbf{X}_{N+1}$ be a closed, balanced, convex, absorbing set, for which 0 is an interior point. Set

$$\rho(\mathbf{K}) = \inf\{\|u\| : u \in \partial\mathbf{K}\}, \tag{4.2.12}$$

where it is explicitly assumed that $\partial\mathbf{K} \neq \varnothing$. Then

$$d_N(\mathbf{K}, \mathbf{X}_{N+1}) = \rho(\mathbf{K}). \tag{4.2.13}$$

Proof: That $\rho(\mathbf{K})$ is a lower bound is an immediate consequence of Proposition 4.2.3. We shall establish that $\rho(\mathbf{K})$ is an upper bound for the N-width. Thus, choose $x_0 \in \partial\mathbf{K}$, such that

$$\|x_0\| = \inf\{\|u\| : u \in \partial\mathbf{K}\} = \rho(\mathbf{K}), \tag{4.2.14}$$

and select the linear functional F, satisfying (4.2.11). Then, defining

$$\mathbf{H} = \{x \in \mathbf{X}_{N+1} : F(x) = 0\},$$

we have $\dim \mathbf{H} = N$, and, by the formula for the distance of a point to a closed subspace, we have, for $x \in \mathbf{K}$,

$$\|x_0\| \geqslant E(\{x_0\}, \mathbf{H}) = F(x_0) \geqslant |F(x)| = E(\{x\}, \mathbf{H}). \tag{4.2.15}$$

The result follows by taking the supremum over $x \in \mathbf{K}$ in (4.2.15). ■

Remark 4.2.3. The major application we shall make of the preceding propositions is to the case where

$$\mathbf{B} = \mathbf{B}_{\mathbf{Y}} = \{y \in \mathbf{Y} : \|y\|_{\mathbf{Y}} < 1\}. \tag{4.2.16}$$

Here $\mathbf{Y}$ is a normed linear space, continuously, though not necessarily densely, embedded in $\mathbf{X}$. For subspaces $\mathbf{Y}_r$ of $\mathbf{Y}$, it is trivial that $\mathbf{B}_r = \mathbf{B}_\mathbf{Y} \cap \mathbf{Y}_r$ satisfies the hypotheses of Proposition 4.2.3, although $\mathbf{B}_\mathbf{Y}$, and even $\mathrm{Cl}_\mathbf{X}\, \mathbf{B}_\mathbf{Y}$, need not have an interior in $\mathbf{X}$. Since

$$\mathbf{B}_\mathbf{Y} \subset \mathbf{U}_\mathbf{Y} := \{y \in \mathbf{Y}: \|y\|_\mathbf{Y} \leqslant 1\} \subset \mathrm{Cl}_\mathbf{X}\, \mathbf{B}_\mathbf{Y}, \tag{4.2.17}$$

$d_N(\mathbf{U}_\mathbf{Y}, X)$ inherits the lower bound (4.2.3). Since $\mathbf{U}_\mathbf{Y}$ is, in some sense, more natural than $\mathbf{B}_\mathbf{Y}$, we shall formulate our statements in terms of this set. Moreover, if $\mathbf{X}_{N+1}$ and $\mathbf{Y}_{N+1} \subset \mathbf{Y}$ coincide as sets, but retain the norm structure induced by $\mathbf{X}$ and $\mathbf{Y}$, respectively, then $\mathbf{K} = \mathbf{U}_\mathbf{Y} \cap \mathbf{Y}_{N+1}$ satisfies the hypotheses of Proposition 4.2.4 and, hence, (4.2.13) holds.

Remark 4.2.4. Note that, if Propositions 4.2.3 and 4.2.4 are applied to $\mathbf{U}_\mathbf{Y}$ and $\mathbf{K}_N = \mathbf{U}_\mathbf{Y} \cap \mathbf{Y}_{N+1}$, then we conclude that the N-width of $\mathbf{K}_N$ does not decrease when $\mathbf{K}_N$ is approximated in $\mathbf{X}$. For $\mathbf{X} = \mathbf{Y}$, this is the famous Gohberg–Krein theorem (see Chapter 1 [14] p. 137). We shall emphasize these remarks in a summarizing proposition.

Proposition 4.2.5. Let $\mathbf{X}$ and $\mathbf{Y}$ be normed linear spaces with $\mathbf{Y}$ continuously embedded in $\mathbf{X}$, and set $\mathbf{U}_\mathbf{Y} = \{y \in \mathbf{Y}: \|y\|_\mathbf{Y} \leqslant 1\}$. Let $\mathbf{Y}_r$ be any fixed subspace of $\mathbf{Y}$ of dimension $r > 0$. Then, for $N < r$,

$$\begin{aligned} d_N(\mathbf{U}_\mathbf{Y}, \mathbf{X}) &= d_N(\mathrm{Cl}_\mathbf{X}\, \mathbf{U}_\mathbf{Y}, \mathbf{X}) \\ &\geqslant \inf\{\|u\|_\mathbf{X}: u \in \mathbf{Y}_r,\ \|u\|_\mathbf{Y} = 1\}. \end{aligned} \tag{4.2.18a}$$

For $r = N + 1$, the final quantity characterizes the N-width of $\mathbf{K}_N = \mathbf{U}_\mathbf{Y} \cap \mathbf{Y}_{N+1}$:

$$d_N(\mathbf{K}_N, \mathbf{X}) = \inf\{\|u\|_\mathbf{X}: u \in \mathbf{Y}_{N+1},\ \|u\|_\mathbf{Y} = 1\}. \tag{4.2.18b}$$

We now pass on to the specific application where $\mathbf{Y}$ is a Sobolev space and $\mathbf{X}$ a Lebesgue space. We state the complete result in a preliminary remark prior to a detailed discussion of certain special cases. The symbol $a(N) \approx b(N)$ below means that $a(N) = O(b(N))$ and $b(N) = O(a(N))$.

Remark 4.2.5. If we denote by $\mathbf{U}^{\ell,p}$ the closed unit ball of $\mathbf{W}^{\ell,p}(Q)$ (see (4.2.26)) on the unit hypercube in $\mathbb{R}^n$, then the numbers $d_N(\mathbf{U}^{\ell,p}, \mathbf{L}^q(Q))$ may be displayed asymptotically in a tableau† as follows, where $1 \leqslant p, q \leqslant \infty$.

† Reproduced with permission from [19], p. 171.

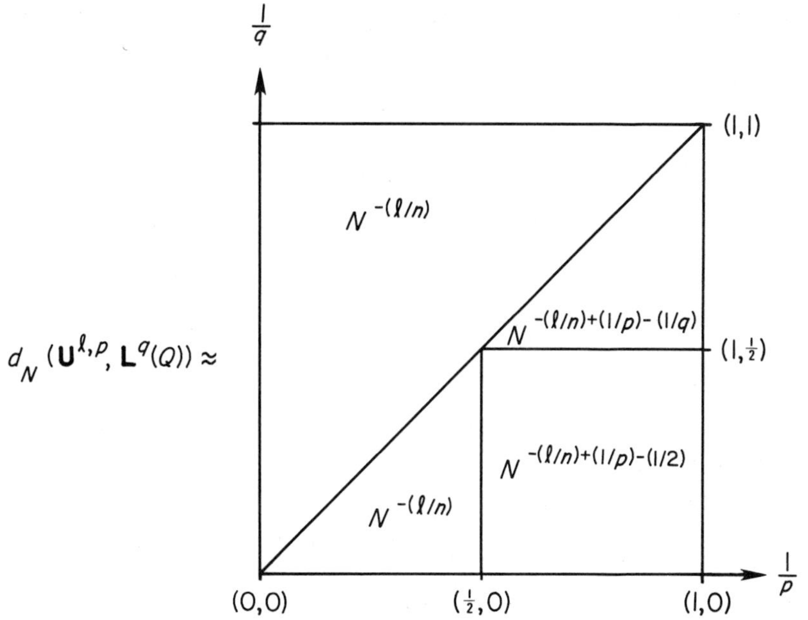

with $\ell - (n/p) + (n/q) > 0$, and $\ell - n/2 > 0$ if $2 \leqslant p < q$, $\ell - n/p > 0$ if $p < 2 < q$. This striking result was obtained by Kašin [23]. The reader will see that only in the ranges $1 \leqslant p \leqslant q \leqslant 2$ and $q \leqslant p$ does the asymptotic estimate match the upper bound given by the projection method of Section 4.1. The fault does not lie entirely with the Sobolev projection, since the values of the Kolmogorov N-width do not correspond asymptotically in all ranges of p and q to the so-called linear N-width defined by operators (see the bibliographical remarks). We shall give detailed proofs only for the cases $1 \leqslant p = q \leqslant \infty$ and $1 < p \leqslant q \leqslant 2$. For these cases, the estimates in the tableau follow by combining Proposition 4.1.5 (upper estimates) with Propositions 4.2.7 and 4.2.10 (lower estimates).

Definition 4.2.3. Let $k \geqslant 1$ be specified, set $h = 1/(2k)$ and construct a uniform partition of Q into $(2k)^n$ cubes $Q_{\boldsymbol{\alpha}}$, defined by (4.1.45), with $0 \leqslant \alpha_i \leqslant 2k - 1$, $i = 1, \ldots, n$. Define the polynomial of degree m^2, $m \geqslant 1$,

$$a(s) = 1 - (1 - s^m)^m, \qquad 0 \leqslant s \leqslant 1, \tag{4.2.19a}$$

and the (deficient) spline $b \in \mathbf{C}^{m-1}(\mathbb{R}^1)$,

$$b(s) = \begin{cases} a(2s), & 0 \leqslant s \leqslant \frac{1}{2}, \\ a(2(1-s)), & \frac{1}{2} \leqslant s \leqslant 1, \\ 0, & \text{otherwise.} \end{cases} \tag{4.2.19b}$$

Further, define the piecewise polynomial function g on $\mathbb{R}^n$ of degree m^2 in each variable $x_1, \ldots, x_n$ by

$$g(\mathbf{x}) = \prod_{i=1}^{n} b(x_i), \tag{4.2.20a}$$

and define the $2^n k^n$ splines $g_{\boldsymbol{\alpha}} \in \mathbf{C}^{m-1}(Q)$ by

$$g_{\boldsymbol{\alpha}}(\mathbf{x}) = \begin{cases} g \circ \tau_{\boldsymbol{\alpha}}^{-1}(\mathbf{x}) = g((\mathbf{x} - \mathbf{x}_{\boldsymbol{\alpha}})/h)), & \mathbf{x} \in Q_{\boldsymbol{\alpha}}, \\ 0, & \mathbf{x} \notin Q_{\boldsymbol{\alpha}}. \end{cases} \tag{4.2.20b}$$

Finally, set $\mathcal{M}_m = \text{linear span}(g_{\boldsymbol{\alpha}})$, and observe that $\mathcal{M}_m \subset \mathbf{W}_0^{m,p}(Q)$, $1 \leqslant p \leqslant \infty$.

Remark 4.2.6. The assertion $b \in \mathbf{C}^{m-1}(\mathbb{R}^1)$ is an easy consequence of the observation that

$$\left[\left(\frac{d}{ds}\right)^j a\right](i) = \begin{cases} 1, & i = 1, \quad j = 0, \\ 0, & (i, j) \in \{0, 1\} \times \{0, \ldots, m-1\} \backslash (1, 0). \end{cases} \tag{4.2.21}$$

Note that $\text{support}(g_{\boldsymbol{\alpha}}) \subset Q_{\boldsymbol{\alpha}}$; in particular, for any $u \in \mathcal{M}_m$, $u = \sum c_{\boldsymbol{\alpha}} g_{\boldsymbol{\alpha}}$, we have

$$\begin{aligned} \|u\|_{\mathbf{L}^p(Q)} &= \left[\int_Q \left|\sum c_{\boldsymbol{\alpha}} g_{\boldsymbol{\alpha}}\right|^p\right]^{1/p} = \left[\sum_{\boldsymbol{\beta}} \int_{Q_{\boldsymbol{\beta}}} \left|\sum_{\boldsymbol{\alpha}} c_{\boldsymbol{\alpha}} g_{\boldsymbol{\alpha}}\right|^p\right]^{1/p} \\ &= \left[\sum_{\boldsymbol{\beta}} \int_{Q_{\boldsymbol{\beta}}} |c_{\boldsymbol{\beta}} g_{\boldsymbol{\beta}}|^p\right]^{1/p} = \left[\sum_{\boldsymbol{\beta}} |c_{\boldsymbol{\beta}}|^p h^n \int_Q |g|^p\right]^{1/p} \\ &= \|\{c_{\boldsymbol{\alpha}}\}\|_{\ell^p} h^{n/p} \|g\|_{\mathbf{L}^p(Q)}, \end{aligned} \tag{4.2.22}$$

for $1 \leqslant p < \infty$, with the interpretation $\|u\|_{\mathbf{L}^\infty(Q)} = \|\{c_{\boldsymbol{\alpha}}\}\|_{\ell^\infty} \|g\|_{\mathbf{L}^\infty(Q)}$, for $p = \infty$.

Proposition 4.2.6. There exist positive constants C_1, C_2, and C_3, such that, for $u \in \mathcal{M}_m$ and $m \geqslant \ell$,

$$\|u\|_{\mathbf{L}^p(Q)} \leqslant C_1 h^{n((1/p)-(1/q))} \|u\|_{\mathbf{L}^q(Q)}, \qquad q \leqslant p, \tag{4.2.23a}$$

$$\|u\|_{\mathbf{W}^{\ell,p}(Q)} \leqslant C_2 h^{-\ell} \|u\|_{\mathbf{L}^p(Q)} \leqslant C_3 \|u\|_{\mathbf{W}^{\ell,p}(Q)}, \qquad 1 \leqslant p \leqslant \infty. \tag{4.2.23b}$$

In fact, C_1, C_2, and C_3 may be chosen explicitly as

$$\begin{aligned} C_1 = \|g\|_{\mathbf{L}^p(Q)} / \|g\|_{\mathbf{L}^q(Q)}, \qquad C_2 = \|g\|_{\mathbf{W}^{\ell,p}(Q)} / \|g\|_{\mathbf{L}^p(Q)}, \\ C_3 = \|g\|_{\mathbf{W}^{\ell,p}(Q)} / |g|_{\mathbf{W}^{\ell,p}(Q)}. \end{aligned} \tag{4.2.23c}$$

Proof: The verification of (4.2.23a) proceeds by using (4.2.22) as a starting point, followed by the inequality

$$\|\{c_{\boldsymbol{\alpha}}\}\|_{\ell^p} \leqslant \|\{c_{\boldsymbol{\alpha}}\}\|_{\ell^q}, \qquad q \leqslant p, \tag{4.2.24}$$

followed once again by (4.2.22), with p replaced by q. In this case, $C_1 = \|g\|_{\mathbf{L}^p(Q)}/\|g\|_{\mathbf{L}^q(Q)}$. Similarly, the equality

$$|u|_{\mathbf{W}^{j,p}(Q)} = h^{-j+(n/p)}|g|_{\mathbf{W}^{j,p}(Q)}\|\{c_{\boldsymbol{\alpha}}\}\|_{\ell^p},$$

when combined with (4.2.22), yields

$$|u|_{\mathbf{W}^{j,p}(Q)} = c_j h^{-j}\|u\|_{\mathbf{L}^p(Q)}, \tag{4.2.25}$$

with $c_j = |g|_{\mathbf{W}^{j,p}(Q)}/\|g\|_{\mathbf{L}^p(Q)}$. This yields the first inequality in (4.2.23b), with $C_2 = \|g\|_{\mathbf{W}^{\ell,p}(Q)}/\|g\|_{\mathbf{L}^p(Q)}$, since $h^{-j} \leqslant h^{-\ell}$ for $j \leqslant \ell$. If (4.2.25) is employed with $j = \ell$, we obtain

$$\|u\|_{\mathbf{L}^p(Q)} = c_\ell^{-1}h^\ell|u|_{\mathbf{W}^{\ell,p}(Q)} \leqslant c_\ell^{-1}h^\ell\|u\|_{\mathbf{W}^{\ell,p}(Q)},$$

which yields the second inequality in (4.2.23b), with $C_3 = C_2 c_\ell^{-1}$. ■

Remark 4.2.7. It is worth mentioning that the constants may be chosen independent of p and q in (4.2.23). Indeed, the choices $C_1 = \|g\|_{\mathbf{L}^\infty(Q)}/\|g\|_{\mathbf{L}^1(Q)}$, $C_2 = \|g\|_{\mathbf{W}^{\ell,\infty}(Q)}/\|g\|_{\mathbf{L}^1(Q)}$, and $C_3 = \|g\|_{\mathbf{W}^{\ell,\infty}(Q)}/|g|_{\mathbf{W}^{\ell,1}(Q)}$, respectively, suffice, since Q is of unit measure. The previous proposition, in conjunction with Proposition 4.2.5, yields lower bounds for the N-widths of the set

$$\mathbf{U}^{\ell,p} = \{u \in \mathbf{W}^{\ell,p}(Q): \|u\|_{\mathbf{W}^{\ell,p}(Q)} \leqslant 1\}, \qquad \ell \geqslant 1, \quad 1 \leqslant p \leqslant \infty, \tag{4.2.26}$$

when embedded in $\mathbf{L}^p(Q)$. We state this as the following proposition.

Proposition 4.2.7. Suppose $1 \leqslant p \leqslant \infty$, and define the sequence

$$N_k = 2^n k^n - 1, \qquad k = 1, 2, \ldots. \tag{4.2.27a}$$

Then, for $N \in \{N_k\}$ and $C = C_2^{-1}$ given in (4.2.23),

$$d_N(\mathbf{U}^{\ell,p}, \mathbf{L}^p(Q)) \geqslant C\left(\frac{1}{2k}\right)^\ell. \tag{4.2.27b}$$

Proof: The embedding $\mathbf{W}^{\ell,p}(Q) \to \mathbf{L}^p(Q)$ is continuous for $1 \leqslant p \leqslant \infty$, though it fails to be dense for $p = \infty$. Thus, Proposition 4.2.5 applies with $r = \dim \mathcal{M}_\ell = N_k + 1$. In particular,

$$d_N(\mathbf{U}^{\ell,p}, \mathbf{L}^p(Q)) \geqslant \inf\left\{\frac{\|u\|_{\mathbf{L}^p(Q)}}{\|u\|_{\mathbf{W}^{\ell,p}(Q)}} : u \in \mathcal{M}_\ell,\, u \neq 0\right\} \geqslant (C_2^{-1}h^\ell), \tag{4.2.28}$$

upon use of (4.2.23). If $h = 1/(2k)$ is substituted into (4.2.28), we immediately obtain (4.2.27b). ■

We pass to the case $1 < p \leqslant q \leqslant 2$. We first prove an auxiliary result in Hilbert space which will serve as a pivoting result.

Proposition 4.2.8. Let **H** be a Hilbert space, and let **V** be an r-dimensional subspace of **H**, with $r > 0$, such that $\{\psi_i\}_1^r$ is a complete orthonormal system in **V**. Suppose that **Y** is a normed linear space, embedded in **H** and containing the space **V** as a subspace. Then,

$$d_N(\mathbf{U_Y}, \mathbf{H}) \geqslant \left(\frac{r - N}{\sum_{i=1}^{r} \|\psi_i\|_{\mathbf{Y}}^2} \right)^{1/2}, \qquad 0 \leqslant N < r. \tag{4.2.29}$$

Proof: Let $\mathcal{N}$ be any N-dimensional subspace of **H**, and set $\mathbf{Z} =$ linear span$(\mathbf{V} \cup \mathcal{N})$. Write

$$\mathcal{N} = \mathcal{N}_1 \oplus \mathcal{N}_2, \qquad \mathcal{N}_1 \subset \mathbf{V}^{\perp}, \quad \mathcal{N}_2 \subset \mathbf{V}.$$

Then $\mathbf{Z} = \mathbf{V} \oplus \mathcal{N}_1$, say $\dim \mathbf{Z} = r + M$. Let $\{\phi_i\}_1^{r+M}$ be a complete orthonormal sequence in **Z**, such that

$$\text{linear span}\{\phi_i\}_1^M = \mathcal{N}_1, \qquad \text{linear span}\{\phi_i\}_{M+1}^N = \mathcal{N}_2.$$

Then, for $1 \leqslant j \leqslant r$,

$$\begin{aligned} 1 = \|\psi_j\|_{\mathbf{H}}^2 &= \sum_{i=1}^{M} |(\psi_j, \phi_i)_{\mathbf{H}}|^2 + \sum_{i=M+1}^{N} |(\psi_j, \phi_i)_{\mathbf{H}}|^2 + \sum_{i=N+1}^{r+M} |(\psi_j, \phi_i)_{\mathbf{H}}|^2 \\ &= \sum_{i=M+1}^{N} |(\psi_j, \phi_i)_{\mathbf{H}}|^2 + \sum_{i=N+1}^{r+M} \left| \left(\frac{\psi_j}{\|\psi_j\|_{\mathbf{Y}}}, \phi_i \right)_{\mathbf{H}} \right|^2 \|\psi_j\|_{\mathbf{Y}}^2. \end{aligned}$$

Summing on j, and using the Fourier projection theorem, we obtain

$$r \leqslant N - M + E_{\mathbf{H}}(\mathbf{U_Y}, \mathcal{N}^{\perp})^2 \sum_{j=1}^{r} \|\psi_j\|_{\mathbf{Y}}^2, \tag{4.2.30}$$

and (4.2.29) is immediate from (4.2.30). ■

Corollary 4.2.9. Suppose $1 \leqslant p \leqslant 2$, $\ell \geqslant 1$, and $((1/2) - (1/p) + (\ell/n)) \geqslant 0$. Define the sequence

$$N_k = 2^{n-1} k^n. \tag{4.2.31a}$$

Then, for $N \in \{N_k\}$ and $C = C_2^{-1}$ given in (4.2.23b),

$$d_N(\mathbf{U}^{\ell,p}, \mathbf{L}^2(Q)) \geqslant 2^{-1/2} C \left(\frac{1}{2k}\right)^{\ell - n((1/p)-(1/2))}. \tag{4.2.31b}$$

It follows that, for general $N = 1, 2, \ldots,$

$$d_N(\mathbf{U}^{\ell,p}, \mathbf{L}^2(Q)) \geqslant C_0 N^{-\ell/n + ((1/p)-(1/2))}, \tag{4.2.32}$$

for some positive constant C_0.

Proof: Define the normalizing factor $\delta_{\boldsymbol{\alpha}}$, so that

$$e_{\boldsymbol{\alpha}} = \delta_{\boldsymbol{\alpha}} g_{\boldsymbol{\alpha}}, \qquad \|e_{\boldsymbol{\alpha}}\|_{\mathbf{L}^2(Q)} = 1. \tag{4.2.33}$$

In particular, for $\mathbf{V} = \mathcal{M}_\ell$, $\{e_{\boldsymbol{\alpha}}\}$ is a complete orthonormal system in the sense of $\mathbf{L}^2(Q)$. By (4.2.23b), we have

$$\left\{\sum_{\boldsymbol{\alpha}} \|e_{\boldsymbol{\alpha}}\|^2_{\mathbf{W}^{\ell,p}(Q)}\right\}^{1/2} \leqslant C_2 h^{-\ell} \left\{\sum_{\boldsymbol{\alpha}} \|e_{\boldsymbol{\alpha}}\|^2_{\mathbf{L}^p(Q)}\right\}^{1/2}. \tag{4.2.34}$$

If the inequality

$$\|e_{\boldsymbol{\alpha}}\|_{\mathbf{L}^p(Q)} \leqslant h^{n((1/p)-(1/q))} \|e_{\boldsymbol{\alpha}}\|_{\mathbf{L}^q(Q)}, \qquad 1 \leqslant p \leqslant q \leqslant \infty, \tag{4.2.35}$$

is applied to (4.2.34), with $q = 2$, we obtain

$$\left\{\sum_{\boldsymbol{\alpha}} \|e_{\boldsymbol{\alpha}}\|^2_{\mathbf{W}^{\ell,p}(Q)}\right\}^{1/2} \leqslant C_2 r^{1/2} h^{-\ell + n((1/p)-(1/2))}, \qquad r = 2^n k^n. \tag{4.2.36}$$

If (4.2.36) is combined with (4.2.29), there results (4.2.31b). Note that we have used the embedding $\mathbf{W}^{\ell,p}(Q) \to \mathbf{L}^2(Q)$. We now verify (4.2.35). Note that the homogeneity property of norms permits the verification of the equivalent statement

$$\|g_{\boldsymbol{\alpha}}\|_{\mathbf{L}^p(Q)} \leqslant h^{n((1/p)-(1/q))} \|g_{\boldsymbol{\alpha}}\|_{\mathbf{L}^q(Q)}, \qquad 1 \leqslant p \leqslant q \leqslant \infty. \tag{4.2.37}$$

Thus,

$$\|g_{\boldsymbol{\alpha}}\|_{\mathbf{L}^p(Q)} = h^{n/p} \|g\|_{\mathbf{L}^p(Q)} \leqslant h^{n/p} \|g\|_{\mathbf{L}^q(Q)} = h^{n((1/p)-(1/q))} \|g_{\boldsymbol{\alpha}}\|_{\mathbf{L}^q(Q)},$$

and (4.2.37) is verified. It remains to verify (4.2.32). For given $N \geqslant 0$, select $k \geqslant 0$, such that

$$2^{n-1} k^n \leqslant N < 2^{n-1}(k+1)^n. \tag{4.2.38}$$

Then, by the second inequality in (4.2.38),

$$d_N(\mathbf{U}^{\ell,p}, \mathbf{L}^2(Q)) \geqslant d_{N_{k+1}}(\mathbf{U}^{\ell,p}, \mathbf{L}^2(Q)) \geqslant 2^{-(1/2)} C \left(\frac{1}{2k+2}\right)^{\ell - n((1/p)-(1/2))} \tag{4.2.39}$$

for $N_{k+1} = 2^{n-1}(k+1)^n$, and, by the first inequality,

$$\left(\frac{1}{2k+2}\right) \geqslant \frac{1}{((2N)^{1/n}+2)}. \tag{4.2.40}$$

Combining (4.2.39) and (4.2.40) yields (4.2.32). ■

This is the Hilbert space result, permitting pivoting, to which we referred. Note that the lower bounds (4.2.31), (4.2.32) may be strengthened to the case where $\mathbf{U}^{\ell,p}$ is replaced by $\mathbf{U}_\pi^{\ell,p}$, the closed unit ball in $\mathbf{W}_\pi^{\ell,p}(Q)$ (see Remark 4.2.8), since $\mathcal{M}_\ell \subset \mathbf{W}_\pi^{\ell,p}(Q)$. The following remark discusses the relationship between N-widths under isomorphism.

Remark 4.2.8. Let J be an isomorphism from a normed linear space **X** onto a normed linear space **Z**. Then,

$$E_{\mathbf{Z}}(\mathbf{JK}, \mathbf{J}\mathcal{M}) \leqslant \|\mathbf{J}\| E_{\mathbf{X}}(\mathbf{K}, \mathcal{M}), \tag{4.2.41}$$

for any set $\mathbf{K} \subset \mathbf{X}$ and any finite dimensional subspace $\mathcal{M} \subset \mathbf{X}$. If the restriction $\tilde{\mathbf{J}}$ of J, to a subspace **W** continuously embedded in **X**, is also an isomorphism onto a subspace **V** continuously embedded in **Z**, then

$$\begin{aligned} d_N(\mathbf{U_V}, \mathbf{Z}) &= d_N(\mathbf{J} \circ \mathbf{J}^{-1}\mathbf{U_V}, \mathbf{JX}) \leqslant \|\mathbf{J}\| d_N(\mathbf{J}^{-1}\mathbf{U_V}, \mathbf{X}) \\ &\leqslant \|\mathbf{J}\| d_N(\|\tilde{\mathbf{J}}^{-1}\| \mathbf{U_W}, \mathbf{X}) \\ &= \|\mathbf{J}\| \, \|\tilde{\mathbf{J}}^{-1}\| d_N(\mathbf{U_W}, \mathbf{X}). \end{aligned} \tag{4.2.42}$$

The principal application we shall make of this result is the case $\mathbf{J}^{-1} = \mathbf{J}_s$ is the Bessel potential operator (see Adams *et al.* [3], Bergh and Löfstrom [6], and El Kolli [15]), which establishes an isomorphism

$$\mathbf{J}_s : \mathbf{W}_\pi^{m,q}(Q) \to \mathbf{W}_\pi^{m-s,q}(Q), \qquad s \geqslant 0, \quad 1 < q < \infty, \tag{4.2.43}$$

between Sobolev spaces. Here the definition of $\mathbf{J}_s$ depends only on s, and, for m $\geqslant 0$, $\mathbf{W}_\pi^{m,q}(Q)$ denotes the subspace of $\mathbf{W}^{m,q}(Q)$ of functions extendible to $\mathbb{R}^n$ as periodic functions of period one in each variable, such that the extended functions are in $\mathbf{W}_{\mathrm{loc}}^{m,q}(\mathbb{R}^n)$. For $m < 0$, $\mathbf{W}_\pi^{m,q}(Q) = \mathbf{W}^{m,q}(Q) = [\mathbf{W}_0^{-m,q'}(Q)]^*$. The isomorphism is proved in [15], by use of the Mihlin multiplier theorem. The classical results on Bessel potential operators establish the corresponding isomorphism result on $\mathbb{R}^n$ (see Adams *et al.* [3], and Bergh and Löfstrom [6]). We can now handle the case $1 < p \leqslant q \leqslant 2$.

Proposition 4.2.10. Suppose that $1 < p \leqslant q \leqslant 2, \ell \geq 1$, and $((1/q) - (1/p) + (\ell/n)) \geqslant 0$. Then, there is a constant C, depending on ℓ, n, p, and q, such that, for $N_\nu = \nu^n$ and $N \in \{N_\nu\}$,

$$d_N(\mathbf{U}^{\ell,p}, \mathbf{L}^q(Q)) \geqslant C\nu^{-\ell + n((1/p)-(1/q))}. \tag{4.2.44}$$

It follows that, for some constant C_0, and $N = 1, 2, \ldots,$

$$d_N(\mathbf{U}^{\ell,p}, \mathbf{L}^q(Q)) \geqslant C_0 N^{-\ell/n + ((1/p)-(1/q))}. \tag{4.2.45}$$

Lemma 4.2.11.† Let $\mathbf{Y}$ be a normed linear space continuously embedded in a normed linear space $\mathbf{X}$, and let $\mathbf{A}$ be a subset of $\mathbf{Y}$. Then,

$$d_{N_1+N_2}(\mathbf{A}, \mathbf{X}) \leqslant d_{N_1}(\mathbf{A}, \mathbf{Y}) d_{N_2}(\mathbf{U}_{\mathbf{Y}}, \mathbf{X}), \qquad N_1 \geqslant 0, \quad N_2 \geqslant 0, \tag{4.2.46}$$

when these numbers are finite.

Proof of Lemma 4.2.11: Suppose $\delta_1 > 0$ and $\delta_2 > 0$ are such that

$$d_{N_1}(\mathbf{A}, \mathbf{Y}) < \delta_1, \qquad d_{N_2}(\mathbf{U}_{\mathbf{Y}}, \mathbf{X}) < \delta_2,$$

and choose $\mathbf{M}_i$, $\dim \mathbf{M}_i = N_i$, such that

$$E_{\mathbf{Y}}(\mathbf{A}, \mathbf{M}_1) < \delta_1, \qquad E_{\mathbf{X}}(\mathbf{U}_{\mathbf{Y}}, \mathbf{M}_2) < \delta_2.$$

Then, given $x \in \mathbf{A}$, there exists $y \in \mathbf{M}_1$, such that $\|x - y\|_{\mathbf{Y}} < \delta_1$, and there exists $z \in \mathbf{M}_2$, such that $\|((x - y)/\delta_1) - z\|_{\mathbf{X}} < \delta_2$. Altogether,

$$\|x - (y + \delta_1 z)\|_{\mathbf{X}} < \delta_1 \delta_2.$$

Since $y + \delta_1 z \in \mathbf{M}_1 + \mathbf{M}_2$, $\dim(\mathbf{M}_1 + \mathbf{M}_2) \leqslant N_1 + N_2$, we have

$$d_{N_1+N_2}(\mathbf{A}, \mathbf{X}) \leqslant E_{\mathbf{X}}(\mathbf{A}, \mathbf{M}_1 + \mathbf{M}_2) \leqslant \delta_1 \delta_2.$$

The result follows by letting δ_i tend to $d_{N_i}(\cdot, \cdot)$. ■

Proof of Proposition 4.2.10: Choose an integer $j > 0$, such that $((1/2) - (1/q) + (j/n)) > 0$; in particular, $\mathbf{W}^{j,q}(Q)$ is continuously embedded in $\mathbf{L}^2(Q)$. If we apply Lemma 4.2.11, with $\mathbf{A} = \mathbf{U}_\pi^{\ell+j,p}$, $\mathbf{Y} = \mathbf{W}_\pi^{j,q}(Q)$, and $\mathbf{X} = \mathbf{L}_\pi^2(Q)$, we obtain

$$d_{N_1}(\mathbf{U}_\pi^{\ell+j,p}, \mathbf{W}_\pi^{j,q}(Q)) \geqslant \frac{d_{N_1+N_2}(\mathbf{U}_\pi^{\ell+j,p}, \mathbf{L}_\pi^2(Q))}{d_{N_2}(\mathbf{U}_\pi^{j,q}, \mathbf{L}_\pi^2(Q))}. \tag{4.2.47}$$

Now, choose an integer μ, such that $\mu^n \geqslant 2^j$, and set $N_2 = v^n \binom{n+j-1}{j-1}$, and $N_1 + N_2 = 2^{n-1}\mu^n v^n$. This gives

$$N_1 = \left[2^{n-1}\mu^n - \binom{n+j-1}{j-1}\right] v^n \geqslant v^n.$$

Thus, applying Proposition 4.1.5, with $k = v$, to the denominator of (4.2.47), and a strengthened form of Corollary 4.2.9, with $k = \mu v$ and $\mathbf{U}^{\ell+j,p}$ replaced

† Due to El Kolli [15].

by $\mathbf{U}_\pi^{\ell+j,p}$, to the numerator, we obtain

$$d_{N_1}(\mathbf{U}_\pi^{\ell+j,p}, \mathbf{W}_\pi^{j,q}(Q)) \geqslant C\nu^{-\ell-j-n((1/2)-(1/p))}\nu^{j+n((1/2)-(1/q))}$$

for some $C > 0$, from which it follows that

$$d_{N_1}(\mathbf{U}_\pi^{\ell+j,p}, \mathbf{W}_\pi^{j,q}(Q)) \geqslant C\nu^{-\ell+n((1/p)-(1/q))}.$$

An application of the inequality (4.2.42), with $\mathbf{W} = \mathbf{W}_\pi^{\ell,p}(Q)$, $\mathbf{V} = \mathbf{W}_\pi^{\ell+j,p}(Q)$, $\mathbf{X} = \mathbf{L}_\pi^q(Q)$, $\mathbf{Z} = \mathbf{W}_\pi^{j,q}(Q)$, and $\mathrm{J} = \mathrm{J}_j^{-1}$, yields

$$d_{N_1}(\mathbf{U}_\pi^{\ell,p}, \mathbf{L}_\pi^q(Q)) \geqslant C\nu^{-\ell+n((1/p)-(1/q))}$$

for some $C > 0$, and the result (4.2.44) follows by the monotonicity properties of N-widths. Inequality (4.2.45) follows in analogy with (4.2.32). ■

Remark 4.2.9. Note that we have now completed the details for the N-width orders in the cases $1 \leqslant p = q \leqslant \infty$ and $1 < p \leqslant q \leqslant 2$ (see Remark 4.2.5.) In the case $p = q$, Proposition 4.2.7 suggests that the largest constant C in (4.2.27b) may be obtained from a nonlinear extremal problem of the form

$$\begin{aligned} C &= \sup\{\|g\|_{\mathbf{L}^p(Q)}/\|g\|_{\mathbf{W}_0^{\ell,p}(Q)} : 0 \neq g \in \mathbf{W}_0^{\ell,p}(Q)\} \\ &= [\inf\{\|g\|_{\mathbf{W}_0^{\ell,p}(Q)}/\|g\|_{\mathbf{L}^p(Q)} : 0 \neq g \in \mathbf{W}_0^{\ell,p}(Q)\}]^{-1}, \end{aligned} \tag{4.2.48}$$

which, when $p = 2$, is the reciprocal square root of the smallest (positive) eigenvalue of the Dirichlet problem on Q, when the appropriate norm is defined on $\mathbf{W}_0^{\ell,p}(Q)$. The reader may justly inquire whether this value of C gives the exact value of d_N in (4.2.27b). This is, in fact, the case, and we shall give the proof using the famous Weinstein–Aronszajn theory for intermediate eigenvalue problems. We shall actually present a much more general result.

Definition 4.2.4. Let $\mathbf{H}$ be a real or complex Hilbert space with norm $\|u\| = (u, u)^{1/2}$. We shall denote by $\mathscr{S}$ the class of self-adjoint linear operators A which are positive, that is,

$$(\mathrm{A}u, u) \geqslant 0$$

on the domain $\mathbf{D} = \mathbf{D}_\mathrm{A}$ of A, such that the lower part of the spectrum of A consists of a finite or infinite number of isolated eigenvalues $0 \leqslant \lambda_1 \leqslant \lambda_2 \leqslant \cdots$ repeated according to their finite multiplicity. Thus, $\lambda_j < \lambda_\infty$ for these eigenvalues, where λ_∞ is the lowest point in the essential spectrum (if non-empty) of A, or $\lambda_\infty = \infty$ otherwise. We define, for $\mathrm{A} \in \mathscr{S}$, the (extended) integer

$$M = \sup\{j : \lambda_j < \lambda_\infty\}, \tag{4.2.49}$$

and the ellipsoid

$$\mathscr{A} = \{u \in \mathbf{D}_{\mathrm{A}} : (\mathrm{A}u, u) \leqslant 1\}. \tag{4.2.50}$$

Proposition 4.2.12. If $0 \leqslant N < M$, then the N-width of the ellipsoid $\mathscr{A}$ is given by

$$d_N(\mathscr{A}, \mathbf{H}) = E(\mathscr{A}, \mathscr{M}_N) = \lambda_{N+1}^{-1/2}. \tag{4.2.51}$$

Here, $\mathscr{M}_N$ is the linear span of the first N eigenvectors of A, and $d_N(\mathscr{A}, \mathbf{H}) = \infty$ if $\lambda_{N+1} = 0$.

Proof: By taking orthogonal complements, we may assume $\lambda_1 > 0$. Let $\mathscr{M}$ be any N-dimensional subspace of $\mathbf{H}$. Then, if P is the orthogonal projector onto $\mathscr{M}$, and $\mathrm{Q} = \mathrm{I} - \mathrm{P}$ is its complement,

$$\begin{aligned} E(\mathscr{A}, \mathscr{M})^{-2} &= [\sup\{\|\mathrm{Q}u\|^2 : (\mathrm{A}u, u) \leqslant 1\}]^{-1} \\ &= [\sup\{(\mathrm{Q}u, \mathrm{Q}u)/(\mathrm{A}u, u) : \mathrm{P}u = 0, u \neq 0\}]^{-1} \\ &= \inf\{(\mathrm{A}u, u)/(\mathrm{Q}u, \mathrm{Q}u) : \mathrm{P}u = 0, u \neq 0\} \\ &= \inf\{(\mathrm{AQ}u, \mathrm{Q}u)/(\mathrm{Q}u, \mathrm{Q}u) : \mathrm{P}u = 0, u \neq 0\} \\ &= \inf\{(\mathrm{QAQ}u, u)/(u, u) : \mathrm{P}u = 0, u \neq 0\}. \end{aligned}$$

According to the Weinstein–Aronszajn theory of intermediate problems for eigenvalues (see Weinstein and Stenger [43] Theorem 1, p. 28) and Stenger [39]), the last infimum is the $(N + 1)$st eigenvalue of QAQ (note that 0 is an eigenvalue of multiplicity at least N) and is attained by an eigenvector of QAQ; moreover,

$$\inf\{(\mathrm{QAQ}u, u)/(u, u) : \mathrm{P}u = 0, u \neq 0\} = \inf\{(\mathrm{A}u, u)/(u, u) : \mathrm{P}u = 0, u \neq 0\}.$$

We, thus, have, by the sup-inf characterization of eigenvalues,

$$E(\mathscr{A}, \mathscr{M})^{-2} \leqslant \lambda_{N+1}, \qquad E(\mathscr{A}, \mathscr{M}_N)^{-2} = \lambda_{N+1},$$

so that (4.2.51) follows and the theorem is proved.

Remark 4.2.10. Proposition 4.2.12 can also be proved by using results of Golomb [17] on the approximation of ellipsoids of general positive self-adjoint operators. The class $\mathscr{S}$ is far richer than the set of inverses of compact self-adjoint operators, and includes, for example, the Schrödinger operator in $\mathbb{R}^3$ for the potential induced by the helium atom (see [43] pp. 96–101).

Two applications, one in general Hilbert space and the other to the computation of $d_N(\mathbf{U}^{\ell,2}, \mathbf{L}^2(\Omega))$, with explicit asymptotic constants, are given now.

Corollary 4.2.13. Let $\mathbf{H}$ be a Hilbert space, and let $\mathbf{V}$ be a dense linear subspace of $\mathbf{H}$, such that $\mathbf{V}$ is a Hilbert space under the inner product

$$(u, v)_{\mathbf{V}} = [u, v] + (u, v)_{\mathbf{H}}, \tag{4.2.52}$$

where $[\cdot,\cdot]$ is a nonnegative, Hermitian bilinear form defined on $\mathbf{V} \times \mathbf{V}$. Let $\mathbf{V}_0$ be a linear manifold dense in $\mathbf{V}$, that is, $\mathrm{Cl}_{\mathbf{V}}\mathbf{V}_0 = \mathbf{V}$, and let $\mathscr{B}_0 = \{u \in \mathbf{V}_0 : [u, u] \leqslant 1\}$. If $\mathscr{B} = \{u \in \mathbf{V} : [u, u] \leqslant 1\}$, then $\mathrm{Cl}_{\mathbf{V}}\mathscr{B}_0 = \mathrm{Cl}_{\mathbf{V}}\mathscr{B} = \mathscr{B}$ and, hence, $\mathrm{Cl}_{\mathbf{H}}\mathscr{B}_0 = \mathrm{Cl}_{\mathbf{H}}\mathscr{B}$. Moreover, there exists a unique positive self-adjoint operator A, with $\mathrm{Cl}_{\mathbf{V}}\mathbf{D}_{\mathrm{A}} = \mathrm{V}$, satisfying

$$[u, v] = (\mathrm{A}u, v)_{\mathbf{H}} \qquad \text{for all} \quad u \in \mathbf{D}_{\mathrm{A}}, \qquad v \in \mathbf{V}, \tag{4.2.53}$$

with $\mathbf{D}_{\mathrm{A}}$ characterized by (4.2.53). In particular, if $\mathscr{A}$ denotes the ellipsoid determined by A, then

$$d_N(\mathscr{A}, \mathbf{H}) = d_N(\mathscr{B}, \mathbf{H}), \qquad N \geqslant 0. \tag{4.2.54}$$

Proof: The operator A is the well-known Friedrichs' operator (see Riesz and Sz-Nagy [30] pp. 332–333), and (4.2.54) follows from the earlier closure results, upon identifying $\mathbf{D}_{\mathrm{A}}$ with $\mathbf{V}_0$ and $\mathscr{A}$ with $\mathscr{B}_0$. The closure statement, $\mathrm{Cl}_{\mathbf{V}}\mathscr{B}_0 = \mathscr{B}$, follows from normalizing any sequence convergent to a nonzero element of $\mathscr{B}$, and the closure statement in $\mathbf{H}$ follows from

$$\mathrm{Cl}_{\mathbf{H}}\mathscr{B} = \mathrm{Cl}_{\mathbf{H}}(\mathrm{Cl}_{\mathbf{V}}\mathscr{B}_0) = \mathrm{Cl}_{\mathbf{H}}\mathscr{B}_0,$$

where the last equality results from the continuity of the embedding of $\mathbf{V}$ into $\mathbf{H}$. The N-widths are of course invariant under closure. ■

We have the following sharp theorem concerning the $\mathbf{L}^2$ N-width of the unit ball of $\mathbf{W}^{\ell,2}(\Omega)$. Note the adjustment in the definition of the norm of $\mathbf{W}^{\ell,2}(\Omega)$ in the following theorem to reflect the number of times a particular derivative appears in a tensor product reformulation (see List of Symbols/Definitions).

Theorem 4.2.14. Let Ω be a domain with $\mathbf{C}^\infty$ boundary. Then

$$d_N(\mathbf{U}^{\ell,2}, \mathbf{L}^2(\Omega)) \sim cN^{-\ell/n}, \qquad N \to \infty, \tag{4.2.55a}$$

where c has the value

$$c = (2\pi)^{-n}\rho_n \operatorname{meas}(\Omega), \tag{4.2.55b}$$

and ρ_n is the volume of the unit ball in $\mathbb{R}^n$, given by

$$\rho_n = \pi^{n/2}/\Gamma((n/2) + 1). \tag{4.2.55c}$$

Lemma 4.2.15. If the self-adjoint Friedrichs' operator A in $\mathbf{L}^2(\Omega)$ is defined by

$$(u, v)_{\mathbf{W}^{\ell,2}(\Omega)} = (\mathrm{A}u, v)_{\mathbf{L}^2(\Omega)}, \qquad \text{for all} \quad v \in \mathbf{W}^{\ell,2}(\Omega), \tag{4.2.56}$$

then,

$$\mathbf{D}_{\mathrm{A}} \subset \mathbf{W}^{2\ell,2}(\Omega). \tag{4.2.57}$$

Here, $\ell \geqslant 1$.

Proof of Lemma 4.2.15: Define self-adjoint operators A_j on $\mathbf{L}^2(\Omega)$ by

$$((u, v))_{\mathbf{W}^{j,2}(\Omega)} = (\mathrm{A}_j u, v)_{\mathbf{L}^2(\Omega)}, \qquad \text{for all} \quad v \in \mathbf{W}^{j,2}(\Omega),$$

for $j = 1, \ldots, \ell$, where the double bracket notation is defined by

$$((u, u))_{\mathbf{W}^{j,2}(\Omega)} = |u|^2_{\mathbf{W}^{j,2}(\Omega)}.$$

An elementary argument shows that $\mathbf{C}_0^\infty(\Omega) \subset \mathbf{D}_{\mathrm{A}_j}$, and

$$\mathrm{A}_j\phi = (-1)^j\Delta^j\phi, \qquad \phi \in \mathbf{C}_0^\infty(\Omega).$$

It follows from this and the self-adjointness of A_j that

$$(\mathrm{A}_j u, \phi)_{\mathbf{L}^2(\Omega)} = (u, (-1)^j\Delta^j\phi)_{\mathbf{L}^2(\Omega)}, \qquad \text{for all} \quad \phi \in \mathbf{C}_0^\infty(\Omega),$$

for $u \in \mathbf{D}_{\mathrm{A}}$. Thus, u is a weak solution of

$$(-1)^j\Delta^j u = \mathrm{A}_j u = f \in \mathbf{L}^2(\Omega),$$

and, hence (see Agmon [1] Theorem 11.10, p. 165), $\mathbf{D}_{\mathrm{A}_j} \subset \mathbf{W}^{2j,2}(\Omega)$. It now follows easily that

$$\mathrm{A}_j = \mathrm{A}_1^j = (-1)^j\Delta^j, \qquad j = 1, \ldots, \ell,$$

and, hence, $\mathbf{D}_{\mathrm{A}_j} \supset \mathbf{D}_{\mathrm{A}_{j+1}}$, $j = 1, \ldots, \ell - 1$. In particular, the self-adjoint operator

$$\hat{\mathrm{A}} = \sum_{j=1}^{\ell} \mathrm{A}_j + I, \qquad \mathbf{D}_{\hat{\mathrm{A}}} = \mathbf{D}_{\mathrm{A}^\ell},$$

satisfies

$$\mathbf{D}_{\hat{\mathrm{A}}} \subset \mathbf{W}^{2\ell,2}(\Omega), \tag{4.2.58}$$

and also satisfies the relation (4.2.56) defining A. However, by the uniqueness in $\mathbf{W}^{\ell,2}(\Omega)$ of A, it follows that $\mathrm{A} = \hat{\mathrm{A}}$, and (4.2.57) is immediate from (4.2.58). ■

Proof of Theorem 4.2.14: Setting $\mathbf{V} = \mathbf{W}^{\ell,2}(\Omega)$, $[\cdot,\cdot] = (\cdot,\cdot)_{\mathbf{W}^{\ell,2}(\Omega)}$ and $\mathbf{H} = \mathbf{L}^2(\Omega)$, we see that Corollary 4.2.13 is applicable. In particular, the N-widths of $\mathbf{U}^{\ell,2}$ are given by $\lambda_{N+1}^{-1/2}$, that is, in terms of the eigenvalues of the operator A satisfying (4.2.56). It is known (see Agmon [1] Theorem 14.6, p. 250) that the asymptotic distribution of the eigenvalues of such self-adjoint operators is given by

$$N(\lambda) = c\lambda^{n/2\ell} + o(\lambda^{n/2\ell}), \qquad \lambda \to \infty, \tag{4.2.59}$$

where $N(\lambda)$ is the number of eigenvalues $\leqslant \lambda$ and

$$c = (2\pi)^{-n} \int_\Omega w(\mathbf{x})\, d\mathbf{x}, \qquad w(\mathbf{x}) \equiv \text{meas}\left\{\xi : 0 < \left(\sum_{i=1}^{n} \xi_i^2\right)^{\ell} < 1\right\}. \tag{4.2.60}$$

Since $N(\lambda_{j+1}) = N(\lambda_j)(1 + O(1))$ (see the proof of Lemma 3.2 of Jerome [20]), a routine argument shows that

$$\lambda_j = (j/c)^{2\ell/n} + o(j^{2\ell/n}), \qquad j \to \infty. \tag{4.2.61}$$

Equations (4.2.51) and (4.2.61) immediately give (4.2.55a). Equation (4.2.55b) follows directly from (4.2.60). ■

4.3 CONVERGENCE RATES FOR THE CONTINUOUS GALERKIN METHOD

We shall describe briefly the linear elliptic theory of finite-element approximations, in a fairly general setting, prior to nonlinear applications to degenerate parabolic problems. Thus, let $\mathbf{G}$ be a Hilbert space with dual $\mathbf{G}^*$ and suppose $a(\cdot,\cdot)$ is a continuous, strictly coercive bilinear form defined on $\mathbf{G} \times \mathbf{G}$. If T is the isomorphism of $\mathbf{G}^*$ onto $\mathbf{G}$, defined by

$$a(\mathrm{T}\ell, v) = \langle \ell, v \rangle, \qquad \text{for all} \quad v \in \mathbf{G}, \tag{4.3.1}$$

and guaranteed by the Lax–Milgram theorem, we seek approximations for $u = \mathrm{T}\ell$ and estimates of that approximation. The operator T is, of course, a generalization of the operators D_0, N_σ ($\sigma \geqslant 0$), and R_ω of Chapter 1. We shall explicitly assume that $a(\cdot,\cdot)$ is symmetric so that T is self-adjoint.

Definition 4.3.1. For $m \geqslant 1$, consider a family $\mathbf{G}_s$ of Hilbert spaces for $s = -m, -m+1, \ldots, 2m-1, 2m$, decreasing as functions of s with dense, continuous inclusions. We identify $\mathbf{G}$ with $\mathbf{G}_m$. These spaces are to have the property

$$\mathbf{G}_k^* = \mathbf{G}_{-k}, \qquad 0 \leqslant k \leqslant m, \tag{4.3.2}$$

and the restriction mappings are required to satisfy

$$\mathrm{T}:\mathbf{G}_k \to \mathbf{G}_{k+2m}, \qquad -m \leqslant k \leqslant 0, \tag{4.3.3}$$

and to be isomorphisms. Assume that $\{\mathbf{S}_h\}_h$ is a family of finite-dimensional subspaces of $\mathbf{G}_m$, and let E_h be the projection onto $\mathbf{S}_h$, defined by

$$a(u - \mathrm{E}_h u, v) = 0, \qquad \text{for all} \quad v \in \mathbf{S}_h. \tag{4.3.4}$$

Here, h is a positive parameter related to $\dim \mathbf{S}_h$ by a simple rational function. We say that $\mathbf{S}_h$ is of degree $k - 1$ for E_h, for some integer $m < k \leqslant 2m$, if

$$\mathbf{S}_h \subset \mathbf{G}_{k-1}, \tag{4.3.5a}$$

and if, for each $s = 0, \ldots, m - 1$, there exists a positive constant $C_{m,k,s}$, such that

$$\|u - \mathrm{E}_h u\|_{\mathbf{G}_m} \leqslant C_{m,k,s} \begin{cases} h^{m-s}\|u\|_{\mathbf{G}_{2m-s}}, & k \geqslant 2m - s, \\ h^{k-m}\|u\|_{\mathbf{G}_k}, & k \leqslant 2m - s, \end{cases} \tag{4.3.5b}$$

for $u \in \mathbf{G}_{2m-s}$.

The approximation order of $u - \mathrm{E}_h u$ in $\mathbf{G}_s$, $s < m$, is now described.

Proposition 4.3.1.[†] Let $f \in \mathbf{G}_{k-2m}$ be given, and suppose $u \in \mathbf{G}_k$ satisfies $u = \mathrm{T}f$. Then, for $s = 0, \ldots, m - 1$, there are constants $C_{m,k,s}$, such that

$$\|u - \mathrm{E}_h u\|_{\mathbf{G}_s} \leqslant C_{m,k,s} \begin{cases} h^{k-s}\|f\|_{\mathbf{G}_{k-2m}}, & s \geqslant 2m - k, \\ h^{2(k-m)}\|f\|_{\mathbf{G}_{k-2m}}, & s \leqslant 2m - k, \end{cases} \tag{4.3.6}$$

if $\mathbf{S}_h$ is of degree $k - 1$ for E_h.

Proof: One proceeds by standard duality. Thus, for $g \in \mathbf{G}_{-s}$, define $v = \mathrm{T}g \in \mathbf{G}_{-s+2m}$. Then,

$$\|u - \mathrm{E}_h u\|_{\mathbf{G}_s} = \sup\{|\langle g, u - \mathrm{E}_h u\rangle| / \|g\|_{\mathbf{G}_{-s}} : 0 \neq g \in \mathbf{G}_{-s}\}, \tag{4.3.7}$$

and

$$\begin{aligned} |\langle g, u - \mathrm{E}_h u\rangle| &= |a(v - \mathrm{E}_h v, u - \mathrm{E}_h u)| \\ &\leqslant C\|v - \mathrm{E}_h v\|_{\mathbf{G}_m}\|u - \mathrm{E}_h u\|_{\mathbf{G}_m}. \end{aligned} \tag{4.3.8}$$

Applying (4.3.5b), for the specified s, to $\|v - \mathrm{E}_h v\|_{\mathbf{G}_m}$, and applying the same inequality, with $s = 2m - k$, to $\|u - \mathrm{E}_h u\|_{\mathbf{G}_m}$, we deduce, from (4.3.8),

$$|\langle g, u - \mathrm{E}_h u\rangle| \leqslant C\|v\|_{\mathbf{G}_{2m-s}}\|u\|_{\mathbf{G}_k} \begin{cases} h^{k-s}, & k \geqslant 2m - s, \\ h^{2(k-m)}, & k \leqslant 2m - s. \end{cases} \tag{4.3.9}$$

[†] This is a generalized form of the Nitsche–Aubin lemma.

Using the isomorphism property of T, we obtain, from (4.3.9),

$$|\langle g, u - \mathrm{E}_h u\rangle| \leqslant C_1\|g\|_{\mathbf{G}_{-s}}\|f\|_{\mathbf{G}_{k-2m}}\begin{cases} h^{k-s}, & k \geqslant 2m - s, \\ h^{2(k-m)}, & k \leqslant 2m - s, \end{cases} \tag{4.3.10}$$

and (4.3.6) is immediate from (4.3.7) and (4.3.10). ■

Remark 4.3.1. It is often useful to have the parallel concept that $\mathbf{S}_h$ is of degree $k - 1$ for P_h, where P_h is the orthogonal projection in $\mathbf{G}_0$. In this case we require, for $s = 0, \ldots, 2m$,

$$\|u - \mathrm{P}_h u\|_{\mathbf{G}_0} \leqslant C_{m,s,k}\begin{cases} h^{2m-s}\|u\|_{\mathbf{G}_{2m-s}}, & k \geqslant 2m - s, \\ h^k\|u\|_{\mathbf{G}_k}, & k \leqslant 2m - s, \end{cases} \tag{4.3.11}$$

for $u \in \mathbf{G}_{2m-s}$ as the appropriate modification of (4.3.5b). The result is the following. For $-m \leqslant s < 0$ and $0 \leqslant j \leqslant 2m$, there exist constants $C_{m,j,k,s}$, such that

$$\|u - \mathrm{P}_h u\|_{\mathbf{G}_s} \leqslant C_{m,j,k,s}\begin{cases} h^{2m-j-s}\|u\|_{\mathbf{G}_{2m-j}}, & k \geqslant 2m - j, \\ h^{k-s}\|u\|_{\mathbf{G}_k}, & k \leqslant 2m - j, \end{cases} \tag{4.3.12}$$

for $u \in \mathbf{G}_{2m-j}$. The proof is analogous. Notice that $u \in \mathbf{G}_0$ is essential for this result, since it is derived under the supposition that $\langle v, w\rangle = (v, w)_{\mathbf{G}_0}$ for $v, w \in \mathbf{G}_0$, that is, under the supposition that $\mathbf{G}_0$ is a pivot space.

Remark 4.3.2. The primary application of Proposition 4.3.1 is to the case where $\mathbf{G}_s = \mathbf{H}^s(\Omega)$ or $\mathbf{G}_s = \mathbf{H}_0^s(\Omega)$, and T is one of the operators N_σ, R_ω, or D_0. Here, the spaces $\mathbf{S}_h$ are typically $\mathbf{C}^{k-2}$ spaces of piecewise polynomials of degree at least $k - 1$, for $k \geqslant 2$, defined by means of underlying meshes or triangulations of the domain. Although the reader might expect such trial spaces automatically to satisfy (4.3.5b), in the light of the results of Section 4.1, in fact, additional arguments are needed. This is due to the fact that $\mathrm{E}_h u_{|e}$ need not be the best approximation of $u_{|e}$ from $\mathrm{S}_{h|e}$ in $\mathbf{H}^m(e)$, for e an element of a given finite-element triangulation, although $\mathrm{E}_h u$ is the best approximation, up to equivalent norms, from $\mathbf{S}_h$ in $\mathbf{H}^m(\bigcup e)$. The verification of (4.3.5b) in Sobolev space for finite-element spaces is achieved by the construction of an interpolation operator, or, in the case $m \leqslant (n/2)$, the composition of a smoothing operator and an interpolation operator (see Strang [40] and Strang and Fix [41] for complete details), and employs the notion of a nodal basis spanning $\mathbf{S}_h$. The argument does use, however, the decomposition $u = P_k + R$ in each element e, where P_k is the polynomial of degree $k - 1$, defined, say, by the Sobolev representation theorem on e. The interpolation operator $u \to u_I$ reproduces P_k, so that the final estimation of $u - u_I$ can, somewhat crudely, estimate R and its interpolation polynomial R_I, separately, on the element e. Of course, the estimation of R was already carried out in Section 4.1 and yields $|R|_{\mathbf{H}^m(e)} \leqslant ch^{k-m}|u|_{\mathbf{H}^k(e)}$, for example, for $u \in \mathbf{H}^{2m}(\Omega)$. The estimation of R_I is a simple application of Sobolev's inequality, applied to terms such as $\mathrm{D}^{\beta}R(z_j)$, when R_I is represented, via a nodal basis $\{\phi_j\}$, in terms of the values of R and its derivatives at nodal points $\mathbf{z}_j \in \Omega$; here, $\mathbf{z}_j = \mathbf{z}_{j'}$ is possible only in the case where derivatives of R appear in the expansion of R_I. For example, for $u \in \mathbf{H}^{2m}(\Omega)$ and distinct $\mathbf{z}_j$, we have, under uniform basis assumptions on $\{\phi_j\}$,

$$|\mathrm{D}^{\alpha}R_I(\mathbf{x})| \leqslant ch^{k-m-(n/2)}|u|_{\mathbf{H}^k(e)}, \qquad |\alpha| = m, \quad \mathbf{x} \in e,$$

which also gives $|R_I|_{\mathbf{H}^m(e)} \leqslant ch^{k-m}|u|_{\mathbf{H}^k(e)}$. These computations are described in Strang and Fix [41] p. 146.

Definition 4.3.2. Given three Hilbert spaces $\mathbf{G}_{-m}$, $\mathbf{G}_0$, and $\mathbf{G}_m$, with $m \geqslant 1$, $\mathbf{G}_m^* = \mathbf{G}_{-m}$, and $\mathbf{G}_m \subset \mathbf{G}_0 \subset \mathbf{G}_{-m}$, where the latter embeddings are dense and continuous, and $\mathbf{G}_0$ is a pivot space, let $a(\cdot,\cdot)$ be a symmetric, continuous, positive definite bilinear form on $\mathbf{G}_m$. Denote by T the isomorphism of $\mathbf{G}_m^*$ onto $\mathbf{G}_m$, defined by (4.3.1). We agree to identify $a(\cdot,\cdot)$ notationally with the inner product on $\mathbf{G}_m$, and we select, for the inner product on $\mathbf{G}_{-m}$,

$$(f,g)_{\mathbf{G}_{-m}} = \langle f, \mathrm{T}g\rangle. \tag{4.3.13}$$

Let $\mathbf{S}_h$ be a finite-dimensional subspace of $\mathbf{G}_m$, where the dimension of $\mathbf{S}_h$ is a simple rational function of h. We say that $\mathbf{S}_h$ satisfies the inverse hypothesis in $\mathbf{G}_m$ if there exists a constant C, such that

$$\|\chi\|_{\mathbf{G}_m} \leqslant \frac{C}{h^m}\|\chi\|_{\mathbf{G}_0}, \qquad \text{for all} \quad \chi \in \mathbf{S}_h. \tag{4.3.14}$$

We denote by E_h the orthogonal projection of $\mathbf{G}_m$ onto $\mathbf{S}_h$, defined by (4.3.4), and, finally, we set $\mathrm{T}_h = \mathrm{E}_h\mathrm{T}$.

Remark 4.3.3. The mappings D_0, N_σ $(\sigma \geqslant 0)$ and R_ω, introduced in Chapter 1, are examples of the mapping T. From the discussion of those examples, which carries over to the general case, we see that T is positive definite and self-adjoint on the pivot space $\mathbf{G}_0$, and (4.3.13) determines a norm on $\mathbf{G}_{-m}$ equivalent to the duality norm. Moreover, T_h may be characterized by

$$\langle f, \chi\rangle = (\mathrm{T}_h f, \chi)_{\mathbf{G}_m}, \qquad \text{for all} \quad \chi \in \mathbf{S}_h, \tag{4.3.15}$$

if f is given in $\mathbf{G}_{-m}$. The mapping T_h is self-adjoint and nonnegative definite on $\mathbf{G}_0$, and positive definite on $\mathbf{S}_h$. Note that the invertibility of T and the identity $\|\mathrm{T}u\|_{\mathbf{G}_m} = \|u\|_{\mathbf{G}_{-m}}$ permit the usual pivot space inequality

$$|(u, v)_{\mathbf{G}_0}| \leqslant \|u\|_{\mathbf{G}_m}\|v\|_{\mathbf{G}_{-m}}, \qquad u, v \in \mathbf{G}_m. \tag{4.3.16}$$

Definition 4.3.3. For ϕ, ψ in $\mathbf{S}_h$, set

$$(\phi, \psi)_{\mathbf{G}_{-m,h}} = (\phi, \mathrm{T}_h\psi)_{\mathbf{G}_0}, \tag{4.3.17}$$

and extend this inner product on $\mathbf{S}_h$ to a semi-inner product on $\mathbf{G}_{-m}$, by permitting ϕ, ψ in (4.3.17) to be arbitrary elements of $\mathbf{G}_{-m}$.

Remark 4.3.4. If, as earlier, P_h denotes the orthogonal projection in $\mathbf{G}_0$ onto $\mathbf{S}_h$, then, for $\phi, \psi \in \mathbf{G}_0$,

$$(\phi, \psi)_{\mathbf{G}_{-m,h}} = (\mathrm{T}_h\phi, \psi)_{\mathbf{G}_0} = (\mathrm{T}_h\phi, \mathrm{P}_h\psi)_{\mathbf{G}_0} = (\mathrm{T}_h\mathrm{P}_h\phi, \mathrm{P}_h\psi)_{\mathbf{G}_0}. \tag{4.3.18}$$

Note also the natural duality,

$$|(\phi, \psi)_{\mathbf{G}_0}| \leqslant \|\phi\|_{\mathbf{G}_{-m,h}}\|\psi\|_{\mathbf{G}_m}, \qquad \phi, \psi \in \mathbf{S}_h, \tag{4.3.19}$$

which follows, since $\mathrm{T}_{h|\mathbf{S}_h}$ is invertible.

Proposition 4.3.2.[†] The inequality

$$\|\chi\|_{\mathbf{G}_{-m,h}} \leqslant \|\chi\|_{\mathbf{G}_{-m}}, \qquad \text{for all} \quad \chi \in \mathbf{S}_h, \tag{4.3.20a}$$

is valid. If $\mathbf{S}_h$ is of degree $2m - 1$ for E_h, and if the inverse hypothesis (4.3.14) holds, then there is a constant C, not depending on h, such that

$$\|\chi\|_{\mathbf{G}_{-m}} \leqslant C\|\chi\|_{\mathbf{G}_{-m,h}}, \qquad \text{for all} \quad \chi \in \mathbf{S}_h. \tag{4.3.20b}$$

Proof: We first establish (4.3.20a). Thus, for $\chi \in \mathbf{S}_h$,

$$\begin{aligned}\|\chi\|_{\mathbf{G}_{-m,h}} &= (\mathrm{T}_h\chi, \chi)_{\mathbf{G}_0}^{1/2} = \|\mathrm{T}_h\chi\|_{\mathbf{G}_m} \\ &= \|\mathrm{E}_h\mathrm{T}\chi\|_{\mathbf{G}_m} \leqslant \|\mathrm{T}\chi\|_{\mathbf{G}_m} \\ &= (\mathrm{T}\chi, \chi)_{\mathbf{G}_0}^{1/2} = \|\chi\|_{\mathbf{G}_{-m}},\end{aligned}$$

where we have used (4.3.1), (4.3.13), and (4.3.17), as well as the fact that $\|\mathrm{E}_h\| \leqslant 1$ as an operator on $\mathbf{G}_m$. To establish (4.3.20b), we first note that

$$\|\chi\|_{\mathbf{G}_0} \leqslant \frac{C}{h^m}\|\chi\|_{\mathbf{G}_{-m,h}}, \tag{4.3.21}$$

[†] Propositions 4.3.2–4.3.4 are adapted from the author's paper with Rose (see Chapter 2, Ref. [9]).

where C is the same constant appearing in (4.3.14). Indeed, by the latter and by (4.3.19), we have

$$\begin{aligned}\|\chi\|_{\mathbf{G}_0} &= \sup\{|(\chi,\psi)_{\mathbf{G}_0}|:\psi \in \mathbf{S}_h, \|\psi\|_{\mathbf{G}_0} \leqslant 1\}\\ &\leqslant \sup\{\|\chi\|_{\mathbf{G}_{-m,h}}\|\psi\|_{\mathbf{G}_m}:\psi \in \mathbf{S}_h, \|\psi\|_{\mathbf{G}_0} \leqslant 1\}\\ &\leqslant \sup\left\{\|\chi\|_{\mathbf{G}_{-m,h}}\frac{C}{h^m}\|\psi\|_{\mathbf{G}_0}:\psi \in \mathbf{S}_h, \|\psi\|_{\mathbf{G}_0} \leqslant 1\right\}\\ &\leqslant \frac{C}{h^m}\|\chi\|_{\mathbf{G}_{-m,h}},\end{aligned}$$

which establishes (4.3.21). We now estimate $\|\chi\|_{\mathbf{G}_{-m}}$:

$$\begin{aligned}\|\chi\|^2_{\mathbf{G}_{-m}} &= (\mathrm{T}\chi,\chi)_{\mathbf{G}_0} = (\mathrm{E}_h\mathrm{T}\chi,\chi)_{\mathbf{G}_0} + ((\mathrm{I}-\mathrm{E}_h)\mathrm{T}\chi,\chi)_{\mathbf{G}_0}\\ &\leqslant \|\chi\|^2_{\mathbf{G}_{-m,h}} + \|(\mathrm{T}-\mathrm{T}_h)\chi\|_{\mathbf{G}_0}\|\chi\|_{\mathbf{G}_0}\\ &\leqslant \|\chi\|^2_{\mathbf{G}_{-m,h}} + Ch^{2m}\|\chi\|^2_{\mathbf{G}_0}\\ &\leqslant \|\chi\|^2_{\mathbf{G}_{-m,h}} + C_1\|\chi\|^2_{\mathbf{G}_{-m,h}},\end{aligned}$$

where we have applied (4.3.21) and the inequality

$$\|(\mathrm{T}-\mathrm{T}_h)f\|_{\mathbf{G}_0} \leqslant Ch^{2m}\|f\|_{\mathbf{G}_0},$$

which follows from (4.3.6), with $k = 2m$ and $u = \mathrm{T}f$. This concludes the verification of (4.3.20b). ■

We introduce now the specific finite-element space of piecewise linear trial functions.

Definition 4.3.4. Let $\{\Delta_h\}$ be a family of triangulations or simplicial decompositions of Ω. Thus, Δ_h consists of closed simplicial elements e. The positive parameter h will be specified shortly. The boundary elements are permitted to have a curvilinear edge coinciding with $\partial\Omega$, and it is required that $\bar{\Omega} = \Omega_h = \bigcup_{e\in\Delta_h} e$. For each $e \in \Delta_h$, we define $\rho(e)$ (respectively $\sigma(e)$) to be the radius of the smallest ball containing e (respectively the largest ball contained in e). Set

$$h = \sup\{\rho(e):e \in \Delta_h\}, \tag{4.3.22}$$

and

$$\mathbf{M}_h = \{\chi \in \mathbf{C}(\bar{\Omega}):\chi_{|e} \quad \text{is affine,} \qquad \text{for all} \quad e \in \Delta_h\}. \tag{4.3.23}$$

For simplicity, we assume $h < 1$ in the sequel.

Remark 4.3.5. We shall assume that $\mathbf{M}_h$ has the approximation property

$$\inf\{\|u-\chi\|_{\mathbf{H}^j(\Omega)}:\chi \in \mathbf{M}_h\} \leqslant Ch^{\ell-j}\|u\|_{\mathbf{H}^\ell(\Omega)}, \qquad \text{for all} \quad u \in \mathbf{H}^\ell(\Omega), \tag{4.3.24}$$

for $\ell = 1, 2$ and $j = 0, \ldots, \ell - 1$. This implies that $\mathbf{M}_h$ is of degree 1 for E_h and P_h (see (4.3.5b) and (4.3.11)). In the framework of the earlier abstract structure, we are taking $m = 1$, $k = 2$, $\mathbf{G}_{-1} = [\mathbf{H}^1(\Omega)]^*$, and $\mathbf{G}_j = \mathbf{H}^j(\Omega)$ for $j \geqslant 0$. We have, thus, chosen to consider explicitly only the Neumann and Robin boundary-value approximations. However, trivial modifications, for Ω a convex, polyhedral domain permit the handling of the Dirichlet problem, where the trial functions in $\mathbf{M}_h$ are required to vanish on the boundary of Ω. The Dirichlet problem in the case of general (smooth) boundary geometry requires interior and exterior approximation by polyhedral domains to order h^2. Even when such approximation is carried out, the convergence rate deteriorates (see Strang and Fix [41] pp. 192–196) due to boundary-layer phenomena (see, however, [33]).

Remark 4.3.6. As noted in Remark 4.3.2, the estimate (4.3.24) is valid whenever the elements $e \in \Delta_h$ permit a polynomial projection locally in e, defined by the Sobolev representation formula (4.1.4) for $\ell = 1, 2$. This places a restriction on the boundary elements e of Δ_h, such as e is the union of star-shaped domains. However (see Remark 4.1.3), this covers a broad class of domains. In order to utilize (4.3.24), via Proposition 4.3.1 and Remark 4.3.1, it is necessary that the appropriate smoothing operator T satisfy

$$\mathrm{T}:\mathbf{L}^2(\Omega) \to \mathbf{H}^2(\Omega), \tag{4.3.25}$$

as an isomorphism. In order to utilize Proposition 4.3.2, we shall require the analog of (4.3.14):

$$\|\chi\|_{\mathbf{H}^1(\Omega)} \leqslant \frac{C}{h} \|\chi\|_{\mathbf{L}^2(\Omega)}, \qquad \text{for all} \quad \chi \in \mathbf{M}_h. \tag{4.3.26}$$

The inverse hypothesis (4.3.26) is known to be implied by the hypothesis that the underlying triangulations are quasi-uniform, that is, for all $h > 0$,

$$\inf\left\{\frac{\sigma(e)}{h} : \sigma \in \Delta_h\right\} \geqslant \gamma_0 > 0, \qquad \text{all} \quad h. \tag{4.3.27}$$

Thus, the results of Propositions 4.3.1 and 4.3.2 and Remark 4.3.1 are valid under the hypotheses associated with (4.3.24), (4.3.25), and (4.3.27).

Remark 4.3.7. One of the remarkable properties of the projection E_h, taken in Hilbert space, is its stability, or near stability, in the pointwise norm spaces. For our purposes, this simply means that, for piecewise linear trial functions, an estimate of the form

$$\|(\mathrm{T} - \mathrm{T}_h)f\|_{\mathbf{L}^\infty(\Omega)} \leqslant Ch^2(\ln(1/h))^{d(n)}\|f\|_{\mathbf{L}^\infty(\Omega)} \tag{4.3.28}$$

can be proved, where $d(n)$ depends upon the Euclidean dimension n. It is known that $d(1) = 0$ and $d(n) = 2$, $n \geqslant 2$, for the standard Dirichlet and Neumann homogeneous boundary-value problems (see Douglas *et al.* [12] and Schatz and Wahlbin [33] and Remark 4.3.10 to follow). Many other results similar to (4.3.28), in which $\|f\|_{\mathbf{L}^\infty(\Omega)}$ is replaced by various second-order norm differential expressions of $\mathrm{T}f$, e.g., $\|\mathrm{T}f\|_{\mathbf{W}^{2,\infty}(\Omega)}$, have been obtained by Scott [36], Nitsche [28], and others. However, for the applications of this section, the estimate (4.3.28) is the crucial one. Frehse and Rannacher [16], obtained a result slightly weaker than that of (4.3.28), with higher powers of the logarithm appearing. It is of interest that, for many higher-order differential problems, involving piecewise polynomials at least of quadratic degree, the exponent $d(n)$ is zero, at least when (4.3.28) is interpreted in the relaxed sense, in which $\|f\|_{\mathbf{L}^\infty(\Omega)}$ is replaced by a more comprehensive expression involving f (see Nitsche [28] and Scott [36]), and, of course, in which the exponent of h is accordingly increased.

Definition 4.3.5. We denote by $\mathbf{M}(\bar{\Omega})$ the space of finite regular Baire measures on $\bar{\Omega}$, normed with the total variation norm. It is well known that $\mathbf{M}(\bar{\Omega}) = [\mathbf{C}(\bar{\Omega})]^*$. We make use of the norm-preserving Lebesgue extension $\tilde{\mu}$ of μ, so that

$$\left|\int_\Omega f\, d\tilde{\mu}\right| \leqslant \|f\|_{\mathbf{L}^\infty(\Omega)} \|\mu\|_{\mathbf{M}(\bar{\Omega})}, \qquad \text{for all} \quad f \in \mathbf{L}^\infty(\Omega), \tag{4.3.29}$$

holds for each $\mu \in \mathbf{M}(\bar{\Omega})$, and we identify μ with $\tilde{\mu}$. Finally, T is an isomorphism of $\mathbf{F} = [\mathbf{H}^1(\Omega)]^*$ onto $\mathbf{H}^1(\Omega)$, defined, via (4.3.1), by a symmetric, positive definite bilinear form $a(\cdot,\cdot)$, $\mathrm{T}_h = \mathrm{E}_h\mathrm{T}$, where E_h is defined by (4.3.4), with $\mathbf{S}_h = \mathbf{M}_h$, and $\mathbf{F}_h$ is the (semi) inner product space described by (4.3.17), that is, $\mathbf{F}_h = \mathbf{G}_{-1,h}$.

Proposition 4.3.3. Suppose that T and T_h satisfy (4.3.28). Then, for $\mu \in [\mathbf{H}^1(\Omega)]^* \cap \mathbf{M}(\bar{\Omega})$, the estimate

$$\|(\mathrm{T} - \mathrm{T}_h)\mu\|_{\mathbf{M}(\bar{\Omega})} \leqslant Ch^2 \left(\ln \frac{1}{h}\right)^{d(n)} \|\mu\|_{\mathbf{M}(\bar{\Omega})} \tag{4.3.30}$$

holds for the same constant C.

Proof: The argument proceeds by standard duality:

$$\begin{aligned}
\|(\mathrm{T} - \mathrm{T}_h)\mu\|_{\mathbf{M}(\bar{\Omega})} &= \sup\{|\langle(\mathrm{T} - \mathrm{T}_h)\mu, \psi\rangle| : \psi \in \mathbf{C}(\bar{\Omega}) : \|\psi\|_{\mathbf{C}(\bar{\Omega})} \leqslant 1\} \\
&= \sup\{|\langle\mu, (\mathrm{T} - \mathrm{T}_h)\psi\rangle| : \psi \in \mathbf{C}(\bar{\Omega}) : \|\psi\|_{\mathbf{C}(\bar{\Omega})} \leqslant 1\} \\
&\leqslant \sup\left\{Ch^2\left(\ln\frac{1}{h}\right)^{d(n)} \|\psi\|_{\mathbf{L}^\infty(\Omega)}\|\mu\|_{\mathbf{M}(\bar{\Omega})} : \psi \in \mathbf{C}(\bar{\Omega}) : \|\psi\|_{\mathbf{C}(\bar{\Omega})} \leqslant 1\right\},
\end{aligned}$$

and (4.3.30) is immediate. Note that we have used the self-adjointness of $\mathrm{T} - \mathrm{T}_h$ with respect to the duality pairing. ■

We turn now to an application of these ideas to nonlinear parabolic evolutions.

Definition 4.3.6. Consider the differential equation in distribution form,

$$\frac{\partial v}{\partial t} - \Delta u + f_0(u) = 0, \qquad v = H(u)^{\dagger}, \tag{4.3.31}$$

on a space–time domain $\mathscr{D} = \Omega \times (0, T_0)$, with initial datum $v_{|_{t=0}} \in \mathbf{L}^{\infty}(\Omega)$ specified. It is explicitly assumed that

$$u \in \mathbf{L}^2((0, T_0); \mathbf{H}^1(\Omega)) \cap \mathbf{H}^1([0, T_0]; \mathbf{L}^2(\Omega)) \cap \mathbf{L}^{\infty}(\mathscr{D}), \tag{4.3.32a}$$

$$v \in \mathbf{L}^{\infty}(\mathscr{D}), \qquad v_t \in \mathbf{L}^{\infty}((0, T_0); \mathbf{M}(\bar{\Omega})), \tag{4.3.32b}$$

f_0 is Lipschitz continuous, (4.3.32c)

$H(\cdot)$ defines a surjective, maximal, monotone graph in $\mathbb{R}^2$ with bounded sections. (4.3.32d)

We also assume that (4.3.31) has a pointwise lifting, via an isomorphic, self-adjoint mapping T onto $\mathbf{H}^1(\Omega)$, so that the pair u, v satisfies, for $f = f_0 - \sigma id$, some $\sigma \geqslant 0$,

$$\frac{\partial \mathrm{T}v}{\partial t} + u + \mathrm{T}f(u) = 0, \qquad \text{almost everywhere in} \quad \Omega, \quad 0 < t < T_0, \tag{4.3.33}$$

and that there exist positive monotone increasing functions c_1 and c_2, defined on $[0, \infty)$, such that, for all $w_1, w_2 \in \mathbf{L}^{\infty}(\Omega)$,

$$|H(w_1(\mathbf{x})) - H(w_2(\mathbf{x}))| \geqslant [c_1(\|w_1\|_{\mathbf{L}^{\infty}(\Omega)})c_2(\|w_2\|_{\mathbf{L}^{\infty}(\Omega)})]^{-1}|w_1(\mathbf{x}) - w_2(\mathbf{x})|, \tag{4.3.34}$$

for almost all $\mathbf{x} \in \Omega$.

Remark 4.3.8. This model includes, with homogeneous Neumann or Robin boundary conditions, the two-phase Stefan problem, the porous-medium equation (see (2.3.21)), and the nondegenerate reaction–diffusion system, if (4.3.31) is understood in vector format. However, the latter can be

† Uniqueness is anticipated here.

analyzed directly in a more straightforward manner and with a sharper estimate, with $\mathbf{v} = \mathbf{u}$, once the maximum principle $\mathbf{u} \in \mathbf{L}^\infty(\mathscr{D})$ is utilized to permit the assumption (4.3.32c). The assumption $v_t \in \mathbf{L}^\infty((0, T_0); \mathbf{M}(\bar{\Omega}))$ is the really distinguishing feature† in (4.3.32); this was not directly required in defining the solution classes in Sections 1.1 and 1.2, but will be demonstrated in Chapter 5 (see Theorem 5.2.1).

Definition 4.3.7. We shall define the finite-element approximation U_h: $[0, T_0] \to \mathbf{M}_h$ as the solution of the differential equation

$$\left\langle \frac{\partial H(U_h)}{\partial t}, \chi \right\rangle + a(U_h, \chi) + (f_0(U_h), \chi)_{\mathbf{L}^2(\Omega)} = 0, \tag{4.3.35a}$$

for all $\chi \in \mathbf{M}_h$, $0 < t < T_0$, subject to the initial condition $U_h(0) \in \mathbf{M}_h$, defined by

$$\mathrm{P}_h H(U_h(0)) = \mathrm{P}_h v|_{t=0}. \tag{4.3.35b}$$

Note that the system (4.3.35a), of $\dim \mathbf{M}_h$ ordinary differential equations in t, must be understood in the sense of distributions on $(0, T_0)$, and that P_h is the orthogonal projection in $\mathbf{L}^2(\Omega)$ onto $\mathbf{M}_h$. Concerning the regularity of U_h, which must be considered in tandem with $H(U_h)$, we require

$$U_h \in \mathbf{L}^2((0, T_0); \mathbf{H}^1(\Omega)) \cap \mathbf{H}^1([0, T_0]; \mathbf{L}^2(\Omega)), \tag{4.3.35c}$$

and

$$H(U_h) \in \mathbf{C}([0, T_0]; \mathbf{L}^2(\Omega)). \tag{4.3.35d}$$

Remark 4.3.9. The existence of a solution pair satisfying (4.3.35) follows from the existence theory developed in Chapter 5. The uniqueness is a consequence of the pointwise lifting relation, where $f = f_0 - \sigma id$, as in Definition 4.3.6,

$$\frac{\partial}{\partial t} \mathrm{T}_h H(U_h) + U_h + \mathrm{T}_h f(U_h) = 0, \qquad \text{almost everywhere in} \quad \Omega, \quad 0 < t < T_0, \tag{4.3.36}$$

which follows from (4.3.35a), by setting $\chi = \mathrm{T}_h \psi$, $\psi \in \mathbf{C}_0^\infty(\Omega)$, and employing the self-adjointness of T_h. The relations (4.3.33) and (4.3.36) are the central relations for the convergence estimates to follow.

† This analytical property makes possible the application of $\mathbf{L}^\infty$ finite-element estimates via extended duality.

Proposition 4.3.4. Suppose that the continuous Galerkin approximations U_h, satisfying (4.3.35) and (4.3.36), are bounded, say,

$$\|U_h\|_{\mathbf{L}^\infty(\mathscr{D})} \leqslant C_0, \tag{4.3.37}$$

for the system satisfying Definition 4.3.6. Then, under the assumption that (4.3.30) holds, there exists a positive constant C, such that

$$\|H(u) - H(U_h)\|^2_{\mathbf{L}^\infty((0,T_0);\mathbf{F}_h)} + \|u - U_h\|^2_{\mathbf{L}^2(\mathscr{D})} \leqslant Ch^2\left[\ln\left(\frac{1}{h}\right)\right]^{d(n)}. \tag{4.3.38}$$

If (4.3.24), (4.3.25), and (4.3.27) hold, then,

$$\|H(u) - H(U_h)\|^2_{\mathbf{L}^\infty((0,T_0);\mathbf{F})} \leqslant Ch^2\left[\ln\left(\frac{1}{h}\right)\right]^{d(n)}. \tag{4.3.39}$$

Here $\mathbf{F}_h = \mathbf{F}$ as a set, with semi-norm defined by (4.3.17).

Proof: Subtracting (4.3.36) from (4.3.33), multiplying the resultant by $H(u) - H(U_h)$, and integrating over Ω, we obtain, for $0 < t < T_0$,

$$\begin{aligned}
&\left(\frac{\partial}{\partial t}\mathrm{T}_h[H(u) - H(U_h)], H(u) - H(U_h)\right)_{\mathbf{L}^2(\Omega)} + (H(u) - H(U_h), u - U_h)_{\mathbf{L}^2(\Omega)} \\
&= \left((\mathrm{T}_h - \mathrm{T})\frac{\partial H(u)}{\partial t}, H(u) - H(U_h)\right)_{\mathbf{L}^2(\Omega)} \\
&\quad + ((\mathrm{T}_h - \mathrm{T})f(u), H(u) - H(U_h))_{\mathbf{L}^2(\Omega)} \\
&\quad + (\mathrm{T}_h(f(U_h) - f(u)), H(u) - H(U_h))_{\mathbf{L}^2(\Omega)}.
\end{aligned} \tag{4.3.40}$$

Suppose the first term on the left-hand side of (4.3.40) is rewritten in $\mathbf{F}_h$ seminorm differentiated terms, and the final term on the right-hand side is estimated by the Cauchy–Schwarz inequality in $\mathbf{F}_h$, and by the domination of the $\mathbf{F}_h$ seminorm by the $\mathbf{F}$ norm as

$$\begin{aligned}
&|(\mathrm{T}_h(f(U_h) - f(u)), H(u) - H(U_h))_{\mathbf{L}^2(\Omega)}| \\
&\quad \leqslant \|f(U_h) - f(u)\|_{\mathbf{F}}\|H(u) - H(U_h)\|_{\mathbf{F}_h} \\
&\quad \leqslant \tfrac{1}{2}C\eta\|U_h - u\|^2_{\mathbf{L}^2(\Omega)} + \tfrac{1}{2}\eta^{-1}\|H(u) - H(U_h)\|^2_{\mathbf{F}_h},
\end{aligned} \tag{4.3.41}$$

for $\sqrt{C}$ the product of the Lipschitz constant of f and the embedding of $\mathbf{L}^2(\Omega)$ into $\mathbf{F}$, and where η is chosen satisfying

$$\eta = C^{-1}[c_1(\|u\|_{\mathbf{L}^\infty(\mathrm{D})})c_2(C_0)]^{-1}, \tag{4.3.42}$$

with C_0 specified in assumption (4.3.37). The estimate (4.3.41) is valid for almost all t, $0 < t < T_0$, by virtue of (4.3.32a,b) and the continuity of U_h on $\mathscr{D}$. Here, the functions c_1 and c_2 are given by (4.3.34). Then, we may rewrite (4.3.40), by the use of (4.3.41) and (4.3.34), to obtain

$$\tfrac{1}{2}\frac{d}{dt}\|H(u)-H(U_h)\|^2_{\mathbf{F}_h}+\tfrac{1}{2}(H(u)-H(U_h),u-U_h)_{\mathbf{L}^2(\Omega)}$$

$$\leqslant \tfrac{1}{2}\eta^{-1}\|H(u)-H(U_h)\|^2_{\mathbf{F}_h}+\left\|(\mathrm{T}-\mathrm{T}_h)\frac{\partial H(u)}{\partial t}\right\|_{\mathbf{M}(\bar\Omega)}\|H(u)-H(U_h)\|_{\mathbf{L}^\infty(\Omega)}$$

$$+\|(\mathrm{T}-\mathrm{T}_h)f(u)\|_{\mathbf{M}(\bar\Omega)}\|H(u)-H(U_h)\|_{\mathbf{L}^\infty(\Omega)}.$$

Integration from $\tau=0$ to $\tau=t$ and application of (4.3.30) yields, for $0<t<T_0$,

$$\|H(u)-H(U_h)\|^2_{\mathbf{F}_h}+\int_0^t (H(u)-H(U_h),u-U_h)_{\mathbf{L}^2(\Omega)}\,d\tau$$

$$\leqslant CT_0h^2\left[\ln\left(\frac{1}{h}\right)\right]^{d(n)}\left\{\|f(u)\|_{\mathbf{L}^\infty((0,T_0);\mathbf{M}(\bar\Omega))}+\left\|\frac{\partial H(u)}{\partial t}\right\|_{\mathbf{L}^\infty((0,T_0);\mathbf{M}(\bar\Omega))}\right\}$$

$$+\eta^{-1}\int_0^t\|H(u)-H(U_h)\|^2_{\mathbf{F}_h}\,d\tau. \tag{4.3.43}$$

Here, we used (4.3.37) and (4.3.32a,b) and have observed that, by (4.3.35b),

$$\|H(u_0)-H(U_h(0))\|_{\mathbf{F}_h}=\|\mathrm{P}_h[H(u_0)-H(U_h(0))]\|_{\mathbf{F}_h}=0.$$

An application of the Gronwall inequality to (4.3.43) yields

$$\|H(u)-H(U_h)\|^2_{\mathbf{L}^\infty((0,T_0);\mathbf{F}_h)}+(H(u)-H(U_h),u-U_h)_{\mathbf{L}^2(\mathscr{D})}$$

$$\leqslant C_1h^2\left[\ln\left(\frac{1}{h}\right)\right]^{d(n)}, \tag{4.3.44}$$

where we have applied (4.3.32a,b). Combining (4.3.44) with (4.3.34) and (4.3.37), we obtain

$$\|H(u)-H(U_h)\|^2_{\mathbf{L}^\infty((0,T_0);\mathbf{F}_h)}+[c_1(\|u\|_{\mathbf{L}^\infty(\mathscr{D})})c_2(C_0)]^{-1}\|u-U_h\|^2_{\mathbf{L}^2(\mathscr{D})}$$

$$\leqslant C_1h^2\left[\ln\left(\frac{1}{h}\right)\right]^{d(n)}, \tag{4.3.45}$$

which implies (4.3.38).

To complete the proof, we must estimate

$$\|(\mathrm{I}-\mathrm{P}_h)[H(u)-H(U_h)]\|_{\mathbf{L}^\infty((0,T_0);\mathbf{F})}.$$

Thus, by Remark 4.3.1,

$$\|(\mathrm{I}-\mathrm{P}_h)[H(u)-H(U_h)]\|_{\mathbf{L}^\infty((0,T_0);\mathbf{F})}$$
$$\leqslant Ch\|H(u)-H(U_h)\|_{\mathbf{L}^\infty((0,T_0);\mathbf{L}^2(\Omega))}$$
$$\leqslant C_1h, \tag{4.3.46}$$

where we have used (4.3.32a) and (4.3.37). Combining the estimate for $\|P_h[H(u) - H(U_h)]\|^2_{F_h}$, given by (4.3.38), with (4.3.20b) and (4.3.46), we obtain (4.3.39). ∎

Remark 4.3.10. We shall elaborate here on the derivation of estimate (4.3.28),† as given in Schatz and Wahlbin [33], for $n \geq 2$. If $p < \infty$ is arbitrary, the estimate

$$\|(T - T_h)f\|_{L^\infty(\Omega)} \leq C \ln\left(\frac{1}{h}\right) h^{2-(n/p)} \|Tf\|_{W^{2,p}(\Omega)} \tag{4.3.47}$$

holds, by combining standard approximation theory with the estimate,

$$\|(T - T_h)f\|_{L^\infty(\Omega)} \leq C_0 \ln\left(\frac{1}{h}\right) \min_{\chi \in M_h} \|Tf - \chi\|_{L^\infty(\Omega)},$$

which has been obtained in Schatz and Wahlbin [33] for Dirichlet problems, and in Scott [36] for Neumann problems. Now the Calderón–Zygmund lemma [11], as employed by Agmon *et al.* [2], gives, for C_1 independent of p,

$$\|(T - T_h)f\|_{L^\infty(\Omega)} \leq C_1 (p h^{-(n/p)}) h^2 \ln\left(\frac{1}{h}\right) \|f\|_{L^p(\Omega)}, \tag{4.3.48}$$

where the explicit linear dependence on p was noted by Johnson and Thomée [22]. A simple calculus minimization of $g(p) = p h^{-(n/p)}$, for $p \geq 1$ and $0 < h < 1$, yields the unique minimum, $en \ln(1/h)$ at $p = n \ln(1/h)$, from which (4.3.28) easily follows, with $d(n) = 2$, $n \geq 2$.

4.4 CONVERGENCE RATES FOR SEMIDISCRETE APPROXIMATIONS‡

We shall derive convergence rates for the fully implicit method applied to a class of degenerate parabolic equations of the form

$$\frac{\partial v}{\partial t} - \Delta u = 0, \qquad v = H(u), \tag{4.4.1}$$

in a format including the porous-medium equation and certain forms of the two-phase Stefan problem.

† The existence of this estimate was brought to the author's attention by Ridgway Scott.

‡ The convergence of the horizontal line method is discussed here.

Definition 4.4.1. Consider the differential equation (4.4.1) in distribution form on a space–time domain $\mathscr{D} = \Omega \times (0, T_0)$. We shall assume that (4.4.1) has a pointwise lifting induced by an isomorphic self-adjoint map, $\mathrm{T}:\mathbf{G}^* \to \mathbf{G}$, onto $\mathbf{G} \subset \mathbf{H}^1(\Omega)$:

$$\frac{\partial \mathrm{T}v}{\partial t} + u = 0, \qquad \text{almost everywhere in } \Omega, \quad 0 < t < T_0, \tag{4.4.2}$$

where we assume explicitly that there is a pair (u, v) satisfying (4.4.2), with $v = H(u)$,[†] and

$$u \in \mathbf{L}^2((0, T_0); \mathbf{H}^1(\Omega)) \cap \mathbf{H}^1([0, T_0]; \mathbf{L}^2(\Omega)), \tag{4.4.3a}$$

$$v \in \mathbf{L}^2(\mathscr{D}), \qquad v:[0, T_0] \to \mathbf{L}^2(\Omega) \quad \text{weakly continuous}, \tag{4.4.3b}$$

H defines a surjective maximal monotone graph with continuous function left inverse, (4.4.3c)

$\mathrm{T}v_{|t=0}$ is specified in the range of T, (4.4.3d)

$$|\xi_1 - \xi_2| \geqslant C|H(\xi_1) - H(\xi_2)|^\gamma, \qquad \text{some} \quad \gamma \geqslant 1; \tag{4.4.3e}$$

here the inequality holds for all elements in $H(\xi_1)$, $H(\xi_2)$, and all $\xi_1, \xi_2 \in \mathbb{R}^1$. Finally, let $\{\mathscr{P}^N\}$ be a sequence of partitions of $[0, T_0]$, and define the fully implicit approximations $\{u_k^N, H(u_k^N)\}_{k=1}^{M(N)}$ in analogy with (2.4.1)–(2.4.2), with $H(u_0^N) = v_{|t=0}$. It is required that $H(u_k^N) \in \mathbf{L}^2(\Omega)$.

Proposition 4.4.1. Let $d_k^N = u(t_k^N) - u_k^N$ and $e_k^N = H(u)(t_k^N) - H(u_k^N)$. Then, the estimate

$$\|e_m^N\|_{\mathbf{G}^*}^2 + C\frac{\gamma}{\gamma+1}\sum_{k=1}^{M(N)} \|e_k^N\|_{\mathbf{L}^{\gamma+1}(\Omega)}^{\gamma+1}(t_k^N - t_{k-1}^N) \leqslant C_1 \int_{\mathscr{D}} \left|\frac{\partial u}{\partial t}\right|^{1+(1/\gamma)} d\mathbf{x}\,dt\, \|\mathscr{P}^N\|^{1+(1/\gamma)} \tag{4.4.4}$$

holds for some positive constant C_1, and C given by (4.4.3e). Under the additional restrictions that $u, v \in \mathbf{L}^\infty(\mathscr{D})$, $v_{|t=0} \in \mathbf{L}^\infty(\Omega)$, and that the pointwise estimates $\|u_k^N\|_{\mathbf{L}^\infty(\Omega)} \leqslant \|H^{-1}v_{|t=0}\|_{\mathbf{L}^\infty(\Omega)}$ hold, as well as the inequality (4.3.34), then we have the inequality

$$\sum_{k=1}^{M(N)} \|d_k^N\|_{\mathbf{L}^2(\Omega)}^2(t_k^N - t_{k-1}^N) \leqslant C_2 \int_{\mathscr{D}} \left|\frac{\partial u}{\partial t}\right|^{1+(1/\gamma)} d\mathbf{x}\,dt\, \|\mathscr{P}^N\|^{1+(1/\gamma)}. \tag{4.4.5}$$

The quantity $\|\mathscr{P}^N\|$, above, denotes the maximal mesh spacing, and the constants C_1 and C_2 contain $C^{-(1/\gamma)}$ as a factor.[‡]

[†] Again, uniqueness is utilized.

[‡] The latter statement has particular relevance for the Stefan problem.

Proof: Using the lifted format, we obtain for the implicit approximations and the evolution equation, respectively, for $k = 1, \ldots, M(N)$,

$$\mathrm{T}[H(u_k^N) - H(u_{k-1}^N)](t_k^N - t_{k-1}^N)^{-1} + u_k^N = 0, \tag{4.4.6a}$$

$$\begin{aligned}
&[\mathrm{T}H(u)(t_k) - \mathrm{T}H(u)(t_{k-1})](t_k^N - t_{k-1}^N)^{-1} + u(t_k^N) \\
&= [\mathrm{T}H(u)(t_k) - \mathrm{T}H(u)(t_{k-1})](t_k^N - t_{k-1}^N)^{-1} - \frac{\partial \mathrm{T}H(u)}{\partial t}(t_k^N) \\
&= (t_k^N - t_{k-1}^N)^{-1} \int_{t_{k-1}^N}^{t_k^N} \left[\frac{\partial \mathrm{T}H(u)}{\partial t}(s) - \frac{\partial \mathrm{T}H(u)}{\partial t}(t_k^N) \right] ds \\
&= (t_k^N - t_{k-1}^N)^{-1} \int_{t_{k-1}^N}^{t_k^N} (t_{k-1}^N - s) \frac{\partial^2 \mathrm{T}H(u)}{\partial t^2}(s)\, ds.
\end{aligned} \tag{4.4.6b}$$

Subtracting (4.4.6a) from (4.4.6b), multiplying by e_k^N, and integrating over Ω, we obtain

$$\begin{aligned}
&(t_k^N - t_{k-1}^N)^{-1} \|e_k^N\|_{\mathbf{G}^*}^2 - (t_k^N - t_{k-1}^N)^{-1} (e_k^N, e_{k-1}^N)_{\mathbf{G}^*} + (e_k^N, d_k^N)_{\mathbf{L}^2(\Omega)} \\
&= (t_k^N - t_{k-1}^N)^{-1} \left(e_k^N, \int_{t_{k-1}^N}^{t_k^N} (s - t_{k-1}^N) \frac{\partial u}{\partial t}(s)\, ds \right)_{\mathbf{L}^2(\Omega)} \\
&= (e_k^N, f_k^N)_{\mathbf{L}^2(\Omega)},
\end{aligned} \tag{4.4.7}$$

where we have noted that $(\partial u)/(\partial t) = -(\partial^2/\partial t^2)\mathrm{T}H(u)$, and where f_k^N is defined by (4.4.7). Now, if C is the constant described in (4.4.3v), then

$$|(e_k^N, f_k^N)_{\mathbf{L}^2(\Omega)}| \leqslant \frac{C}{\gamma + 1} \|e_k^N\|_{\mathbf{L}^{\gamma+1}(\Omega)}^{\gamma+1} + C^{-(1/\gamma)} \frac{\gamma}{\gamma + 1} \|f_k^N\|_{\mathbf{L}^{1+(1/\gamma)}(\Omega)}^{1+(1/\gamma)}, \tag{4.4.8}$$

which follows from the Hölder inequality with $p = \gamma + 1$ and $q = 1 + (1/\gamma)$, followed by the inequality

$$ab \leqslant \frac{a^p}{p} + \frac{b^q}{q}, \qquad a \geqslant 0, \quad b \geqslant 0, \quad \frac{1}{p} + \frac{1}{q} = 1. \tag{4.4.9}$$

Applying (4.4.3e) and (4.4.8) to (4.4.7), and summing over $k = 1, \ldots, m$, we obtain

$$\begin{aligned}
&\tfrac{1}{2} \|e_m^N\|_{\mathbf{G}^*}^2 + C \frac{\gamma}{\gamma + 1} \sum_{k=1}^{m} \|e_k^N\|_{\mathbf{L}^{\gamma+1}(\Omega)}^{\gamma+1} (t_k^N - t_{k-1}^N) \\
&\leqslant C^{-(1/\gamma)} \frac{\gamma}{\gamma + 1} \sum_{k=1}^{m} \|f_k^N\|_{\mathbf{L}^{1+(1/\gamma)}(\Omega)}^{1+(1/\gamma)} (t_k^N - t_{k-1}^N),
\end{aligned} \tag{4.4.10}$$

after multiplication by $(t_k^N - t_{k-1}^N)$. To estimate this right-hand side, we have

$$\begin{aligned}
\|f_k^N\|_{\mathbf{L}^{1+(1/\gamma)}(\Omega)}^{1+(1/\gamma)} &\leqslant \left[(t_k^N - t_{k-1}^N)^{-1} \int_{t_{k-1}^N}^{t_k^N} (s - t_{k-1}^N) \left\|\frac{\partial u}{\partial t}\right\|_{\mathbf{L}^{1+(1/\gamma)}(\Omega)} ds\right]^{1+(1/\gamma)} \\
&\leqslant \left[\left(\int_{t_{k-1}^N}^{t_k^N} ds\right)^{1/(\gamma+1)} \left(\int_{t_{k-1}^N}^{t_k^N} \left\|\frac{\partial u}{\partial t}\right\|_{\mathbf{L}^{1+(1/\gamma)}(\Omega)}^{1+(1/\gamma)} ds\right)^{\gamma/(\gamma+1)}\right]^{1+(1/\gamma)} \\
&\leqslant (t_k^N - t_{k-1}^N)^{(1/\gamma)} \int_{t_{k-1}^N}^{t_k^N} \left\|\frac{\partial u}{\partial t}\right\|_{\mathbf{L}^{1+(1/\gamma)}(\Omega)}^{1+(1/\gamma)} ds. \qquad (4.4.11)
\end{aligned}$$

Applying (4.4.11) to (4.4.10), we obtain (4.4.4). Combining (4.3.34) and the pointwise boundedness hypotheses, we may begin with (4.4.7) and proceed in a similar manner to obtain (4.4.5). ∎

An immediate corollary is the following.

Corollary 4.4.2. Let $H(\cdot)$ and u_0 be given, as in Section 1.2, for the case of the porous-medium equation. Then (4.4.4) and (4.4.5) hold, with $\mathbf{G} = \mathbf{H}_0^1(\Omega)$.

Remark 4.4.1. Note that the rates in $\mathbf{L}^{\gamma+1}(\mathscr{D})$ and $\mathbf{L}^2(\mathscr{D})$, given by (4.4.4) and (4.4.5) for the step-function approximation of $H(u)$ and u, respectively, are $\|\mathscr{P}^N\|^{1/\gamma}$ and $\|\mathscr{P}^N\|^{(1/2)(1+(1/\gamma))}$. Note that Proposition 4.4.1 does not apply directly to the Stefan problem because (4.4.3e), with $\gamma = 1$, fails. It does hold, however, if H is replaced by the smoothed function H_ε of Section 2.1, and, in this case, the constant C of (4.4.3e) satisfies $C = \varepsilon/\lambda$ and $\gamma = 1$. If the line method is applied to the smoothed problem with $u_0^\varepsilon = u_0$ (see Remark 2.1.3), then estimate (4.4.5) yields (under hypothesis (2.1.25b))

$$\|U^{\varepsilon,N} - u^\varepsilon\|_{\mathbf{L}^2(\mathscr{D})} \leqslant C\varepsilon^{-1/2}\|\mathscr{P}^N\| \, \|u_t^\varepsilon\|_{\mathbf{L}^2(\mathscr{D})} \qquad (4.4.12)$$

for the step function $U^{\varepsilon,N}$. If the choice $\varepsilon = \|\mathscr{P}^N\|$ is made in (4.4.12), and it is noted that $\|u_t^\varepsilon\|_{\mathbf{L}^2(\mathscr{D})}$ may be bounded, independently of ε, by standard arguments (see Chapter 2 [9]), there results the estimate

$$\|U^{\varepsilon,N} - u^\varepsilon\|_{\mathbf{L}^2(\mathscr{D})} \leqslant C_0\|\mathscr{P}^N\|^{1/2}. \qquad (4.4.13)$$

If (4.4.13) is combined with (2.1.15b) (under hypothesis (2.1.25b)), we have the estimate, for some constant C,

$$\|U^{\varepsilon,N} - u\|_{\mathbf{L}^2(\mathscr{D})} \leqslant C\|\mathscr{P}^N\|^{1/2}, \qquad \varepsilon = \|\mathscr{P}^N\|, \qquad (4.4.14)$$

for the convergence of the fully implicit method.

4.5 BIBLIOGRAPHICAL REMARKS

The exposition of Section 4.1 is based upon the paper of Dupont and Scott [13]. In the proof of Proposition 4.1.3, the case of equality in (4.1.21) rests upon arguments found in Stein [38, pp. 120–121]. In Dupont and Scott [13], the authors are interested in versions of Proposition 4.1.3, in which the bound does not depend upon p and q. Moreover, elsewhere in the paper, they prove theorems where approximation by complete multidimensional polynomials is relaxed to situations including tensor products and other situations involving mixed indices.

An extremely readable account of an alternative proof of Kaŝin's result, together with companion results on the linear N-width and entropy, has been given by Höllig [19]. The linear N-width is defined with respect to linear operators only. It is larger than the Kolmogorov N-width in certain cases, and perhaps represents a fairer yardstick of comparison for linear approximation processes. A tableau for the linear N-widths differs in two of the four regions from the tableau displayed in Remark 4.2.5, namely, when $p \leqslant q$ and $q \geqslant 2$. For example, in the range $2 \leqslant p \leqslant q$, the linear N-width is of order $N^{-(\ell/n)+(1/p)-(1/q)}$. Though this differs from the Kolmogorov N-width, it corresponds to the upper bound determined by the Sobolev representation formula for piecewise polynomial approximation. For additional details see Höllig [19]. Schumaker [35, Section 6.6] gives independent proofs for the sharpness of the order of piecewise polynomial approximation.

Proposition 4.2.4 is due to Brown [9], who has also shown [10] that the Borsuk theorem, used in the lower-bound estimates of Proposition 4.2.3, is logically equivalent to these estimates. Lemma 4.2.11 has been proved by El Kolli [15]. A variant of Proposition 4.2.8 has been used by Scholz [34], who also basically used the spline space introduced in Definition 4.2.3. The arguments here are the author's, however. The author is indebted to David Fox† for the argument used in the proof of Proposition 4.2.12.

The key result of Section 4.3, Proposition 4.3.4, is proved in the author's paper with Rose (Chapter 2 [9]). The author is indebted to Scott for bringing the paper of Johnson and Thomée [22] to his attention, and to Lars Wahlbin for sketching the argument of Remark 4.3.10. Various forms of Proposition 4.3.1 are due to Aubin [4] and Nitsche [27]. Rannacher and Scott [29] show that the logarithm term is not present in $\mathbf{L}^\infty$ first-derivative estimation, if piecewise-linear elements are used.

A version of Proposition 4.4.1 was proved by the author [21]. A general linear theory for the convergence of the line method for the generators of

† Private communication.

C_0 semigroups and groups has been given by Hersh and Kato [18], and Brenner and Thomée [8].

We close with a very brief description of previous numerical work on the two-phase Stefan problem and the porous-medium equation. For the former, we refer to the existence proof of Kamenomostskaja (Chapter 5 [28]), the paper of Samarskiĭ and Moiseenko [32], and the paper of Meyer [24] as being representative, though far from exhaustive (see the bibliography of (Chapter 2 [9]) for additional references). For the porous-medium equation, we refer to the paper of Rose [31]. The unified error analysis of Proposition 4.3.4, applicable to both the two-phase Stefan problem and the porous-medium equation (see Remark 4.3.8), appears to be a new observation.

REFERENCES

[1] S. Agmon, "Lectures on Elliptic Boundary Value Problems." Van Nostrand-Reinhold, New York, 1965.

[2] S. Agmon, A. Douglis, and L. Nirenberg, Estimates near the boundary for solutions of elliptic partial differential equations satisfying general boundary conditions, *Comm. Pure Appl. Math.* **12**, 623–727 (1959).

[3] R. A. Adams, N. Aronszajn, and K. T. Smith, Theory of Bessel potentials, part II, *Ann. Inst. Fourier (Grenoble)* **17**, 1–135 (1967).

[4] J.-P. Aubin, Approximation des espaces de distributions et des opérateurs differéntiels, *Bull. Soc. Math. France Suppl. Mém.* **12** (1967).

[5] E. F. Beckenbach and R. Bellman, "Inequalities." Cambridge Univ. Press, London and New York, 1961.

[6] J. Bergh and J. Löfstrom, "Interpolation Spaces." Springer-Verlag, Berlin and New York, 1976.

[7] J. H. Bramble and S. R. Hilbert, Estimation of linear functionals on Sobolev spaces with application to Fourier transforms and spline interpolation, *SIAM J. Numer. Anal.* **7**, 112–124 (1970).

[8] P. Brenner and V. Thomée, On rational approximations of semigroups, *SIAM J. Numer. Anal.* **16**, 683–694 (1979); On rational approximations of groups of operators, *ibid.* **17**, 119–125 (1980).

[9] A. Brown, Best *n*-dimensional approximation to sets of functions, *Proc. London Math. Soc.* **14**, 577–594 (1964).

[10] A. Brown, The Borsuk–Ulam theorem and orthogonality in normed linear spaces, *Amer. Math. Monthly* **86**, 766–767 (1979).

[11] A. Calderón and A. Zygmund, On the existence of certain singular integrals, *Acta Math.* **88**, 85–139 (1952).

[12] J. Douglas, T. Dupont, and L. Wahlbin, Optimal L_∞ error estimates for Galerkin approximations to solutions of two-point boundary value problems, *Math. Comp.* **29**, 475–483 (1975).

[13] T. Dupont and R. Scott, Polynomial approximations of functions in Sobolev spaces, *Math. Comp.* **34**, 441–463 (1980).

[14] R. E. Edwards, "Functional Analysis, Theory and Applications." Holt, New York, 1965.

[15] A. El Kolli, Niéme épaisseur dans les espaces de Sobolev, *J. Approx. Theory* **10**, 268–294 (1974).

[16] J. Frehse and R. Rannacher, Eine L^1-Fehlerabschätzung für diskrete Grundlösungen in der Methode der finiten Elemente, *in* Finite Elemente," pp. 92–114. Tagung, Inst. für Angew. Math., Univ. Bonn, 1975.

[17] M. Golomb, Optimal approximating manifolds in L_2-spaces, *J. Math. Anal. Appl.* **12**, 505–512 (1965).

[18] R. Hersh and T. Kato, High-accuracy stable difference schemes for well-posed initial-value problems, *SIAM J. Numer. Anal.* **16**, 670–682 (1979).

[19] K. Höllig, Approximationszahlen von Sobolev-Einbettungen, *Math. Ann.* **242**, 273–281 (1979); Diameters of classes of smooth functions, *in* "Quantitative Approximation," pp. 163–175. Academic Press, New York, 1980.

[20] J. Jerome, On *n*-widths in Sobolev spaces with applications to elliptic boundary value problems, *J. Math. Anal. Appl.* **29**, 201–215 (1970).

[21] J. Jerome, Horizontal line analysis of the multidimensional porous medium equation, *in* "Numerical Analysis" (G. A. Watson, ed.), Lecture Series in Mathematics 773, pp. 64–82. Springer-Verlag, Berlin and New York, 1980.

[22] C. Johnson and V. Thomée, Error estimates for some mixed finite element methods for parabolic type problems, *R.A.I.R.O. Numer. Anal.* **15**, 41–78 (1981).

[23] B. S. Kašin, The widths of certain finite dimensional sets and classes of smooth functions, *Izv. Akad. Nauk SSSR* **41**, 334–351 (1977); *Math. USSR Izv.* **11**, 317–333 (1977).

[24] G. H. Meyer, Multidimensional Stefan problems, *SIAM J. Numer. Anal.* **10**, 522–538 (1973).

[25] C. B. Morrey, "Multiple Integrals in the Calculus of Variations." Springer-Verlag, Berlin and New York, 1966.

[26] L. Nirenberg, "Topics in Nonlinear Functional Analysis." Courant Institute of Mathematical Sciences, New York Univ., New York, 1973–74.

[27] J. Nitsche, Ein Kriterium für die Quasi-Optimalität des Ritzschen Verfahrens, *Numer. Math.* **11**, 346–348 (1968).

[28] J. Nitsche, L_∞ convergence of finite element approximation, *Rennes Conf. Finite Elements, 2nd, Rennes, France* (1975); "Lecture Notes in Mathematics 606," pp. 261–274. Springer-Verlag, Berlin and New York, 1977.

[29] R. Rannacher and R. Scott, Some optimal error estimates for piecewise linear finite element approximations, *Math. Comp.* **38**, 437–445 (1982).

[30] F. Riesz and B. Sz-Nagy, "Functional Analysis." Ungar, New York, 1955.

[31] M. Rose, Numerical methods for flows through porous media. Report ANL-78-80, Argonne National Laboratory, 1978.

[32] A. A. Samarskiĭ and B. D. Moiseenko, An efficient scheme for (the) thorough computation in a many dimensional Stefan problem, *Ž. Vyčisl. Mat. i Mat. Fiz.* **5**, 816–827 (1965) (Russian).

[33] A. H. Schatz and L. B. Wahlbin, On the quasi-optimality in L_∞ of the $\mathring{H}^1$ projection into finite element spaces, *Math. Comp.* **38**, 1–22 (1982).

[34] R. Scholz, Abschätzungen linearer Durchmesser in Sobolev und Besov Räumen, *Manuscripta Math.* **11**, 1–14 (1974).

[35] L. L. Schumaker, "Spline Functions: Basic Theory." Wiley (Interscience), New York, 1980.

[36] R. Scott, Optimal L^∞ estimates for the finite element method on irregular meshes, *Math. Comp.* **30**, 681–697 (1976).

[37] H. S. Shapiro, On some Fourier and distribution theoretic methods in approximation theory, *in* "Approximation Theory III" (E. W. Cheney, ed.), pp. 87–124. Academic Press, New York, 1980.
[38] E. M. Stein, "Singular Integrals and Differentiability Properties of Functions." Princeton University Press, Princeton, New Jersey, 1970.
[39] W. Stenger, On the variational principles for eigenvalues for a class of unbounded operators, *J. Math. Mech.* **17**, 641–648 (1968).
[40] G. Strang, Approximation in the finite element method, *Numer. Math.* **19**, 81–98 (1972).
[41] G. Strang and G. Fix, "An Analysis of the Finite Element Method." Prentice-Hall, Englewood Cliffs, New Jersey, 1973.
[42] A. E. Taylor, "Introduction to Functional Analysis." Wiley, New York, 1961.
[43] A. Weinstein and W. Stenger, "Methods of Intermediate Problems for Eigenvalues." Academic Press, New York, 1972.

5 EXISTENCE ANALYSIS VIA THE STABILITY OF CONSISTENT SEMIDISCRETE APPROXIMATIONS

5.0 INTRODUCTION

In Chapter 2, we examined the pointwise stability of semidiscrete approximation schemes, fully implicit in the monotone part, and fully explicit in the Lipschitz perturbation. These pointwise estimates permit a unified existence analysis (see Theorem 5.2.1) for the two-phase Stefan problem and the porous-medium equation which is carried out in Sections 5.1 and 5.2, via the method of horizontal lines. The appropriate auxiliary estimates are obtained in Section 5.1. For the two-phase Stefan problem, an existence analysis is also presented (see Theorem 5.2.9) in the case when the initial datum, $H(u_0)$, is not necessarily essentially bounded. In fact, convergence of semidiscrete schemes is carried out in Theorem 5.2.9, in the case of convex linear combinations of the fully implicit and fully explicit schemes; the Crank–Nicolson scheme is included as a special case. Pointwise estimates are neither derived nor utilized for this result.

One of the key estimates of this chapter is contained in Proposition 5.1.5. The proof is an original one, though the result itself is a restatement of known $\mathbf{L}^1$ contraction principles in a somewhat different format. This estimate leads directly to the properties of $\partial H(u)/\partial t$ in terms of measures. In Chapter 2, we had occasion to introduce regularizations for the two-phase

Stefan problem involving initial data u_0^ε, not necessarily in $\mathbf{H}^1$. Although the existence theory for such problems is well known, an independent proof could be assembled along the lines of the proof of Proposition 5.2.12, where $u_0 \in \mathbf{L}^2(\Omega)$ is assumed.

Sections 5.3 and 5.4 are sparing in detail. These contain, respectively, the existence analysis for reaction–diffusion systems, and the Navier–Stokes system for incompressible viscous fluids. The only major technical hurdle in Section 5.3 involves the relaxation of the vector field conditions on the invariant region boundary from strict inequality to conditions which permit equality as well.

We mention, finally, that the existence theory of Section 5.2 is sufficiently general and flexible so as to cover models not explicitly described in Chapter 1. Thus, for example, porous media horizontal water flow and certain chemical engineering problems involve formulations for which H' may vanish. Theorem 5.2.1 to follow covers cases such as these and other degenerate cases involving a mixing of various types of singular behavior. We do not discuss the additional regularity of $H(u)$ for $1 < \gamma < 3$.

5.1 STABILITY IN SOBOLEV NORMS FOR SEMIDISCRETIZATIONS OF DEGENERATE PARABOLIC EQUATIONS

In this section we shall derive the stability properties required for an existence theory and an approximation theory for solutions of a class of degenerate parabolic equations, including the two-phase Stefan problem and the porous-medium equation. We shall find it advantageous to correlate, as precisely as possible, the derived stability properties, with the associated features of the degenerate parabolic class, including the regularity of the initial datum, and the particular semidiscrete method chosen for analysis. The existence theory of the next section requires only a stability analysis for one (mixed) implicit method, and, in this case, the results of Chapter 2 lead to the conclusion that the respective analyses of the two-phase Stefan problem and the porous-medium equation essentially merge, except for technical questions associated with the multivalued enthalpy function in the former case. The reason for this merger is that the pointwise stability of the (mixed) implicit scheme leads, in the latter case, to the conclusion that $J = H^{-1}$ is a *de facto* globally Lipschitz continuous monotone function, despite the fact that, for the porous-medium equation, J has superlinear growth at infinity. Here J represents a left inverse.

We shall, in fact, examine those semidiscrete methods which are convex combinations of the fully implicit method and the (purely) explicit method, including the Crank–Nicolson method. For these methods, we prove stability in familiar energy, or Sobolev norms for both $\mathbf{L}^2$ and $\mathbf{H}^1$ initial data, although we do not prove pointwise stability, as in Chapter 2. For the estimates of the divided differences in time, however, we require global Lipschitz properties of J, or an equivalent substitute, in terms of pointwise stability, together with $\mathbf{H}^1$ initial data. Note that, for most of this section, we assume that the function f, in equation (5.1.1) to follow, is Lipschitz continuous. We later relax this assumption and sketch the modified estimates.

Definition 5.1.1. Let $H(\cdot)$ be a surjective multivalued function on $\mathbb{R}^1$, defined so that a unique left inverse $J = H^{-1}$ exists and is continuous, with $0 \in \operatorname{int}(\operatorname{Dom} H)$ and $0 \in H(0)$, such that $H(\cdot)$ defines a maximal monotone graph in $\mathbb{R}^1 \times \mathbb{R}^1$. Suppose, also, that the induced maximal monotone operator on $\mathbf{L}^2(\Omega)$ (Example 3.2.2) is bounded. Let T denote one of the operators $\mathrm{N}_\sigma(\sigma > 0)$ or D_0. Let f be a Lipschitz continuous function, and let u_0 and $H(u_0)$ be given functions in $\mathbf{L}^2(\Omega)$. Consider the model equation

$$\frac{\partial H(u)}{\partial t} - \Delta u + f(u) = 0 \qquad \text{on} \quad \mathscr{D} = \Omega \times (0, T_0), \tag{5.1.1}$$

in distribution form, with (weakly) prescribed homogeneous Dirichlet or Neumann boundary conditions and initial data in terms of u_0 and $H(u_0)$. In the lifted formalism, we write

$$\frac{\partial \mathrm{T}H(u)}{\partial t} + u + \mathrm{T}f_\sigma(u) = 0 \qquad \text{in} \quad \mathscr{D}, \tag{5.1.2a}$$

$$\mathrm{T}H(u)|_{t=0} = H(u_0), \tag{5.1.2b}$$

where $f_\sigma = f - \sigma \mathrm{id}$ if $\mathrm{T} = \mathrm{N}_\sigma\ (\sigma > 0)$, and $f_\sigma = f$ if $\mathrm{T} = \mathrm{D}_0$, and, for the moment, we understand (5.1.2a) in the distribution differentiated sense. For

$$\tfrac{1}{2} \leqslant \theta_1 \leqslant 1, \qquad 0 \leqslant \theta_2 \leqslant 1, \tag{5.1.3}$$

and $\{\mathscr{P}^N\}$ a sequence of partitions of $[0, T_0]$, we define the admissible semidiscretization schemes of (5.1.2) (and, hence, of (5.1.1)) by $H(u_0^N) = H(u_0)$, and

$$\begin{aligned}(t_k^N - t_{k-1}^N)^{-1}\mathrm{T}[H(u_k^N) - H(u_{k-1}^N)] + \theta_1 u_k^N + (1 - \theta_1)u_{k-1}^N \\ + \theta_2 \mathrm{T}f_\sigma(u_k^N) + (1 - \theta_2)\mathrm{T}f_\sigma(u_{k-1}^N) = 0.\end{aligned} \tag{5.1.4}$$

The following properties are explicitly assumed. If (5.1.5b) to follow holds, then

$$(1 - \varepsilon)\lambda(t_k^N - t_{k-1}^N)^{-1} \geqslant \theta_2(\|f\|_{\mathbf{Lip}} + \sigma), \tag{5.1.5a}$$

for some fixed ε, $0 < \varepsilon \leqslant 1$; for if $\theta_2\|f_\sigma\|_{\mathbf{Lip}} \neq 0$, $H(\cdot)$ is assumed to satisfy

$$[H(\xi) - H(\eta)](\xi - \eta) \geqslant \lambda(\xi - \eta)^2, \qquad \lambda > 0, \tag{5.1.5b}$$

for all $\xi, \eta \in \mathbb{R}^1$. In (5.1.5b), we have identified sets and elements.

Remark 5.1.1. It follows from Proposition 3.2.1 that unique solutions u_k^N of (5.1.4) exist in $\mathbf{D}_{\mathrm{T}^{-1}} \subset \mathbf{H}^1(\Omega)$, with $H(u_k^N) \in \mathbf{L}^2(\Omega)$, under assumption (5.1.5a). Note that $H(u_k^N)$ is uniquely determined.

Remark 5.1.2. The stability analysis proceeds by showing, first, that $\{u_k^N\}_{k,N}$ and, hence, $\{H(u_k^N)\}$ lies in a fixed ball of $\mathbf{L}^2(\Omega)$. A significant technical obstacle is overcome by using the format of (5.1.4), rather than (the weak form of) the equivalent elliptic boundary-value problem, to derive these latter estimates on $\{H(u_k^N)\}$. We shall use the elliptic problems to obtain the gradient estimates, however. The reason for the technical simplification revolves around the fact that $H(w) \notin \mathbf{H}^1(\Omega)$, in general, if $w \in \mathbf{H}^1(\Omega)$. Thus, the standard technique of theoretical numerical analysis, of obtaining an estimate on $H(u_k^N)$ by using this function as a test function in the weak formulation, breaks down. Of course, the standard resolution of this dilemma is a smoothing of H, such as introduced in Chapter 2, and realized for general H by the Yosida approximation. This smoothing technique gives satisfactory bounds for $\{H(u_k^N)\}$ in the case of the fully implicit scheme $\theta_1 = 1$, but apparently not otherwise without further restrictions on H. The use of the lifted format, thus, provides greater generality as well as greater simplicity.

Proposition 5.1.1. Consider the semidiscrete scheme defined by (5.1.4). Under the hypotheses of Definition 5.1.1, solutions u_k^N exist in $\mathbf{D}_{\mathrm{T}^{-1}} \subset \mathbf{H}^1(\Omega)$, and are stable in $\mathbf{L}^2(\Omega)$. More precisely, there is a constant C, not depending on k or N, such that

$$\|u_k^N\|_{\mathbf{L}^2(\Omega)} \leqslant C, \qquad k = 1, \ldots, M(N), \qquad \text{all} \quad N. \tag{5.1.6a}$$

If $\theta_1 > \frac{1}{2}$, then the estimate

$$\sum_{k=1}^{M(N)} \|u_k^N - u_{k-1}^N\|_{\mathbf{L}^2(\Omega)}^2 \leqslant C, \qquad \text{all} \quad N, \tag{5.1.6b}$$

holds for some C.

Proof: We multiply (5.1.4) by $u_k^N - u_{k-1}^N$ and integrate over Ω to obtain

$$\theta_1 \|u_k^N\|^2_{\mathbf{L}^2(\Omega)} - (1-\theta_1)\|u_{k-1}^N\|^2_{\mathbf{L}^2(\Omega)} - (2\theta_1 - 1)(u_k^N, u_{k-1}^N)_{\mathbf{L}^2(\Omega)}$$
$$+ \left(\mathrm{T}\left[\frac{H(u_k^N) - H(u_{k-1}^N)}{t_k^N - t_{k-1}^N} + \theta_2(f_\sigma(u_k^N) - f_\sigma(u_{k-1}^N)) \right], u_k^N - u_{k-1}^N \right)_{\mathbf{L}^2(\Omega)}$$
$$= -(\mathrm{T}f_\sigma(u_{k-1}^N), u_k^N - u_{k-1}^N)_{\mathbf{L}^2(\Omega)}. \tag{5.1.7}$$

The fourth term may be seen to dominate the term

$$\varepsilon\lambda(t_k^N - t_{k-1}^N)^{-1}(\mathrm{T}(u_k^N - u_{k-1}^N), u_k^N - u_{k-1}^N)_{\mathbf{L}^2(\Omega)}, \tag{5.1.8}$$

if the nonnegativity of T is combined with the monotonicity of $H(\cdot)$, and with (5.1.5a), when $\theta_2 f_\sigma \neq 0$. If the inequality

$$|(v, w)| \leqslant \frac{\delta}{2}\|u\|^2 + \frac{\delta^{-1}}{2}\|v\|^2, \qquad \delta > 0, \tag{5.1.9}$$

is applied to the third term in (5.1.7), with $\delta = 1$, and if the dominance of (5.1.8) by the fourth term is utilized, we obtain

$$\tfrac{1}{2}\|u_k^N\|^2_{\mathbf{L}^2(\Omega)} - \tfrac{1}{2}\|u_{k-1}^N\|^2_{\mathbf{L}^2(\Omega)} + \lambda\varepsilon(t_k^N - t_{k-1}^N)^{-1}(T(u_k^N - u_{k-1}^N), u_k^N - u_{k-1}^N)_{\mathbf{L}^2(\Omega)}$$
$$\leqslant |(\mathrm{T}f_\sigma(u_{k-1}^N), u_k^N - u_{k-1}^N)_{\mathbf{L}^2(\Omega)}|, \tag{5.1.10a}$$

when $\theta_2 f_\sigma \neq 0$. By formally setting $\varepsilon = 0$, we see that the corresponding inequality holds when $\theta_2 = 0$ and $f_\sigma \neq 0$. Also, we have the simpler inequality

$$\tfrac{1}{2}\|u_k^N\|^2_{\mathbf{L}^2(\Omega)} - \tfrac{1}{2}\|u_{k-1}^N\|^2_{\mathbf{L}^2(\Omega)} \leqslant 0, \tag{5.1.10b}$$

when $f_\sigma = 0$. If (5.1.9) is applied to the right-hand side of (5.1.10a), in the "negative" norm, $\|v\| = (\mathrm{T}v, v)^{1/2}_{\mathbf{L}^2(\Omega)}$, with the choice $\delta^{-1} = \lambda\varepsilon(t_k^N - t_{k-1}^N)^{-1}$, we obtain

$$\tfrac{1}{2}\|u_k^N\|^2_{\mathbf{L}^2(\Omega)} - \tfrac{1}{2}\|u_{k-1}^N\|^2_{\mathbf{L}^2(\Omega)} + (\lambda\varepsilon(t_k^N - t_{k-1}^N)^{-1}/2)(\mathrm{T}(u_k^N - u_{k-1}^N), u_k^N - u_{k-1}^N)_{\mathbf{L}^2(\Omega)}$$
$$\leqslant C(t_k^N - t_{k-1}^N)[\|f_\sigma\|^2_{\mathbf{Lip}}\|u_{k-1}^N\|^2_{\mathbf{L}^2(\Omega)} + |f_\sigma(0)|^2|\Omega|], \tag{5.1.11}$$

where $C = (\lambda\varepsilon)^{-1}C_0^2$ and C_0 is the norm of the injection of $\mathbf{L}^2(\Omega)$ into the ("negatively" normed) dual space. Summing (5.1.11) over $k = 1, \ldots, m$, we obtain

$$\|u_m^N\|^2_{\mathbf{L}^2(\Omega)} \leqslant C_1 + C_2 \sum_{k=0}^{m-1} \|u_k^N\|^2_{\mathbf{L}^2(\Omega)}(t_{k+1}^N - t_k^N), \tag{5.1.12}$$

where $C_1 = 2CT_0|f_\sigma(0)|^2|\Omega| + \|u_0\|^2_{\mathbf{L}^2(\Omega)}$, and $C_2 = 2C\|f_\sigma\|^2_{\mathbf{Lip}}$, for $m = 1, \ldots, M(N)$. If the version of the discrete Gronwall inequality contained in Remark 2.2.1 is applied to (5.1.12) (cf. (2.2.11b)), we obtain

$$\|u_k^N\|^2_{\mathbf{L}^2(\Omega)} \leqslant (C_1 + C_2\|u_0\|^2_{\mathbf{L}^2(\Omega)}T_0)\exp(C_2 T_0), \tag{5.1.13}$$

for $k = 1, \ldots, M(N)$ and all N. This yields (5.1.6a) if $f_\sigma \neq 0$. If $f_\sigma = 0$, we sum (5.1.10b), instead of (5.1.11), and the result (5.1.6a) is clearly immediate with $C = \|u_0\|_{\mathbf{L}^2(\Omega)}$. The proof of (5.1.6b) proceeds similarly. In this case, the left-hand side of (5.1.10) is augmented by the term

$$(\theta_1 - \tfrac{1}{2})\|u_k^N - u_{k-1}^N\|^2_{\mathbf{L}^2(\Omega)}. \quad \blacksquare$$

The hypothesis that H defines a bounded operator on $\mathbf{L}^2(\Omega)$, in conjunction with the previous proposition, yields the following.

Corollary 5.1.2. Under the hypotheses of Definition 5.1.1, there exists a constant C, such that

$$\|H(u_k^N)\|_{\mathbf{L}^2(\Omega)} \leqslant C, \qquad k = 1, \ldots, M(N), \qquad \text{all} \quad N. \tag{5.1.14}$$

Remark 5.1.3. Suppose f is permitted the more general representation $f = g + h$, where h is a Lipschitz continuous function, and g is a monotone Lipschitz continuous function, vanishing at 0 and satisfying

$$[g(\xi) - g(\eta)][\xi - \eta] \geqslant c(\xi - \eta)^2, \qquad c > 0. \tag{5.1.15}$$

The analog of (5.1.4) is now given by

$$\begin{aligned}(t_k^N - t_{k-1}^N)^{-1}\mathrm{T}[H(u_k^N) - H(u_{k-1}^N)] + \theta_1 u_k^N + (1 - \theta_1)u_{k-1}^N \\ + \mathrm{T}g(u_k^N) + \theta_2 \mathrm{T}h_\sigma(u_k^N) + (1 - \theta_2)\mathrm{T}h_\sigma(u_{k-1}^N) = 0,\end{aligned} \tag{5.1.16}$$

where we may conveniently choose σ in (5.1.16) to have the value c in (5.1.15). The estimates (5.1.6) and (5.1.14) are valid in this case also, provided $\|f\|_{\mathbf{Lip}} + \sigma$ in the statement of (5.1.5a) is replaced by $\|h\|_{\mathbf{Lip}}$, and provided (5.1.5b) holds when $\theta_2\|h\|_{\mathbf{Lip}} \neq 0$, though $\|f_\sigma\|_{\mathbf{Lip}}$ is retained in the estimates (5.1.11), etc. The motivation for (5.1.16) is the semidiscretization (2.2.18), which is the differential equation corresponding to (5.1.16) with $\theta_1 = 1$ and $\theta_2 = 0$. Although g was only required to be locally Lipschitz continuous in the earlier format, the case of essentially bounded initial data leads, via the estimates of Section 2.2, to a *de facto* Lipschitz continuous function.

The next step is the derivation of gradient estimates.

Proposition 5.1.3. Under the hypotheses of Definition 5.1.1 and the additional hypothesis $u_0 \in \mathbf{D}_{\mathrm{T}^{-1}}$, there is a constant C, such that, for the solutions u_k^N of the semidiscrete scheme (5.1.4),

$$\|u_k^N\|_{\mathbf{H}^1(\Omega)} \leqslant C, \qquad k = 1, \ldots, M(N), \qquad \text{all} \quad N. \tag{5.1.17}$$

Proof: We use the weak differential equation formulation corresponding to (5.1.4): $H(u_0^N) = H(u_0)$, and

$$(t_k^N - t_{k-1}^N)^{-1}(H(u_k^N) - H(u_{k-1}^N), v)_{\mathbf{L}^2(\Omega)} + \theta_1(\nabla u_k^N, \nabla v)_{\mathbf{L}^2(\Omega)} \\ + (1-\theta_1)(\nabla u_{k-1}^N, \nabla v)_{\mathbf{L}^2(\Omega)} + \theta_2(f(u_k^N), v)_{\mathbf{L}^2(\Omega)} + (1-\theta_2)(f(u_{k-1}^N), v)_{\mathbf{L}^2(\Omega)} \\ = 0, \tag{5.1.18}$$

for all $v \in \mathbf{D}_{\mathrm{T}^{-1}}$. Setting $v = u_k^N - u_{k-1}^N$ in (5.1.18), and using an analog of (5.1.7) and (5.1.10a), we obtain, in analogy with (5.1.11),

$$\tfrac{1}{2}\|\nabla u_k^N\|_{\mathbf{L}^2(\Omega)}^2 - \tfrac{1}{2}\|\nabla u_{k-1}^N\|_{\mathbf{L}^2(\Omega)}^2 + (\lambda\varepsilon(t_k^N - t_{k-1}^N)^{-1}/2)\|u_k^N - u_{k-1}^N\|_{\mathbf{L}^2(\Omega)}^2 \\ \leqslant C(t_k^N - t_{k-1}^N)[\|f\|_{\mathbf{Lip}}^2\|u_{k-1}^N\|_{\mathbf{L}^2(\Omega)}^2 + |f(0)|^2|\Omega|], \tag{5.1.19}$$

where now C has the value $C = (\lambda\varepsilon)^{-1}$. Note that f_σ is replaced by f in the nonlifted formalism, and that a weaker form of (5.1.5a) has been used. If (5.1.19) is summed from $k = 1$ to $k = m$, the right-hand side is seen to be bounded, independent of m and N, by (5.1.6), and there is obtained immediately the gradient estimate

$$\|\nabla u_m^N\|_{\mathbf{L}^2(\Omega)}^2 \leqslant \|\nabla u_0\|_{\mathbf{L}^2(\Omega)}^2 + C, \tag{5.1.20}$$

for some positive constant C. The gradient estimate (5.1.20), in conjunction with the $\mathbf{L}^2$ estimate (5.1.6), yields (5.1.17). ■

Corollary 5.1.4. Under the hypotheses of Definition 5.1.1 and the additional hypothesis $u_0 \in \mathbf{D}_{\mathrm{T}^{-1}}$, there exists a constant C, such that

$$\sum_{k=1}^{M(N)} (t_k^N - t_{k-1}^N)^{-1}\|u_k^N - u_{k-1}^N\|_{\mathbf{L}^2(\Omega)}^2 \leqslant C, \tag{5.1.21}$$

for all N.

Proof: Use (5.1.19) as a starting point, and sum from $k = 1$ to $k = m$, as in the proof of the preceding proposition. The left-hand side of (5.1.20) is then augmented by a constant multiple $\varepsilon\lambda/2$ of the left-hand side of (5.1.21). ■

Remark 5.1.4. The content of Remark 5.1.3 applies to the results of Proposition 5.1.3 and Corollary 5.1.4, if (5.1.18) is replaced by

$$(t_k^N - t_{k-1}^N)^{-1}(H(u_k^N) - H(u_{k-1}^N), v)_{\mathbf{L}^2(\Omega)} + \theta_1(\nabla u_k^N, \nabla v)_{\mathbf{L}^2(\Omega)} \\ + (1-\theta_1)(\nabla u_{k-1}^N, \nabla v)_{\mathbf{L}^2(\Omega)} + (g(u_k^N), v)_{\mathbf{L}^2(\Omega)} + \theta_2(h(u_k^N), v)_{\mathbf{L}^2(\Omega)} \\ + (1-\theta_2)(h(u_{k-1}^N), v)_{\mathbf{L}^2(\Omega)} = 0, \quad \text{for all} \quad v \in \mathbf{D}_{\mathrm{T}^{-1}}. \tag{5.1.22}$$

Specifically, the estimates (5.1.17) and (5.1.21) hold for the solutions of (5.1.22), with the same provisions as in Remark 5.1.3.

There is an analog of (5.1.21) in the $\mathbf{L}^1$ norm, applied to difference quotients involving $H(\cdot)$. For simplicity, we shall derive these somewhat complicated estimates only in the case of the fully implicit scheme. We first make an important preliminary remark.

Remark 5.1.5. In Chapter 2, we proved the pointwise stability of the mixed scheme corresponding to (5.1.18), with $\theta_1 = 1$ and $\theta_2 = 0$, as *adjusted* by (5.1.22). The proofs given in the special cases described in Sections 2.2 and 2.4 may be unified into a single proof, valid for equations described in Definition 5.1.1. Thus, in the following proposition, we shall assume that there exists a constant $c = c(\|u_0\|_{\mathbf{L}^\infty(\Omega)})$, such that, for $\theta_1 = 1$ and $\theta_2 = 0$,

$$\|u_k^N\|_{\mathbf{L}^\infty(\Omega)} \leqslant c, \tag{5.1.23}$$

$k = 1, \ldots, M(N)$, $N \geqslant 1$. Now, (5.1.5b) may be relaxed, in this case, to the local condition $\|J_{|H([-c,c])}\|_{\mathbf{Lip}} \leqslant 1/\lambda$, described by

$$[H(\xi) - H(\eta)](\xi - \eta) \geqslant \lambda(\xi - \eta)^2, \qquad |\xi|,|\eta| \leqslant c, \tag{5.1.24}$$

where c is prescribed by (5.1.23), and $J = H^{-1}$. This has the important implication, noted at the outset of this section, that (5.1.21) holds, when (5.1.5b) is replaced by (5.1.24). Thus, under (5.1.24), Proposition 5.1.5, to follow, applies to semidiscretizations of both the two-phase Stefan problem and the porous-medium equation. Some additional regularity is required of u_0 in the following proposition, viz., $u_0 \in \mathbf{W}^{2,1}(\Omega)$. We also require $H(u_0)$ to be the $\mathbf{L}^2$ limit of admissible smoothings to preclude further assumptions on u_0.

Proposition 5.1.5. Consider the equation (5.1.2), and consider the corresponding semidiscretization (5.1.4), with $\theta_1 = 1$ and $\theta_2 = 0$. Suppose the assumptions of Definition 5.1.1, excluding (5.1.5b), hold, with the following additional assumptions:

Hypotheses (5.1.23) and (5.1.24) hold; (5.1.25a)

$$u_0 \in \mathbf{D}_{\mathbf{T}^{-1}} \cap \mathbf{W}^{2,1}(\Omega); \tag{5.1.25b}$$

$H(u_0) \in \mathbf{L}^\infty(\Omega)$ and $H(u_0) = (\mathbf{L}^2) \lim_{\varepsilon \to 0} H_\varepsilon(u_0)$, for an admissible smoothing H_ε (see Definition 5.1.2); (5.1.25c)

$$(t_k^N - t_{k-1}^N) = \delta t^N,\ k = 1, \ldots, M(N) \text{ (uniform partition).} \tag{5.1.25d}$$

Then, there is a constant C, such that

$$(\delta t^N)^{-1}\|H(u_k^N) - H(u_{k-1}^N)\|_{\mathbf{L}^1(\Omega)} \leqslant C, \tag{5.1.26}$$

for all N and $k = 1, \ldots, M(N)$.

Proof: Define, for $1 \leqslant k < M(N)$,

$$\omega_k^N = \frac{H(u_k^N) - H(u_{k+1}^N)}{u_k^N - u_{k+1}^N}, \qquad \{u_k^N \neq u_{k+1}^N\}. \tag{5.1.27}$$

Certain technical aspects of the proof require the assumptions that ω_k^N has an essential pointwise bound, independent of k and N, and $\Delta u_0 \in \mathbf{L}^{p_0}(\Omega)$, some $p_0 > 1$. Since there is no *a priori* reason why the former must be the case, and since we wish to minimize the assumptions on Δu_0, we are led to a smoothing argument, in which u_0 is approximated in $\mathbf{D}_{\mathrm{T}^{-1}} \cap \mathbf{W}^{2,1}(\Omega)$ by a function u_0^ε in $\mathbf{C}^\infty(\bar{\Omega})$, and H is replaced by a smoothing, say, its Yosida approximation

$$H_\varepsilon = \frac{1}{\varepsilon}(I - K_\varepsilon) = \frac{1}{\varepsilon}[I - (I + \varepsilon H)^{-1}], \qquad \varepsilon > 0, \tag{5.1.28}$$

in which we may unambiguously interpret the inversion in (5.1.28) either as the inversion of a function, or the inversion of an $\mathbf{L}^2$-operator. The function H_ε is an admissible approximation in the sense of Definition 5.1.2, to follow, and is monotone and Lipschitz continuous, with Lipschitz constant $1/\varepsilon$. It is now a standard fact (see Lions [33] or Lemma 5.1.6, to follow) that, if $u_k^{\varepsilon,N}$ denote the solutions of (5.1.4) ($\theta_1 = 1$, $\theta_2 = 0$), with H replaced by H_ε and u_0 by u_0^ε, then sequences $\{u_k^{\varepsilon_\nu,N}\}$ and $\{H_{\varepsilon_\nu}(u_k^{\varepsilon_\nu,N})\}$ exist, such that $u_k^{\varepsilon_\nu,N} \to u_k^N$ (in $\mathbf{L}^2(\Omega)$) and $H_{\varepsilon_\nu}(u^{\varepsilon_\nu,N}) \rightharpoonup H(u_k^N)$ (weakly in $\mathbf{L}^2(\Omega)$) as $\varepsilon_\nu \to 0$. Clearly, the latter sequence also converges weakly in $\mathbf{L}^1(\Omega)$. Strictly speaking, we have complicated the usual argument by introducing u_0^ε. However, this smoothing sequence can be chosen, so that $H_\varepsilon(u_0^\varepsilon) \to H(u_0)$ in $\mathbf{L}^2(\Omega)$, by the Lipschitz property of H_ε (choose $\|u_0^\varepsilon - u_0\|_{\mathbf{L}^2(\Omega)} < \varepsilon^2$) and the assumed property $H_\varepsilon(u_0) \to H(u_0)$ in $\mathbf{L}^2(\Omega)$ (see Brézis, Chapter 3 [8] in the case of the Yosida approximation), where $H(u_0)$, here, must be defined uniquely in the case of the Yosida approximation by choosing the real number of minimal modulus in the closed convex set $H(u_0(x))$, $x \in \Omega$. As noted in (5.1.25c), other choices of admissible smoothings correspondingly determine $H(u_0)$. It follows, finally, that, if estimates of the form

$$(\delta t^N)^{-1}\|H_\varepsilon(u_k^{\varepsilon,N}) - H_\varepsilon(u_{k-1}^{\varepsilon,N})\|_{\mathbf{L}^1(\Omega)} \leqslant C \tag{5.1.29}$$

can be derived, in which C is independent of ε, k, and N, then the lower semicontinuity of the $\mathbf{L}^1$ norm, with respect to weak convergence, implies (5.1.26), when the limit infimum of (5.1.29) is taken. Note that C is actually permitted to depend upon quantities, such as u_0^ε, which are strongly convergent as $\varepsilon \to 0$, since slight adjustments yield a constant independent of ε. The point of introducing the smoothing is the obvious inequality $\omega_k^{\varepsilon,N} \leqslant 1/\varepsilon$ for all k, N, and the relation $\Delta u_0^\varepsilon \in \mathbf{L}^2(\Omega)$. For ε sufficiently small, one sees that $\omega_k^{\varepsilon,N} \geqslant \lambda/2$,† where λ is given in (5.1.24), and

$$\|u_0 - u_0^\varepsilon\|_{\mathbf{D}_{\mathrm{T}}^{-1} \cap \mathbf{W}^{2,1}(\Omega)} \leqslant \tfrac{1}{2}\|u_0\|_{\mathbf{D}_{\mathrm{T}}^{-1} \cap \mathbf{W}^{2,1}(\Omega)}.$$

For maximum clarity, we shall simply prove the result in the unsmoothed case, when there exists an upper bound λ_1 for ω_k^N, and when $\Delta u_0 \in \mathbf{L}^2(\Omega)$, i.e., in the cases

$$\lambda \leqslant \omega_k^N \leqslant \lambda_1, \tag{5.1.30a}$$

$$\Delta u_0 \in \mathbf{L}^2(\Omega), \tag{5.1.30b}$$

and we shall derive the estimate (5.1.23), with C independent of λ_1. We may extend the definition of ω_k^N to all of Ω, by defining ω_k^N to have the value λ on $\{u_k^N = u_{k+1}^N\}$.

With the assumption (5.1.30), the proof naturally divides into two parts. In the first part, we prove that the verification of (5.1.26) reduces to the verification of the case $k = 1$. For this part of the proof, we find it convenient to introduce the auxiliary problems, for $k = M(N) - 1, \ldots, 1$, given by

$$\omega_k^N(\delta t^N)^{-1}(\zeta_{k-1}^N - \zeta_k^N) - \Delta\zeta_{k-1}^N = 0, \qquad \zeta_{k-1}^N \in \mathbf{D}_{\mathrm{T}^{-1}}, \tag{5.1.31a}$$

which are fully implicit discretizations of a backward, or dual, parabolic problem, and are understood in the weak sense. We specify the terminal value $\zeta_{M(N)}$ by

$$\zeta_{M(N)}^N = \operatorname{sgn}(u_{M(N)-1}^N - u_{M(N)}^N). \tag{5.1.31b}$$

Prior to utilizing the auxiliary problems (5.1.31), we note the key *a priori* estimate

$$\|\zeta_k^N\|_{\mathbf{L}^\infty(\Omega)} \leqslant 1, \qquad k = 1, \ldots, M(N), \quad N \geqslant 1. \tag{5.1.32}$$

Inequality (5.1.32) follows, as in the proof of Proposition 2.2.4, via the choice $v = \Theta_{j,q-1}(\zeta_{k-1}^N)$ in the weak formulation corresponding to (5.1.31). The analog of (2.2.40) is the inequality

$$\|\omega_k^N\zeta_{k-1}^N\|_{\mathbf{L}^q(\Omega)} \leqslant \|\omega_k^N\zeta_k^N\|_{\mathbf{L}^q(\Omega)}, \tag{5.1.33}$$

where q is an arbitrary even integer. Now, using the bounds (5.1.30a) in (5.1.33), and letting $q \to \infty$, we obtain

$$\|\zeta_{k-1}^N\|_{\mathbf{L}^\infty(\Omega)} \leqslant \|\zeta_k^N\|_{\mathbf{L}^\infty(\Omega)} \leqslant \|\zeta_{M(N)}^N\|_{\mathbf{L}^\infty(\Omega)} = 1, \tag{5.1.34}$$

which implies (5.1.32).

† In the case of the Yosida approximation, a sharper lower bound of λ holds.

For notational convenience, we introduce the backward difference operator

$$\eta_k = \delta_k \xi_k = \xi_k - \xi_{k-1}, \tag{5.1.35}$$

which maps $\{\xi_k\}_0^L$ onto $\{\eta_k\}_1^L$. The familiar summation-by-parts formula takes the form

$$\sum_{k=1}^{L} (\delta_k a_k) b_k + \sum_{k=1}^{L} (\delta_k b_k) a_{k-1} = a_L b_L - a_0 b_0 := a_j b_j \Big]_{j=0}^{j=L}, \tag{5.1.36}$$

so that, with $L = M(N) - 1$,

$$a_k = \zeta_{k+1}^N, \qquad b_k = H(u_k^N) - H(u_{k+1}^N),$$

we obtain from (5.1.36) the identity

$$\begin{aligned}(H(u_j^N) - H(u_{j+1}^N), \zeta_{j+1}^N)_{\mathbf{L}^2(\Omega)}\Big]_{j=0}^{j=M(N)-1} &= \sum_{k=1}^{M(N)-1} (\omega_k^N \delta_k(\zeta_{k+1}^N), u_k^N - u_{k+1}^N)_{\mathbf{L}^2(\Omega)} \\ &\quad + \sum_{k=1}^{M(N)-1} (\delta_k(H(u_k^N) - H(u_{k+1}^N)), \zeta_k^N)_{\mathbf{L}^2(\Omega)}.\end{aligned} \tag{5.1.37}$$

If (5.1.31a) is applied to the first term on the right-hand side of (5.1.37), and the defining equation (5.1.4) is applied to the second term ($\theta_1 = 1, \theta_2 = 0$), we obtain

$$(H(u_j^N) - H(u_{j+1}^N), \zeta_{j+1}^N)_{\mathbf{L}^2(\Omega)}\Big]_{j=0}^{j=M(N)-1} = \delta t^N \sum_{k=1}^{M(N)-1} (\delta_k f(u_k^N), \zeta_k^N)_{\mathbf{L}^2(\Omega)}, \tag{5.1.38}$$

so that

$$\begin{aligned}&(\delta t^N)^{-1} \|\delta_{M(N)} H(u_{M(N)}^N)\|_{\mathbf{L}^1(\Omega)} \\ &\quad \leqslant (\delta t^N)^{-1} \|\delta_1 H(u_1^N)\|_{\mathbf{L}^1(\Omega)} + \sum_{k=1}^{M(N)-1} \|\delta_k f(u_k^N)\|_{\mathbf{L}^1(\Omega)}.\end{aligned} \tag{5.1.39}$$

The second term on the right-hand side of (5.1.39) may be estimated by

$$\begin{aligned}\sum_{k=1}^{M(N)-1} \|\delta_k f(u_k^N)\|_{\mathbf{L}^1(\Omega)} &\leqslant \|f\|_{\mathbf{Lip}} \sum_{k=1}^{M(N)-1} \|\delta_k u_k^N\|_{\mathbf{L}^1(\Omega)} \\ &\leqslant \tfrac{1}{2} \|f\|_{\mathbf{Lip}} \left\{ T_0 + |\Omega| \sum_{k=1}^{M(N)-1} (\delta t^N)^{-1} \|\delta_k u_k^N\|_{\mathbf{L}^2(\Omega)}^2 \right\}.\end{aligned} \tag{5.1.40}$$

If (5.1.21) is applied to (5.1.40), this sum is seen to be bounded by a constant independent of k, N, and λ_1, although dependent on λ. Altogether, we have verified the case $k = M(N)$ of the inequality

$$(\delta t^N)^{-1}\|H(u_{k+1}^N) - H(u_k^N)\|_{\mathbf{L}^1(\Omega)} \leqslant (\delta t^N)^{-1}\|H(u_1^N) - H(u_0^N)\|_{\mathbf{L}^1(\Omega)} + C. \quad (5.1.41)$$

The verification for other values of k involves the obvious modifications in (5.1.31b) and in the choice of L (see (5.1.36)). The derivation of (5.1.41), thus, completes the reduction of the first part of the proof.

We introduce two transformations to facilitate the derivation of the estimate

$$(\delta t^N)^{-1}\|H(u_1^N) - H(u_0^N)\|_{\mathbf{L}^1(\Omega)} \leqslant C, \qquad N \geqslant 1. \quad (5.1.42)$$

For $\delta \leqslant 1$, and for $g \in \mathbf{L}^1(\Omega)$, set $\mathrm{S}_\delta^- g = -\mathrm{S}_\delta^+(-g)$, where

$$\mathrm{S}_\delta^+(g)(x) = \begin{cases} g(x), & g(x) \geqslant \dfrac{1}{\delta}, \\ \dfrac{1}{\delta}, & \text{otherwise.} \end{cases} \quad (5.1.43)$$

For $1 < p \leqslant 2$, and for $v \in \mathbf{L}^p(\Omega)$, $v \neq 0$, set

$$\mathrm{I}_p(v) = \frac{v|v|^{p-2}}{\|v\|_{\mathbf{L}^p(\Omega)}^{p-1}}. \quad (5.1.44)$$

The operator I_p maps the set $\mathbf{L}^p(\Omega)\backslash\{0\}$ into the unit sphere in $\mathbf{L}^{p'}(\Omega)$, where $(1/p) + (1/p') = 1$. Moreover,† $I_p \to \mathrm{sgn}(\cdot)$, $p \downarrow 1$, in the following sense. For $v \in \mathbf{L}^2(\Omega)$, $\mathrm{I}_p(v) \to \mathrm{sgn}(v)$ pointwise, with $\mathrm{I}_p(v)$ dominated by an $\mathbf{L}^2(\Omega)$ function. Indeed, the domination is evident from the inequality

$$\|v\|_{\mathbf{L}^p(\Omega)}^{p-1}|\mathrm{I}_p(v)| \leqslant \begin{cases} 1, & |v| \leqslant 1, \\ |v|, & |v| \geqslant 1. \end{cases}$$

Moreover, $\mathrm{I}_p \circ \mathrm{S}_\delta^\pm$ is a mapping of $\mathbf{H}^1(\Omega)$ into itself as induced by the composition of two Lipschitz continuous functions, with an element of $\mathbf{H}^1(\Omega)$ and a constant multiplier. Note that, here, we use the fact that the function $\xi \mapsto \xi|\xi|^{p-2}$, $|\xi| \geqslant 1/\delta$, is Lipschitz continuous for $p \leqslant 2$. Also, the crucial inequality

$$\nabla v \cdot \nabla(\mathrm{I}_p \circ \mathrm{S}_\delta^\pm \circ v) \geqslant 0 \quad (5.1.45)$$

holds for $v \in \mathbf{H}^1(\Omega)$, $\delta \geqslant 1$, and $1 < p \leqslant 2$.

We are now ready to verify (5.1.42). We begin with (5.1.18), with $k = 1$, $\theta_1 = 1$, and $\theta_2 = 0$:

$$(\delta t^N)^{-1}(H(u_1^N) - H(u_0), v)_{\mathbf{L}^2(\Omega)} + (\nabla u_1^N, \nabla v)_{\mathbf{L}^2(\Omega)} + (f(u_0), v)_{\mathbf{L}^2(\Omega)} = 0. \quad (5.1.46)$$

† We have used the Italic symbol here, since I_p clearly acts as a function in this statement.

We select $v = \mathrm{I}_p \circ \mathrm{S}_\delta^\pm(u_1^N - u_0)$ in (5.1.46), neglect the energy term, via (5.1.45), after addition and subtraction of u_0, and apply Hölder's inequality, in conjunction with integration by parts, to obtain, for $1 < p \leqslant 2$,

$$
\begin{aligned}
(\delta t^N)^{-1}(H(u_1^N) - H(u_0), \mathrm{I}_p \circ \mathrm{S}_\delta^\pm(u_1^N - u_0))_{\mathbf{L}^2(\Omega)} \\
\leqslant |(\Delta u_0 + f(u_0), \mathrm{I}_p \circ \mathrm{S}_\delta^\pm(u_1^N - u_0))_{\mathbf{L}^2(\Omega)}| \\
\leqslant \|\Delta u_0\|_{\mathbf{L}^p(\Omega)} + C\|u_0\|_{\mathbf{L}^p(\Omega)} + C_1,
\end{aligned}
$$

where $C = \|f\|_{\mathbf{Lip}}$ and $C_1 = |f(0)|\,|\Omega|^{1/p}$. Letting $\delta \to 0$ gives

$$
(\delta t^N)^{-1}(H(u_1^N) - H(u_0), \mathrm{I}_p(u_1^N - u_0)^\pm)_{\mathbf{L}^2(\Omega)} \leqslant \|\Delta u_0\|_{\mathbf{L}^p(\Omega)} + C\|u_0\|_{\mathbf{L}^p(\Omega)} + C_1. \tag{5.1.47}
$$

Multiplying (5.1.47) by $|\Omega|^{-1/p}$ and letting $p \downarrow 1$ give

$$
(\delta t^N)^{-1}\|[H(u_1^N) - H(u_0)]^\pm\|_{\mathbf{L}^1(\Omega)} \leqslant \|\Delta u_0\|_{\mathbf{L}^1(\Omega)} + C\|u_0\|_{\mathbf{L}^1(\Omega)} + |f(0)|\,|\Omega|, \tag{5.1.48}
$$

after multiplying by $|\Omega|$. Note that we have used, here, the Lebesgue dominated-convergence theorem, and the fact that $u_1^N - u_0$ and $H(u_1^N) - H(u_0)$ have the same sign. Note that (5.1.42) follows directly from (5.1.48). Estimate (5.1.26) follows directly from (5.1.41) and (5.1.42). ∎

We shall now derive gradient estimates for the case $\theta_1 = 1$, $\theta_2 = 0$, under the strong monotonicity condition (5.1.5b), where the condition $u_0 \in \mathbf{D}_{\mathrm{T}^{-1}} \subset \mathbf{H}^1(\Omega)$ is relaxed to $u_0 \in \mathbf{L}^2(\Omega)$. Since smoothing is an indispensible tool in this analysis, we introduce a class of admissible smoothings.

Definition 5.1.2. Let $H(\cdot)$ be a multivalued surjective function, defining a maximal monotone graph on $\mathbb{R}^1 \times \mathbb{R}^1$, with continuous left inverse J, so that $J \circ H = \mathrm{id}$. We shall say that $\{H_\varepsilon\}_{0 < \varepsilon \leqslant \varepsilon*}$ is an admissible smoothing for H, provided the following conditions hold.

(1) H_ε and $J_\varepsilon = H_\varepsilon^{-1}$ are monotone Lipschitz continuous functions with $H_\varepsilon' \leqslant c/\varepsilon$, $c > 0$.
(2) For all $v \in \mathbf{L}^2(\Omega)$, $J_\varepsilon(v) \to J(v)$ in $\mathbf{L}^2(\Omega)$ as $\varepsilon \to 0$.
(3) If H satisfies (5.1.5b), then $H_\varepsilon' \geqslant \lambda/2$ for ε sufficiently small.
(4) If $0 \in \mathrm{int}(\mathrm{Dom}\, H)$ and $0 \in H(0)$, then $H_\varepsilon(0) = 0$.
(5) For each $v \in \mathbf{L}^2(\Omega)$, there exists a constant $C = C(v)$, such that $\|H_\varepsilon(v)\|_{\mathbf{L}^2(\Omega)} \leqslant (1 + C)\|H(v)\|_{\mathbf{L}^2(\Omega)}$.

Lemma 5.1.6. Under the hypotheses of Definition 5.1.2, suppose that $u^\varepsilon \to u$ in $\mathbf{L}^2(\Omega)$ (strongly), and $H_\varepsilon(u^\varepsilon) \rightharpoonup \chi$ in $\mathbf{L}^2(\Omega)$ (weakly). Then $\chi = H(u)$ for an admissible $\mathbf{L}^2(\Omega)$ selection $H(u)$.

Proof: Since H defines a maximal monotone operator H[†] in $\mathbf{L}^2(\Omega)$, it suffices to show that

$$(\chi - H(v), u - v)_{\mathbf{L}^2(\Omega)} \geqslant 0, \qquad \text{for all} \quad (v, H(v)) \in \textbf{graph}\ (\mathrm{H}), \quad (5.1.49)$$

which shows that (u, χ) is an element of the graph of H. To verify (5.1.49), we write

$$\begin{aligned}(\chi - H(v), u - v)_{\mathbf{L}^2(\Omega)} &= (\chi - H_\varepsilon(u^\varepsilon), u - v)_{\mathbf{L}^2(\Omega)} + (H_\varepsilon(u^\varepsilon) - H(v), u - u^\varepsilon)_{\mathbf{L}^2(\Omega)} \\ &\quad + (H_\varepsilon(u^\varepsilon) - H(v), J_\varepsilon \circ H_\varepsilon(u^\varepsilon) - J_\varepsilon \circ H(v))_{\mathbf{L}^2(\Omega)} \\ &\quad + (H_\varepsilon(u^\varepsilon) - H(v), J_\varepsilon \circ H(v) - J \circ H(v))_{\mathbf{L}^2(\Omega)} \\ &\geqslant (\chi - H_\varepsilon(u^\varepsilon), u - v)_{\mathbf{L}^2(\Omega)} + (H_\varepsilon(u^\varepsilon) - H(v), u - u^\varepsilon)_{\mathbf{L}^2(\Omega)} \\ &\quad + (H_\varepsilon(u^\varepsilon) - H(v), J_\varepsilon \circ H(v) - J \circ H(v))_{\mathbf{L}^2(\Omega)},\end{aligned}$$

for all $\varepsilon > 0$. Letting $\varepsilon \to 0$ gives (5.1.49), if we use (2) of Definition 5.1.2, together with the convergence properties of u^ε and $H_\varepsilon(u^\varepsilon)$. Note that $H_\varepsilon(u^\varepsilon)$ is necessarily bounded in $\mathbf{L}^2(\Omega)$. ■

Remark 5.1.6. The Yosida approximation of (5.1.25) is admissible (see Crandall and Pazy [12]); so also is the special approximation of Section 2.1.

Definition 5.1.3. We shall consider a function $H(\cdot)$ and an operator T, satisfying the hypotheses of Definition 5.1.1. Independent of θ_2, we assume (5.1.5b) or a substitute, such as (5.1.24). Specify the semidiscretization of (5.1.1) ($\theta_1 = 1$, $\theta_2 = 0$), given in weak form by $H(u_0^N) = H(u_0)$, and

$$\begin{aligned}&(t_k^N - t_{k-1}^N)^{-1}(H(u_k^N) - H(u_{k-1}^N), v)_{\mathbf{L}^2(\Omega)} + (\nabla u_k^N, \nabla v)_{\mathbf{L}^2(\Omega)} \\ &\quad + (f(u_{k-1}^N), v)_{\mathbf{L}^2(\Omega)} = 0, \qquad \text{for all} \quad v \in \mathbf{D}_{\mathrm{T}^{-1}}. \qquad (5.1.50)\end{aligned}$$

Here, f is assumed to be a Lipschitz continuous function, and u_0 and $H(u_0)$ are assumed to be in $\mathbf{L}^2(\Omega)$.

Proposition 5.1.7. Under the assumptions of Definition 5.1.3, there exists a constant C, such that

$$\sum_{k=1}^{M(N)} \|u_k^N\|_{\mathbf{H}^1(\Omega)}^2 (t_k^N - t_{k-1}^N) \leqslant C, \qquad (5.1.51)$$

for all N, where the $\{u_k^N\}$ are defined by (5.1.50).

Proof: By employing an admissible smoothing, with $H_\varepsilon(u_0^{\varepsilon,N}) = H(u_0)$, combined with Lemma 5.1.6 and standard lower semicontinuity arguments, it suffices to obtain (5.1.51) for a differentiable H (say H_ε), where C is not to

[†] For notational convenience, we conceive of H as defining function composition rather than operator composition for most of this proof.

depend upon ε. More precisely, we shall obtain the estimate (5.1.51), under the assumption

$$\lambda \leqslant H' \leqslant \lambda_1, \tag{5.1.52}$$

with C independent of λ_1. Using (5.1.50) as a starting point, with H now understood to satisfy (5.1.52), we may set $v = H(u_k^N)$, sum on $k = 1, \ldots, m$, and obtain by standard methods, for $\|\mathscr{P}^N\| \leqslant T_0$,

$$\begin{aligned}\tfrac{1}{4}\|H(u_m^N)\|^2_{\mathbf{L}^2(\Omega)} &+ \lambda \sum_{k=1}^{M(N)} \|\nabla u_k^N\|^2_{\mathbf{L}^2(\Omega)}(t_k^N - t_{k-1}^N)\\ &\leqslant \tfrac{1}{2}\|H(u_0)\|^2_{\mathbf{L}^2(\Omega)} + 2|f(0)|^2|\Omega|^2 T_0^2\\ &\quad + 2C^2 T_0 \sum_{k=0}^{m-1} \|u_k^N\|^2_{\mathbf{L}^2(\Omega)}(t_{k+1}^N - t_k^N)\\ &\quad + \frac{1}{4T_0}\sum_{k=1}^{m-1} \|H(u_k^N)\|^2_{\mathbf{L}^2(\Omega)}(t_k^N - t_{k-1}^N),\end{aligned} \tag{5.1.53}$$

where C is a Lipschitz constant for f. Note that the inequalities (5.1.9) and (5.1.52) have been used here. If the inequality

$$\|H(u_m^N)\|^2_{\mathbf{L}^2(\Omega)} \geqslant \lambda^2 \|u_m^N\|^2_{\mathbf{L}^2(\Omega)}$$

is used in (5.1.53), followed by the discrete Gronwall inequality of Remark 2.2.1, we obtain the estimate (5.1.51). ■

The final estimate of this section deals with the appropriate weakening of (5.1.21).

Proposition 5.1.8. Under the hypotheses of Definition 5.1.3, there is a constant C, such that

$$\sum_{k=1}^{M(N)} (t_k^N - t_{k-1}^N)^{-1} \|u_k^N - u_{k-1}^N\|^2_{\mathbf{D}_\mathrm{T}} \leqslant C, \tag{5.1.54}$$

for all N. Here, $\mathbf{D}_\mathrm{T}$ is the dual space with norm $\|\cdot\|_{\mathbf{D}_\mathrm{T}} = (\mathrm{T}(\cdot), (\cdot))^{1/2}_{\mathbf{L}^2(\Omega)}$ on $\mathbf{L}^2(\Omega)$.

Proof: This result is a corollary of the proof of Proposition 5.1.1. Indeed, the hypothesis that (5.1.5b) holds permits us to use (5.1.11) as a starting point. If we proceed, as in the proof of Proposition 5.1.1, carrying the left-hand side of (5.1.54) from (5.1.11) to (5.1.12), we are led directly to (5.1.54). ■

Remark 5.1.7. Estimates (5.1.26) and (5.1.51) remain intact if $f = g + h$, as in Remark 5.1.3, and (5.1.50) is extended by

$$\begin{aligned}(t_k^N - t_{k-1}^N)^{-1}(H(u_k^N) - H(u_{k-1}^N), v)_{\mathbf{L}^2(\Omega)} + (\nabla u_k^N, \nabla v)_{\mathbf{L}^2(\Omega)}\\ + (g(u_k^N), v)_{\mathbf{L}^2(\Omega)} + (h(u_{k-1}^N), v)_{\mathbf{L}^2(\Omega)}, \qquad \text{for all } v \in \mathbf{D}_{\mathrm{T}^{-1}}.\end{aligned} \tag{5.1.55}$$

5.2 EXISTENCE OF WEAK SOLUTIONS FOR THE STEFAN PROBLEM AND THE POROUS-MEDIUM EQUATION AND APPROXIMATION RESULTS

The stability estimates of the preceding section permit us to prove the existence of weak solutions of degenerate parabolic equations, including the two-phase Stefan problem and the porous-medium equation as special cases.

Definition 5.2.1. Let $H(\cdot)$ be a surjective multivalued function on $\mathbb{R}^1$, defined so that a unique left inverse $J = H^{-1}$ exists and is continuous, with $0 \in \operatorname{int}(\operatorname{Dom} H)$ and $0 \in H(0)$, such that $H(\cdot)$ defines a maximal monotone graph in $\mathbb{R}^1 \times \mathbb{R}^1$, and such that the induced maximal monotone operators on $\mathbf{L}^2(\Omega)$ and $\mathbf{L}^2(\Omega \times (0, T_0))$ are bounded. Let $f = g + h$, where g is a locally Lipschitz continuous, monotone increasing function on $\mathbb{R}^1$, vanishing at zero, and h is a Lipschitz continuous function on $\mathbb{R}^1$, and let u_0 and $H(u_0)$ be specified functions in $\mathbf{L}^\infty(\Omega)$, where $H(u_0)$ is the $\mathbf{L}^2$ limit of an admissible smoothing (see Definition 5.1.2). We assume that there exists a number $\lambda > 0$, such that

$$(x - y)(\xi - \eta) \geqslant \lambda(\xi - \eta)^2, \qquad |\xi|, |\eta| \leqslant c, \tag{5.2.1a}$$

for every $x \in H(\xi)$ and $y \in H(\eta)$, where $0 < c < \infty$ is defined by

$$c = \frac{1}{\lambda}\left[|h(0)|T_0 + (1 + C)(1 + T_0\|h\|_{\mathbf{Lip}}/\lambda)\|H(u_0)\|_{\mathbf{L}^\infty(\Omega)}\right] \exp\{T_0\|h\|_{\mathbf{Lip}}/\lambda\}, \tag{5.2.1b}$$

where C is the constant of Definition 5.1.2(5). Note that (5.2.1) imposes a Lipschitz condition on $J = H^{-1}$ restricted to the interval $H[-c, c]$. Denote by $\mathbf{V}$ either the space $\mathbf{H}_0^1(\Omega)$ or $\mathbf{H}^1(\Omega)$ and assume that $u_0 \in \mathbf{V}$. By a weak solution u of the formal nonlinear distribution equation of evolution

$$\frac{\partial H(u)}{\partial t} - \Delta u + f(u) = 0 \qquad \text{in} \quad \mathscr{D} = \Omega \times (0, T_0), \tag{5.2.2a}$$

$$H(u(\cdot, 0)) = H(u_0), \tag{5.2.2b}$$

$$u(\cdot, t) \in \mathbf{V}, \qquad \text{for almost all} \quad t \in (0, T_0), \tag{5.2.2c}$$

$$\frac{\partial u}{\partial \nu} = 0 \qquad \text{on} \quad \partial\Omega, \quad \text{if} \quad \mathbf{V} = \mathbf{H}^1(\Omega), \tag{5.2.2d}$$

is meant a pair u, v, satisfying the following properties:

$$v \in \mathbf{L}^{\infty}(\mathscr{D}) \cap \mathbf{H}^1([0, T_0]; \mathbf{V}^*) \quad \text{and} \quad v \text{ is a selection of } H(u); \tag{5.2.3a}$$

$$u \in \mathscr{C}, \quad f(u) \in \mathbf{L}^{\infty}(\mathscr{D}); \tag{5.2.3b}$$

and the relation

$$\int_{\mathscr{D}} \left[v \frac{\partial \psi}{\partial t} - \nabla u \cdot \nabla \psi - f(u)\psi \right] d\mathbf{x}\, dt - \int_{\Omega \times \{T_0\}} v\psi \, d\mathbf{x} + \int_{\Omega \times \{0\}} H(u_0)\psi \, d\mathbf{x} = 0 \tag{5.2.3c}$$

holds for all $\psi \in \mathscr{C}_0$, where

$$\mathscr{C} = \mathbf{L}^{\infty}((0, T_0); \mathbf{V}) \cap \mathbf{H}^1([0, T_0]; \mathbf{L}^2(\Omega)) \cap \mathbf{L}^{\infty}(\mathscr{D}), \tag{5.2.3d}$$

and

$$\mathscr{C}_0 = \mathbf{C}([0, T_0]; \mathbf{V}) \cap \mathbf{H}^1([0, T_0]; \mathbf{L}^2(\Omega)). \tag{5.2.3e}$$

Note that the second integral in (5.2.3c) has a genuine interpretation as a distribution since $v \in \mathbf{C}([0, T_0]; V^*)$ is guaranteed. However, (see Chapter 1 [23] p. 263) under the regularity condition (5.2.3a), it follows that v is weakly continuous from $[0, T_0]$ into $\mathbf{L}^2(\Omega)$.

Theorem 5.2.1. Suppose that the hypotheses of Definition 5.2.1 hold. Then, there exists a unique weak solution pair u, v of (5.2.3). If $u_0 \in \mathbf{W}^{2,1}(\Omega)$, then $v = H(u)$ satisfies the property that $\partial v/\partial t \in \mathbf{L}^{\infty}((0, T_0); \mathbf{M}(\bar{\Omega}))$, where $\mathbf{M}(\bar{\Omega})$ denotes the finite regular Baire measures on $\bar{\Omega}$.

The proof will follow a sequence of preliminary lemmas. The first of these is a slight variant of the Rellich compactness theorem. We quote it in the form useful for our applications. The generalized version, to be applied again later in this section (see Lemma 5.2.11), and in Section 5.4, is known as the Aubin lemma (see Lions [33] p. 58, Chapter 1 [23] p. 271]), and its statement is given in Lemma 5.4.3.

Lemma 5.2.2. A set bounded in both $\mathbf{H}^1([0, T_0]; \mathbf{L}^2(\Omega))$ and $\mathbf{L}^2((0, T_0); \mathbf{V})$ is relatively compact in $\mathbf{L}^2(\mathscr{D})$.

Lemma 5.2.3. Let $\{w_k\}$ be a sequence of functions on $\mathscr{D}$, satisfying

$$w_k \to w \quad (\text{in} \quad \mathbf{L}^2(\mathscr{D})), \tag{5.2.4a}$$

$$H(w_k) \rightharpoonup \chi \quad (\text{weakly in} \quad \mathbf{L}^2(\mathscr{D})). \tag{5.2.4b}$$

Then, $\chi = H(w)$ for an admissible selection.

Proof: It suffices to show that

$$(\chi - H(\phi), w - \phi)_{\mathbf{L}^2(\mathscr{D})} \geqslant 0, \qquad \text{for all} \quad (\phi, H(\phi)) \in \mathbf{graph}(\mathrm{H}) \quad (5.2.5)$$

(cf. Lemma 5.1.6). To verify (5.2.5), we write

$$\begin{aligned}(\chi - H(\phi), w - \phi)_{\mathbf{L}^2(\mathscr{D})} &= (\chi - H(w_k), w - \phi)_{\mathbf{L}^2(\mathscr{D})} \\ &\quad + (H(w_k) - H(\phi), w - w_k)_{\mathbf{L}^2(\mathscr{D})} + (H(w_k) - H(\phi), w_k - \phi)_{\mathbf{L}^2(\mathscr{D})} \\ &\geqslant \lim_{k\to\infty} (\chi - H(w_k), w - \phi)_{\mathbf{L}^2(\mathscr{D})} + \limsup_{k\to\infty} (H(w_k) - H(\phi), w - w_k)_{\mathbf{L}^2(\mathscr{D})},\end{aligned}$$

where we have used the monotonicity of H. Using (5.2.4b) and the Schwarz inequality we obtain

$$\begin{aligned}(\chi - H(\phi), w - \phi)_{\mathbf{L}^2(\Omega)} &\geqslant -\liminf_{k\to\infty} \|H(w_k) - H(\phi)\|_{\mathbf{L}^2(\Omega)} \|w - w_k\|_{\mathbf{L}^2(\Omega)} \\ &\geqslant -C \liminf_{k\to\infty} \|w - w_k\|_{\mathbf{L}^2(\Omega)}, \qquad (5.2.6)\end{aligned}$$

where C denotes an upper bound for $\{\|H(w_k) - H(\phi)\|_{\mathbf{L}^2(\Omega)}\}$. The proof is now completed by applying (5.2.4a) to the right-hand side of (5.2.6). ■

Lemma 5.2.4. Consider the semidiscretization, defined by (2.2.18), rewritten as

$$\begin{aligned}&(t_k^N - t_{k-1}^N)^{-1}([H(u_k^N) - H(u_{k-1}^N)], \phi)_{\mathbf{L}^2(\Omega)} + (\nabla u_k^N, \nabla \phi)_{\mathbf{L}^2(\Omega)} \\ &\quad + (g(u_k^N), \phi)_{\mathbf{L}^2(\Omega)} + (h(u_{k-1}^N), \phi)_{\mathbf{L}^2(\Omega)} = 0, \qquad \text{for all} \quad \phi \in \mathbf{V}.\end{aligned} \quad (5.2.7)$$

Let $\{H_\varepsilon\}$ denote an admissible smoothing (see Definition 5.1.2) of H, and let $\{u_k^{\varepsilon,N}\}$ satisfy $u_0^{\varepsilon,N} = u_0$ and, for $1 \leqslant k \leqslant M(N)$,

$$\begin{aligned}&(t_k^N - t_{k-1}^N)^{-1}([H_\varepsilon(u_k^{\varepsilon,N}) - H_\varepsilon(u_{k-1}^{\varepsilon,N})], \phi)_{\mathbf{L}^2(\Omega)} + (\nabla u_k^{\varepsilon,N}, \nabla \phi)_{\mathbf{L}^2(\Omega)} \\ &\quad + (g(u_k^{\varepsilon,N}), \phi)_{\mathbf{L}^2(\Omega)} + (h(u_{k-1}^{\varepsilon,N}), \phi)_{\mathbf{L}^2(\Omega)} = 0, \qquad \text{for all} \quad \phi \in \mathbf{V}.\end{aligned} \quad (5.2.8)$$

Then, $\{u_k^{\varepsilon,N}\}$ and $\{H_\varepsilon(u_k^{\varepsilon,N})\}$ lie in fixed balls of $\mathbf{L}^\infty(\Omega)$, with radii c and λc, respectively, where c is defined by (5.2.1b). Moreover, $\{u_k^N\}$ and $\{H(u_k^N)\}$ are similarly bounded, with the same constants, and the locally Lipschitz function g may be assumed Lipschitz continuous. Finally, subsequences $\{u_k^{\varepsilon_\nu,N}\}_\nu$ and $\{H_{\varepsilon_\nu}(u_k^{\varepsilon_\nu,N})\}_\nu$ exist, such that for fixed N, k (respectively, $k-1$)

$$u_k^{\varepsilon_\nu,N} \to u_k^N \qquad (\text{in } \mathbf{L}^2(\Omega)), \qquad (5.2.9a)$$

$$H_{\varepsilon_\nu}(u_k^{\varepsilon_\nu,N}) \rightharpoonup H(u_k^N) \qquad (\text{weakly in } \mathbf{L}^2(\Omega)). \qquad (5.2.9b)$$

Proof: Although the estimates (2.2.31) and (2.2.32) were derived under the formal assumption that H satisfies (1.1.8), as described in Remark 1.1.1, with H_ε defined via (2.1.5), the estimates remain valid for H described in

Definition 5.2.1, provided the constant C is inserted in (5.2.1b). This reflects the fact that (2.2.23) is replaced by $u_0^{\varepsilon,N} = u_0$. In fact, $C = 0$ for the smoothing defined by (1.1.8). More generally, (5.2.1b) holds because an admissible smoothing (see Definition 5.1.2) exists for H, and no further properties were used in deriving (2.2.31) and (2.2.32). Thus, we have verified the fact that $\{u_k^{\varepsilon,N}\}$ and $\{H_\varepsilon(u_k^{\varepsilon,N})\}$ lie in fixed closed balls of $\mathbf{L}^\infty(\Omega)$, with radii c and λc, respectively. Note that (5.2.7) and (5.2.8) correspond to (2.2.18) and (2.2.22), respectively, and the existence and uniqueness theory of Proposition 3.2.1 applies. We may obtain a gradient estimate on $\{u_k^{\varepsilon,N}\}$ in standard fashion by setting $\phi = u_k^{\varepsilon,N} - u_{k-1}^{\varepsilon,N}$ in (5.2.8), and using the pointwise estimates already derived. Applying this estimate with Lemma 5.1.6, we obtain (5.2.9), though, as yet, we are not able to identify the limits with the solution pair of (5.2.7). We shall take limits in the lifted version of (5.2.8), i.e.,

$$(t_k^N - t_{k-1}^N)^{-1}\mathrm{T}[H_\varepsilon(u_k^{\varepsilon,N}) - H_\varepsilon(u_{k-1}^{\varepsilon,N})] + u_k^{\varepsilon,N} + \mathrm{T}g(u_k^{\varepsilon,N}) + \mathrm{T}h_\sigma(u_{k-1}^{\varepsilon,N}) = 0. \tag{5.2.10}$$

Here, T represents one of the operators N_σ ($\sigma > 0$) or D_0, and $h_\sigma = h - \sigma\mathrm{id}$. Of course, g may be assumed Lipschitz continuous by extending it as such outside the interval $[-c, c]$. Note that we use the compactness of T as an operator on $\mathbf{L}^2(\Omega)$. Altogether, the net result is the lifted version of (5.2.7), namely:

$$(t_k^N - t_{k-1}^N)^{-1}\mathrm{T}[H(u_k^N) - H(u_{k-1}^N)] + u_k^N + \mathrm{T}g(u_k^N) + \mathrm{T}h_\sigma(u_{k-1}^N) = 0.$$

Finally, we use the facts that the limit in (5.2.9a) lies in the closed ball of radius c in $\mathbf{L}^\infty(\Omega)$, and that the limit in (5.2.9b) lies in the closed ball of radius λc in $\mathbf{L}^\infty(\Omega)$. ■

Remark 5.2.1. For most admissible smoothings, the property

$$\|H(u_0)\|_{\mathbf{L}^2(\Omega)} \leqslant \liminf_{\varepsilon\to 0} \|H_\varepsilon(u_0^{\varepsilon,N})\|_{\mathbf{L}^2(\Omega)}$$

holds, so that the limit functions $\{u_k^N\}$ and $\{H(u_k^N)\}$ actually are expected to lie in (contracted) balls, for which C in (5.2.1b) has value zero.

Lemma 5.2.5. For $\psi \in \mathbf{C}^\infty(\bar{D})$, $\psi(\cdot, t) \in \mathbf{V}$, $0 \leqslant t \leqslant T_0$, set

$$\psi_k^N(\mathbf{x}) = (t_k^N - t_{k-1}^N)^{-1} \int_{t_{k-1}^N}^{t_k^N} \psi(\mathbf{x}, t)\, dt, \qquad k = 1, \ldots, M, \tag{5.2.11}$$

and define the step function

$$\psi^N(\mathbf{x}, t) = \sum_{k=1}^{M} \psi_k^N(\mathbf{x})\omega_k^N(t), \tag{5.2.12a}$$

where

$$\omega_k^N(t) = \begin{cases} 1, & t_{k-1}^N \leqslant t < t_k^N, \\ 0, & \text{otherwise}, \end{cases} \tag{5.2.12b}$$

for $k = 1, \ldots, M$. Similarly, define the translate

$$\chi^N(\mathbf{x}, t) = \sum_{k=1}^{M} \chi_k^N(\mathbf{x}, t)\omega_k^N(t), \tag{5.2.13a}$$

where

$$\chi_k^N(\mathbf{x}, t) = \begin{cases} \psi_{k-1}^N(\mathbf{x}), & 2 \leqslant k \leqslant M, \\ \psi(\mathbf{x}, t), & k = 1. \end{cases} \tag{5.2.13b}$$

Finally, define the difference quotient

$$\zeta^N(\mathbf{x}, t) = \sum_{k=1}^{M} \zeta_k^N(\mathbf{x}, t)\omega_k^N(t) \tag{5.2.14a}$$

where

$$\zeta_k^N(\mathbf{x}, t) = \begin{cases} (t_k^N - t_{k-1}^N)^{-1}(\chi_{k+1}^N(\mathbf{x}) - \psi_{k+1}^N(\mathbf{x})), & 1 \leqslant k \leqslant M - 1, \\ -\dfrac{\partial \psi}{\partial t}(\mathbf{x}, t), & k = M. \end{cases} \tag{5.2.14b}$$

Suppose $\|\mathscr{P}^N\| \to 0$ and that the partitions satisfy (2.2.41c) and the bounded ratio hypothesis:

$$\frac{t_{k+1}^N - t_k^N}{t_k^N - t_{k-1}^N} \leqslant 1 + O(\|\mathscr{P}^N\|), \qquad k = 1, \ldots, M(N) - 1, \quad N \geqslant 1. \tag{5.2.15}$$

Then

$$\psi^N \to \psi \qquad (\text{in} \quad \mathbf{L}^2((0, T_0); \mathbf{V})), \quad N \to \infty, \tag{5.2.16a}$$

$$\chi^N \to \psi \qquad (\text{in} \quad \mathbf{L}^2((0, T_0); \mathbf{V})), \quad N \to \infty, \tag{5.2.16b}$$

$$\zeta^N \to -\frac{\partial \psi}{\partial t} \qquad (\text{in} \quad \mathbf{L}^2(\mathscr{D})), \qquad N \to \infty. \tag{5.2.16c}$$

Proof: By the Schwarz inequality,

$$|\psi_k^N(\mathbf{x})|^2 \leqslant \int_{t_{k-1}^N}^{t_k^N} |\psi(\mathbf{x}, t)|^2 \, dt, \tag{5.2.17a}$$

$$|\nabla \psi_k^N(\mathbf{x})|^2 \leqslant \int_{t_{k-1}^N}^{t_k^N} |\nabla \psi(\mathbf{x}, t)|^2 \, dt, \tag{5.2.17b}$$

for $k = 1, \ldots, M$. It follows that the sequence $\{\psi^N\}$ is pointwise dominated by a function in $\mathbf{L}^2((0, T_0); \mathbf{V})$. The limit (5.2.16a) will follow from the Lebesgue dominated convergence theorem, if we can establish the pointwise estimates

$$\psi^N(\mathbf{x}, t) \to \psi(\mathbf{x}, t), \qquad \frac{\partial \psi^N}{\partial x_i}(\mathbf{x}, t) \to \frac{\partial \psi}{\partial x_i}(\mathbf{x}, t), \tag{5.2.18}$$

almost everywhere in $\mathscr{D}$. However, by the mean value theorem for integrals, if $t_{k-1}^N \leqslant t < t_k^N$, and $i = 1, \ldots, n$,

$$\begin{aligned} \frac{\partial \psi^N}{\partial x_i}(\mathbf{x}, t) = \frac{\partial \psi_k^N}{\partial x_i}(\mathbf{x}) &= (t_k^N - t_{k-1}^N)^{-1} \int_{t_{k-1}^N}^{t_k^N} \frac{\partial \psi}{\partial x_i}(\mathbf{x}, s)\, ds \\ &= \frac{\partial \psi}{\partial x_i}(\mathbf{x}, t_*), \end{aligned} \tag{5.2.19}$$

for some $t_* = t_*(\mathbf{x})$ lying on the interval (t_{k-1}^N, t_k^N). Since a similar representation holds for $\psi^N(\mathbf{x}, t)$, we conclude that (5.2.18) and, hence, (5.2.16a), holds. The limit (5.2.16b) holds, since $\{\chi^N - \psi^N\}$ is convergent to 0 in $\mathbf{L}^2((0, T_0); \mathbf{V})$, as a consequence of (5.2.19) and (5.2.17).

In order to verify (5.2.16c), we must argue somewhat more delicately, since the partitions are not assumed uniform, or even asymptotically uniform. First, we prove that $\{\zeta^N\}$ is uniformly pointwise bounded. Indeed, by two applications of Taylor's theorem, if $t \in [t_{k-1}^N, t_k^N]$ for $1 \leqslant k \leqslant M - 1$,

$$\begin{aligned} -\zeta_k^N(\mathbf{x}, t) = {} & (t_k^N - t_{k-1}^N)^{-1} \Bigg[(t_{k+1}^N - t_k^N)^{-1} \int_{t_k^N}^{t_{k+1}^N} \Bigg[\psi(\mathbf{x}, t_k^N) + \frac{\partial \psi}{\partial t}(\mathbf{x}, t_k^N)(s - t_k^N) \\ & + o((s - t_k^N)^2) \Bigg] ds - (t_k^N - t_{k-1}^N)^{-1} \int_{t_{k-1}^N}^{t_k^N} \Bigg[\psi(\mathbf{x}, t_{k-1}^N) \\ & + \frac{\partial \psi}{\partial t}(\mathbf{x}, t_{k-1}^N)(s - t_{k-1}^N) + o((s - t_{k-1}^N)^2) \Bigg] ds \Bigg] \\ = {} & (t_k^N - t_{k-1}^N)^{-1} [\psi(\mathbf{x}, t_k^N) - \psi(\mathbf{x}, t_{k-1}^N)] \\ & + \frac{1}{2} \Bigg[\frac{\partial \psi}{\partial t}(\mathbf{x}, t_k^N)(t_{k+1}^N - t_k^N)(t_k^N - t_{k-1}^N)^{-1} \\ & - \frac{\partial \psi}{\partial t}(\mathbf{x}, t_{k-1}^N) \Bigg] \\ & + o((t_{k+1}^N - t_k^N)^2 (t_k^N - t_{k-1}^N)^{-1} + o((t_k^N - t_{k-1}^N)) \end{aligned} \tag{5.2.20}$$

holds, and this is clearly uniformly bounded pointwise if the bounded mesh ratio hypothesis (2.2.41c) holds. The constants depend only upon ψ and the mesh ratio. The proof will be completed if we can show that

$$\zeta^N \rightharpoonup -\frac{\partial \psi}{\partial t} \qquad \text{(weakly in} \quad \mathbf{L}^2(\mathscr{D})). \tag{5.2.21}$$

Indeed, if $\tilde{\zeta}^N$ denotes the piecewise linear (in time) interpolant of ζ_k^N, then Lemma 5.2.2 shows that an $\mathbf{L}^2(\mathscr{D})$ gradient estimate on $\tilde{\zeta}^N$, and an estimate of the form (5.2.23) on the first differences of ζ_k^N (i.e., an $\mathbf{L}^2(\mathscr{D})$ estimate on the time derivative of ζ^N) will imply that every subsequence of $\tilde{\zeta}^N$ has a subsequence convergent in $\mathbf{L}^2(\mathscr{D})$. By Lemma 5.2.6, to follow, these limits must uniquely coincide with that defined by (5.2.21). Now, the gradient estimate is trivial from the properties of ψ, whereas an even stronger estimate than (5.2.23) holds, viz., the uniform boundedness of the difference quotients. This latter property follows directly from (5.2.20) and the properties of ψ.

To obtain (5.2.21), we construct the functions

$$\theta^N(\cdot, t) = -\int_0^t \zeta^N(\cdot, s)\, ds + \psi_1^N(\cdot),$$

whose restrictions to $[t_1^N, t_{M-1}^N]$ coincide with the piecewise linear functions

$$\tilde{\theta}^N(\cdot, t) = \frac{t_{k+1}^N - t}{t_k^N - t_{k-1}^N}(\psi_k^N - \psi_{k+1}^N) + \psi_{k+1}^N, \qquad t_k^N \leqslant t < t_{k+1}^N, \quad 0 < k < M - 1,$$

perturbed by an expression of order $O(\|\mathscr{P}^N\|)$. By (5.2.16a) and Lemma 5.2.6, to follow, which depends upon (5.2.16a,b) only, the sequence $\{\tilde{\theta}^N\}$ is weakly convergent to ψ in $\mathbf{L}^2$ $(\mathscr{D})$. We immediately conclude that $\{\tilde{\theta}^N\}$ is (strongly) convergent to ψ in $\mathbf{L}^2(\mathscr{D})$ by Lemma 5.2.2, since $\{\tilde{\theta}^N\}$ is appropriately bounded in the proper topology. The convergence of $\{\tilde{\theta}^N\}$ is transferred to $\{\theta^N\}$, since $\theta^N - \tilde{\theta}^N \to 0$ in $\mathbf{L}^2(\mathscr{D})$. Finally, $(\partial\theta^N/\partial t)$ and $(\partial\theta^N/\partial x_i)$ are pointwise bounded, and hence, bounded in $\mathbf{L}^2(\mathscr{D})$, so that $(\partial\theta^N/\partial t)$ is weakly convergent in $\mathbf{L}^2(\mathscr{D})$ to $(\partial\psi/\partial t)$, i.e., (5.2.21) holds. Note that the pointwise boundedness of $(\partial\theta^N/\partial x_i)$ asserted above follows from obvious modifications of (5.2.20). ■

Lemma 5.2.6. Let $\{w^N\}$ and $\{\tilde{w}^N\}$ denote step function and piecewise linear sequences, defined by

$$w^N(\mathbf{x}, t) = \sum_{k=1}^{M} w_k^N(\mathbf{x})\omega_k^N(t), \tag{5.2.22a}$$

where $\{\omega_k^N\}$ are defined by (5.2.12b), and

$$\tilde{w}^N(\mathbf{x}, t) = \frac{t_{k+1}^N - t}{t_{k+1}^N - t_k^N}(w_k^N(\mathbf{x}) - w_{k+1}^N(\mathbf{x})) + w_{k+1}^N(\mathbf{x}),$$

$$t_k \leqslant t < t_{k+1}, \quad 0 \leqslant k \leqslant M - 1. \tag{5.2.22b}$$

Suppose that $\{w_k^N\} \subset V$, and satisfies

$$\sum_{k=1}^{M} (t_k^N - t_{k-1}^N)^{-1} \|w_k^N - w_{k-1}^N\|_{\mathbf{L}^2(\Omega)}^2 \leqslant C, \qquad N \geqslant 1, \tag{5.2.23}$$

for some constant C. Then, if

$$w^N \rightharpoonup w \qquad (\text{weakly in} \quad \mathbf{L}^2(\mathscr{D})), \tag{5.2.24a}$$

we conclude that

$$\tilde{w}^N \rightharpoonup w \qquad (\text{weakly in} \quad \mathbf{L}^2(\mathscr{D})), \tag{5.2.24b}$$

provided $\|\mathscr{P}^N\| \to 0$.

Proof: It is routine to see that the boundedness of $\{w^N\}$ in $\mathbf{L}^2(\mathscr{D})$ implies the corresponding boundedness of $\{\tilde{w}^N\}$. This does not require (5.2.23). It is now enough to show that every weakly convergent subsequence of $\{\tilde{w}^N\}$ has limit w. Suppose that such a subsequence, labeled again by N, has limit $\tilde{w}$. Thus, if $\psi \in \mathbf{C}^\infty(\bar{\mathscr{D}})$, $\psi(\cdot, t) \in \mathbf{V}$, $t \in [0, T_0]$, and ψ^N is defined as in the previous lemma, we have, since the product of a weakly convergent and convergent sequence is weakly convergent on the class of bounded functions,

$$\begin{aligned}
(w, \psi)_{\mathbf{L}^2(\mathscr{D})} &= \lim_{N\to\infty} (w^N, \psi^N)_{\mathbf{L}^2(\mathscr{D})} \\
&= \lim_{N\to\infty} \sum_{k=1}^{M} (w_k^N, \psi_k^N)_{\mathbf{L}^2(\Omega)} (t_k^N - t_{k-1}^N) \\
&= \lim_{N\to\infty} (\tilde{w}^N, \psi^N)_{\mathbf{L}^2(\mathscr{D})} + \lim_{N\to\infty} \tfrac{1}{2} \sum_{k=0}^{M-1} (w_{k+1}^N - w_k^N, \psi_{k+1}^N)_{\mathbf{L}^2(\Omega)} (t_{k+1}^N - t_k^N) \\
&= (\tilde{w}, \psi)_{\mathbf{L}^2(\mathscr{D})}.
\end{aligned}$$

Here, we have used the inequality

$$\begin{aligned}
&\left| \sum_{k=0}^{M-1} (w_{k+1}^N - w_k^N, \psi_{k+1}^N)_{\mathbf{L}^2(\Omega)} (t_{k+1}^N - t_k^N) \right| \\
&\qquad \leqslant \tfrac{1}{2} \|\mathscr{P}^N\| \left[\sum_{k=0}^{M-1} (t_{k+1}^N - t_k^N)^{-1} \|w_{k+1}^N - w_k^N\|_{\mathbf{L}^2(\Omega)}^2 \right. \\
&\qquad\qquad \left. + \sum_{k=0}^{M-1} \|\psi_{k+1}^N\|_{\mathbf{L}^2(\Omega)}^2 (t_{k+1}^N - t_k^N) \right],
\end{aligned}$$

which clearly tends to zero as $N \to \infty$ by (5.2.23). The remaining statements follow similarly. ∎

Proof of Theorem 5.2.1: Let $\psi \in \mathbf{C}^\infty(\bar{\mathscr{D}})$, $\psi(\cdot, t) \in \mathbf{V}$, $t \in (0, T_0)$. In order to obtain a solution pair $[u, v]$ satisfying (5.2.3(a–e)), we first select a sequence

$\mathscr{P}^N$ of partitions satisfying (5.2.15), (2.2.41c), and $\|\mathscr{P}^N\| \to 0$, and we define $[u_k^N, H(u_k^N)]$ via (5.2.7); we then construct the corresponding step function sequences, as defined by (5.2.22a), which we label $\{u_k^N\}$ and $\{v_k^N\}$, respectively. We shall also have need of the corresponding piecewise linear (in time) sequences $\{\tilde{u}^N\}$ and $\{\tilde{v}^N\}$, defined as in (5.2.22b). Here, $u_0^N = u_0$ and $v_0^N = H(u_0)$.

Using Proposition 5.1.3 and Corollary 5.1.4, as modified by Remark 5.1.4, and Lemmas 5.2.2, 5.2.3, 5.2.4, and 5.2.6, we may obtain functions† $u \in \mathscr{C}$ (see (5.2.3d)) and $v = H(u)$, satisfying (5.2.3a), and subsequences $\{u^{N_i}\}$, $\{\tilde{u}^{N_i}\}$, $\{v^{N_i}\}$, and $\{\tilde{v}^{N_i}\}$, satisfying

$$u^{N_i} \rightharpoonup^* u \qquad (\text{weak * in } \mathbf{L}^\infty(\mathscr{D}) \quad \text{and} \quad \mathbf{L}^\infty((0, T_0); \mathbf{V})), \tag{5.2.25a}$$

$$\tilde{u}^{N_i} \rightharpoonup u \qquad (\text{weakly in } \mathbf{H}^1([0, T_0]; \mathbf{L}^2(\Omega))), \tag{5.2.25b}$$

$$u^{N_i} \to u \qquad (\text{in } \mathbf{L}^2(\mathscr{D})), \tag{5.2.25c}$$

$$v^{N_i} \rightharpoonup^* v \qquad (\text{weak * in } \mathbf{L}^\infty(\mathscr{D})), \tag{5.2.25d}$$

$$\tilde{v}^{N_i} \rightharpoonup v \qquad (\text{weakly in } \mathbf{H}^1([0, T_0]; \mathbf{V}^*)). \tag{5.2.25e}$$

Moreover, if $u_0 \in \mathbf{W}^{2,1}(\Omega)$, and the partitions are selected to be uniform, then, by Proposition 5.1.5, we may select u, so that $(\partial v/\partial t) \in \mathbf{L}^\infty((0, T_0); \mathbf{M}(\bar{\Omega}))$, and $\{\tilde{v}^{N_i}\}$, so that

$$\frac{\partial \tilde{v}^{N_i}}{\partial t} \rightharpoonup^* \frac{\partial v}{\partial t} \qquad (\text{weak * in } \mathbf{L}^\infty((0, T_0); \mathbf{M}(\hat{\Omega}))). \tag{5.2.25f}$$

Note that the boundedness estimate needed for (5.2.25e) is obtained directly from the discrete equations in terms of other quantities already estimated. Moreover, a modified form of Lemma 5.2.6, in which $\mathbf{L}^2(\Omega)$ is replaced by $\mathbf{V}^*$, is employed to equate the limits in (5.2.25d) and (5.2.25e).

We are now ready to proceed to the verification of (5.2.3c). Using (5.2.7) as a starting point, we set $\phi = \psi_k^N$, sum on k, and apply summation by parts to obtain

$$\begin{aligned} &\sum_{k=1}^{M-1} \left(v_k^N, \frac{\psi_k^N - \psi_{k+1}^N}{t_k^N - t_{k-1}^N} \right)_{\mathbf{L}^2(\Omega)} (t_k^N - t_{k-1}^N) + (v_M^N, \psi_M^N)_{\mathbf{L}^2(\Omega)} - (v_0^N, \psi_1^N)_{\mathbf{L}^2(\Omega)} \\ &\quad + \sum_{k=1}^{M-1} (\nabla u_k^N, \nabla \psi_k^N)_{\mathbf{L}^2(\Omega)} (t_k^N - t_{k-1}^N) \\ &\quad + \sum_{k=1}^{M-1} (g(u_k^N) + h(u_{k-1}^N), \psi_k^N)_{\mathbf{L}^2(\Omega)} (t_k^N - t_{k-1}^N) = 0. \end{aligned} \tag{5.2.26}$$

† The equality of the appropriate limits in (5.2.25) follows from the results quoted above. Note that (5.2.25c) depends upon (5.2.23) or, alternatively, upon the weaker estimate (5.1.6b). These estimates relate the convergence of $\{u^{N_i}\}$ and $\{\tilde{u}^{N_i}\}$.

Replacing the discrete format (5.2.26) by the corresponding continuous format gives, in the notation of Lemma 5.2.5,

$$\begin{aligned}\int_0^{t_{M-1}^N} (v^N, \zeta^N)_{\mathbf{L}^2(\Omega)}\,dt + (v_M^N, \psi_M^N)_{\mathbf{L}^2(\Omega)} - (v_0, \psi_1^N)_{\mathbf{L}^2(\Omega)} \\ + \int_0^{t_{M-1}^N} (\nabla u^N, \nabla\psi^N)\,dt + \int_0^{t_{M-1}^N} (g(u^N), \psi^N)_{\mathbf{L}^2(\Omega)}\,dt \\ + \int_{t_1^N}^{t_{M-1}^N} (h(u_*^N), \psi^N)_{\mathbf{L}^2(\Omega)}\,dt + (h(u_0), \psi_1^N)_{\mathbf{L}^2(\Omega)} t_1^N = 0, \qquad (5.2.27)\end{aligned}$$

where we have used the notation u_*^N to designate the same translation relation to u^N as χ^N bears to ψ^N.

Now, the following limit relations are valid.

$$\lim_{N_i\to\infty} \int_0^{T_0} (v^{N_i}, \zeta^{N_i})_{\mathbf{L}^2(\Omega)}\,dt = \int_0^{T_0} \left(v, -\frac{\partial\psi}{\partial t}\right)_{\mathbf{L}^2(\Omega)} dt,$$

since the product, when defined, of a weakly convergent and a convergent sequence in $\mathbf{L}^2(\mathscr{D})$ is weakly convergent. Also,

$$\lim_{N_i\to\infty} \int_{t_{M(N_i)-1}^{N_i}}^{T_0} (v^{N_i}, \zeta^{N_i})_{\mathbf{L}^2(\Omega)}\,dt = 0,$$

since $\{v^{N_i}\}$ is bounded in $\mathbf{L}^2(\mathscr{D})$, and $\{\zeta^{N_i}\}$ is pointwise bounded. Thus,

$$\lim_{N_i\to\infty} \int_0^{t_{M(N_i)-1}^{N_i}} (v^{N_i}, \zeta^{N_i})_{\mathbf{L}^2(\mathscr{D})}\,dt = \int_0^{T_0} \left(v, -\frac{\partial\psi}{\partial t}\right)_{\mathbf{L}^2(\mathscr{D})} dt. \qquad (5.2.28)$$

In a similar way,

$$\lim_{N_i\to\infty} \int_0^{t_{M(N_i)-1}^{N_i}} (\nabla u^{N_i}, \nabla\psi^{N_i})_{\mathbf{L}^2(\Omega)}\,dt = \int_0^{T_0} (\nabla u, \nabla\psi)_{\mathbf{L}^2(\Omega)}\,dt, \qquad (5.2.29)$$

and

$$\lim_{N_i\to\infty} \int_0^{t_{M(N_i)-1}^{N_i}} (g(u^{N_i}), \psi^{N_i})_{\mathbf{L}^2(\Omega)} = \int_0^{T_0} (g(u), \psi)_{\mathbf{L}^2(\Omega)}\,dt, \qquad (5.2.30)$$

the latter relation following, since the *a priori* estimates permit g to be assumed Lipschitz. By difference quotient estimates, we have $\{u^{N_i} - u_*^{N_i}\} \to 0$ as $N_i \to \infty$, so that

$$\lim_{N_i\to\infty} \int_{t_1^N}^{t_{M(N_i)-1}^{N_i}} (h(u_*^{N_i}), \psi^{N_i})_{\mathbf{L}^2(\Omega)}\,dt = \int_0^{T_0} (h(u), \psi)_{\mathbf{L}^2(\Omega)}\,dt. \qquad (5.2.31)$$

The final term in (5.2.27) clearly tends to 0, and the third term tends to $-(v_0, \psi(\cdot, 0))_{\mathbf{L}^2(\Omega)}$, by the Lebesgue dominated convergence theorem. It remains to analyze the second term in (5.2.27). By the fundamental theorem of calculus in reflexive Banach spaces,

$$v_{M(N_i)}^{N_i} - v_0 = \tilde{v}^{N_i}(\cdot, T_0) - \tilde{v}^{N_i}(\cdot, 0) = \int_0^{T_0} \left(\frac{\partial \tilde{v}^{N_i}}{\partial t} \right) dt,$$

and the weak limit in $\mathbf{V}$ may be taken in this relation, by (5.2.25e). This shows that

$$v_{M(N_i)}^{N_i} \rightharpoonup v(\cdot, T_0) \qquad \text{(weakly in } \mathbf{V}). \tag{5.2.32}$$

These remarks, together with (5.2.28)–(5.2.31), permit us to let N tend to infinity through values N_i in (5.2.27), thereby obtaining (5.2.3c).

It remains to show that (5.2.3c) remains valid for all $\psi \in \mathscr{C}_0$. If an interpolation sequence is defined, built upon a sequence $\{\mathscr{P}^N\}$, $\|\mathscr{P}^N\| \to 0$, of partitions of $[0, T_0]$, each member of the sequence can be approximated arbitrarily closely in $\mathscr{C}_0$ by a function in $\mathbf{C}^\infty(\bar{\mathscr{D}}) \cap \mathscr{C}_0$. The convergence of the interpolation sequence in $\mathscr{C}_0$ then yields the existence result. Uniqueness follows from Proposition 1.5.1. ■

Corollary 5.2.7. Under the hypotheses of Section 1.1, a unique weak solution pair $[u, H(u)]$ exists for the two-phase Stefan problem, satisfying (1.1.10) and (1.1.11). Pointwise bounds for u and $H(u)$ are given by c and λc, respectively, where c is defined by (5.2.1) ($C = 0$). If $u_0 \in \mathbf{W}^{2,1}(\Omega)$, then $\partial H(u)/\partial t \in \mathbf{L}^\infty((0, T_0); \mathbf{M}(\bar{\Omega}))$.

Corollary 5.2.8. Under the hypotheses of Section 1.2, a unique nonnegative weak solution u exists, satisfying (1.2.15) and (1.2.16). A pointwise bound for the solution is given by $\|u_0\|_{\mathbf{L}^\infty(\Omega)}$. If $u_0 \in \mathbf{W}^{2,1}(\Omega)$, then $\partial H(u)/\partial t \in \mathbf{L}^\infty_{\text{loc}}((0, \infty); \mathbf{M}(\bar{\Omega}))$.

Proof: The additional arguments revolve about the fact that the class $\mathscr{C}$ in (1.2.16a) involves (vector) functions, defined on the time interval $(0, \infty)$. Corresponding adjustments are necessary in the redefinition of the partitions, which now contain countably (infinite) many points, and in the redefinition of $\mathscr{D}$ as $\mathscr{D} = \Omega \times (0, \infty)$. Note that $\mathbf{V} = H_0^1(\Omega)$ now. In the new notation, the sequences selected in (5.2.25) are selected as before, except that (5.2.25e,f) need only hold locally on $(0, \infty)$, and one additional condition is specified. Corresponding estimates are required, and we address these now. Proposition 2.4.1 directly yields a pointwise estimate, and the estimate

$$\|\nabla u_k^N\|_{\mathbf{L}^2(\Omega)} \leqslant \|\nabla u_0\|_{\mathbf{L}^2(\Omega)}$$

is routinely derived by setting $v = u_k^N - u_{k-1}^N$ in (2.4.2). Thus, we may select sequences satisfying (5.2.25a), (5.2.25c), and (5.2.25d). The estimate

(5.1.21), for any M, holds, with C given by $1/(2\lambda)\|\nabla u_0\|^2_{\mathbf{L}^2(\Omega)}$; here, λ satisfies (5.1.24), with $c = \|u_0\|_{\mathbf{L}^\infty(\Omega)}$, and may be chosen explicitly, via (2.3.21), as $c_1(c_1/c_2)^{\gamma-1}/\|u_0\|^{(1-1/\gamma)}_{\mathbf{L}^\infty(\Omega)}$. It remains to obtain an estimate of the form

$$\sum_{k=1}^{M} \|\nabla u_k^N\|^2_{\mathbf{L}^2(\Omega)}(t_k^N - t_{k-1}^N) \leqslant C, \qquad \text{all} \quad M \quad \text{and} \quad N. \tag{5.2.33}$$

This and the preceding estimate discussed permit the choice of sequences, satisfying (5.2.25b) and the additional condition

$$u^{N_i} \rightharpoonup u \qquad (\text{weakly in} \quad \mathbf{L}^2((0,\infty); \mathbf{H}_0^1(\Omega)).$$

Equation (5.2.33) is routinely obtained by selecting v in (2.4.2), as described by (2.4.7), with $\ell = 1$ and j sufficiently large, in which case $\Theta_{j,\gamma}(H(u_k^N)) = [H(u_k^N)]^\gamma$. The simple inequality

$$\gamma[H(t)]^{\gamma-1}H'(t) \geqslant \gamma^\gamma c_1^\gamma$$

shows that (2.4.8) can be strengthened to

$$(\nabla u_k^N, \nabla v)_{\mathbf{L}^2(\Omega)} \geqslant \gamma^\gamma c_1^\gamma \|\nabla u_k^N\|^2_{\mathbf{L}^2(\Omega)},$$

which implies (5.2.33), if the correspondingly strengthened form of (2.4.9) is used. ■

Remark 5.2.2. The following result will achieve two objectives. First, the approximations, defined via the more general format of (5.1.18), will be shown to converge in $\mathbf{L}^2(\mathscr{D})$ to the unique solution of (5.2.3c). Second, we relax the requirement that $H(u_0)$ is essentially bounded. Note that $u_0 \in \mathbf{V}$ is retained, however. Actually, minor modifications of the proof of Theorem 5.2.1 permit the hypothesis $u_0 \in \mathbf{V}$ to be relaxed to $u_0 \in \mathbf{H}^1(\Omega)$, that is compatibility between the initial datum and the boundary datum need not hold. For simplicity, however, we retain $u_0 \in \mathbf{V}$.

Theorem 5.2.9. We assume that the hypotheses stated in Definition 5.2.1 hold with the following qualifications. The function $H(u_0)$ is required to belong to $\mathbf{L}^2(\Omega)$, rather than $\mathbf{L}^\infty(\Omega)$, $c = \infty$ in (5.2.1a), and f in (5.2.2a) is assumed Lipschitz continuous. Suppose a sequence of partitions $\{\mathscr{P}^N\}$ is specified, satisfying (2.2.41c) and (5.2.15), and $\|\mathscr{P}^N\| \to 0$ as $N \to \infty$. Let θ_1 and θ_2 be specified, with $\frac{1}{2} \leqslant \theta_1 \leqslant 1$ and $0 \leqslant \theta_2 \leqslant 1$, and define $\{u_k^N\}$ as the recursively generated semidiscrete solutions of (5.1.18). Let $\{u^N\}$ and $\{\tilde{u}^N\}$ be the step function sequence and piecewise linear (in time) sequence, as defined in Lemma 5.2.6. Then, both sequences converge in $\mathbf{L}^2(\mathscr{D})$ to a function u. If $v_k^N = H(u_k^N)$ are defined by (5.1.18), and $\{v^N\}$ and $\{\tilde{v}^N\}$ denote the corresponding sequences, then both sequences converge weakly in $\mathbf{L}^2(\mathscr{D})$ to a function $v = H(u)$. The pair $[u, v]$ is a solution of (5.2.3c,e), subject to

$$u \in \mathbf{L}^\infty((0, T_0); \mathbf{V}) \cap \mathbf{H}^1([0, T_0]; \mathbf{L}^2(\Omega)), \tag{5.2.34a}$$

$$v \in \mathbf{L}^2(\mathscr{D}) \cap \mathbf{H}^1([0, T_0]; \mathbf{V}^*). \tag{5.2.34b}$$

Proof: No additional arguments are required. Thus, by the estimates of Corollary 5.1.2, Proposition 5.1.3, and Corollary 5.1.4, together with Lemma 5.2.3, Lemma 5.2.6, and the Rellich compactness property of Lemma 5.2.2, we conclude that subsequences can be found satisfying

$$u^{N_i} \rightharpoonup {}^* u \qquad (\text{weak* in} \quad \mathbf{L}^\infty((0, T_0); \mathbf{V}), \tag{5.2.35a}$$

$$\tilde{u}^{N_i} \rightharpoonup u \qquad (\text{weakly in} \quad \mathbf{H}^1([0, T_0]; \mathbf{L}^2(\Omega)), \tag{5.2.35b}$$

$$u^{N_i} \to u \qquad (\text{in} \quad \mathbf{L}^2(\mathscr{D})), \tag{5.2.35c}$$

$$v^{N_i} \rightharpoonup v \qquad (\text{weakly in} \quad \mathbf{L}^2(\mathscr{D})), \tag{5.2.35d}$$

$$\tilde{v}^{N_i} \rightharpoonup v \qquad (\text{weakly in} \quad \mathbf{H}^1([0, T_0]; \mathbf{V}^*), \tag{5.2.35e}$$

where u and $v = H(u)$ satisfy (5.2.34). The remainder of the proof proceeds as Theorem 5.2.1. Note that the uniqueness of the solution pair $[u, v]$ implies that the subsequences displayed in (5.2.35) may be selected to be the entire sequences. ■

Remark 5.2.3. Note that Theorem 5.2.9 does not cover the porous-medium equation, since (5.2.1a) does not hold with $c = \infty$. Note also that f is required to be Lipschitz continuous, whereas Theorem 5.2.1 permitted a locally Lipschitz monotone increasing perturbation. Moreover, it is worth making explicit the convergence theory used to prove Theorem 5.2.1. We prefer to view the following result as a corollary of both theorems.

Corollary 5.2.10. We assume the hypotheses of Definition 5.2.1 and a sequence of partitions, as defined in Theorem 5.2.9. If $\{u_k^N\}$ are defined via (5.2.7), and $\{u^N\}$, $\{\tilde{u}^N\}$, $\{v^N\}$, $\{\tilde{v}^N\}$ have the meaning of Theorem 5.2.9, then (5.2.35) holds, with $\{N_i\} = \{N\}$. In particular, the solution pair, described in Theorem 5.2.1, is realized as the unique limit of this construction.

Our final result relaxes the requirement that $u_0 \in \mathbf{V}$. We also relax the pointwise boundedness hypothesis as well. However, to achieve this, a stronger compactness criterion than that of Lemma 5.2.2 is needed. We shall state the result in the form necessary for our purposes (see Lemma 5.4.3, to follow).

Lemma 5.2.11. A set bounded in both $\mathbf{H}^1([0, T_0]; \mathbf{V}^*)$ and $\mathbf{L}^2((0, T_0); \mathbf{V})$ is relatively compact in $\mathbf{L}^2(\mathscr{D})$.

Proposition 5.2.12. We assume the hypotheses of Definition 5.2.1 with the following qualifications. The function u_0 is not required to belong to $\mathbf{V}(u_0 \in \mathbf{L}^2(\Omega)$ only), and f is required to be Lipschitz continuous. Then a unique solution pair $[u, v]$, $v = H(u)$, exists for (5.2.3c,e), satisfying

$$u \in \mathbf{L}^2((0, T_0); \mathbf{V}) \cap \mathbf{H}^1([0, T_0]; \mathbf{V}^*), \tag{5.2.36a}$$

$$v \in \mathbf{L}^2(\mathscr{D}) \cap \mathbf{H}^1([0, T_0]; \mathbf{V}^*). \tag{5.2.36b}$$

Proof: We select a sequence of partitions, as in Theorem 5.2.9, and define the sequences of functions, as in Corollary 5.2.10. For the estimates, we use Propositions 5.1.7 and 5.1.8. We also use Lemmas 5.2.3, 5.2.4, and 5.2.11. In addition we use a modified version of Lemma 5.2.6 with $\mathbf{L}^2(\Omega)$ replaced by $\mathbf{V}^*$. Altogether, we may select subsequences, satisfying

$$u^{N_i} \rightharpoonup u \quad \text{(weakly in } \mathbf{L}^2((0, T_0); \mathbf{V})), \tag{5.2.37a}$$

$$\tilde{u}^{N_i} \rightharpoonup u \quad \text{(weakly in } \mathbf{H}^1([0, T_0]; \mathbf{V}^*) \text{ and } \mathbf{L}^2((0, T_0); \mathbf{V})), \tag{5.2.37b}$$

$$\tilde{u}^{N_i} \to u \quad \text{(in } \mathbf{L}^2(\mathscr{D})), \tag{5.2.37c}$$

$$v^{N_i} \rightharpoonup v \quad \text{(weakly in } \mathbf{L}^2(\mathscr{D})), \tag{5.2.37d}$$

$$\tilde{v}^{N_i} \rightharpoonup v \quad \text{(weakly in } \mathbf{H}^1([0, T_0]; \mathbf{V}^*)). \tag{5.2.37e}$$

The remainder of the proof follows that of Theorem 5.2.1 once it has been shown that (5.2.37c) implies the $\mathbf{L}^2(\mathscr{D})$ convergence of the step function sequence. This follows, as has been observed earlier, from estimate (5.1.6b). ■

5.3 EXISTENCE FOR REACTION–DIFFUSION SYSTEMS

The task of this section is to obtain gradient estimates and divided difference estimates in time for the concentration and potential variables, and to relax the strict inequalities in (2.5.6) to simple inequalities. We note that pointwise estimates have been obtained in Section 2.5 (see Proposition 2.5.1) and do not depend upon (2.5.6). They do require that f_i be a bounded, locally Lipschitz continuous function for $i = 1, \ldots, i_0$, and that g_i be a locally Lipschitz continuous, monotone increasing function for $i = i_0 + 1, \ldots, m$, with h_i Lipschitz continuous. The pointwise estimates permit us to assume that these functions are Lipschitz continuous.

Proposition 5.3.1. There exists a constant C, such that, for $i = 1, \ldots, m$,

$$\|\nabla u_{k,i}^N\|_{\mathbf{L}^2(\Omega)}^2 \leqslant C, \qquad \text{for all} \quad N, \quad k = 1, \ldots, M(N), \tag{5.3.1a}$$

and

$$\sum_{k=1}^{M(N)} \|u_{k,i}^N - u_{k-1,i}^N\|_{\mathbf{L}^2(\Omega)}^2 (t_k^N - t_{k-1}^N)^{-1} \leqslant C, \qquad \text{for all} \quad N. \tag{5.3.1b}$$

Here, $\{\mathbf{u}_k^N\}$ denote the (vector) solutions of (2.5.2), guaranteed by Proposition 3.3.3.

Proof: We use (2.5.2) as a starting point, and set $v_i = u_{k,i}^N - u_{k-1,i}^N$. After some estimation, this leads to, for $i = 1, \ldots, i_0$,

$$\begin{aligned}
&\tfrac{1}{2}\|u_{k,i}^N - u_{k-1,i}^N\|_{\mathbf{L}^2(\Omega)}^2 (t_k^N - t_{k-1}^N)^{-1} + \tfrac{1}{2}D_i\|\nabla u_{k,i}^N\|_{\mathbf{L}^2(\Omega)}^2 - \tfrac{1}{2}D_i\|\nabla u_{k-1,i}^N\|_{\mathbf{L}^2(\Omega)}^2 \\
&\quad + \tfrac{1}{2}D_i\|\sqrt{\omega_i}u_{k,i}^N\|_{\mathbf{L}^2(\partial\Omega)}^2 - \tfrac{1}{2}D_i\|\sqrt{\omega_i}u_{k-1,i}^N\|_{\mathbf{L}^2(\partial\Omega)}^2 \\
&\leqslant \tfrac{1}{2}|f_i(0)|^2|\Omega|(t_k^N - t_{k-1}^N) + 2^{m-2}\sum_{j\neq i} c_{j,i}^2\|u_{k-1,j}^N\|_{\mathbf{L}^2(\Omega)}^2 (t_k^N - t_{k-1}^N) \\
&\quad + 2^{m-2}c_{i,i}^2\|u_{k,i}^N\|_{\mathbf{L}^2(\Omega)}^2 (t_k^N - t_{k-1}^N),
\end{aligned} \tag{5.3.2}$$

where the constants $c_{j,i}$ are defined from

$$|f_i(y_1, \ldots, y_m) - f_i(x_1, \ldots, x_m)| \leqslant \sum_{j=1}^{m} c_{j,i}|y_j - x_j|.$$

The corresponding inequality, for $i = i_0 + 1, \ldots, m$, is

$$\begin{aligned}
&\tfrac{1}{2}\|u_{k,i}^N - u_{k-1,i}^N\|_{\mathbf{L}^2(\Omega)}^2 (t_k^N - t_{k-1}^N)^{-1} + \tfrac{1}{2}D_i\|\nabla u_{k,i}^N\|_{\mathbf{L}^2(\Omega)}^2 - \tfrac{1}{2}D_i\|\nabla u_{k-1,i}^N\|_{\mathbf{L}^2(\Omega)}^2 \\
&\quad + \tfrac{1}{2}D_i\|\sqrt{\omega_i}u_{k,i}^N\|_{\mathbf{L}^2(\partial\Omega)}^2 - \tfrac{1}{2}D_i\|\sqrt{\omega_i}u_{k-1,i}^N\|_{\mathbf{L}^2(\partial\Omega)}^2 \\
&\leqslant \tfrac{1}{2}|h_i(0)|^2|\Omega|(t_k^N - t_{k-1}^N) + 2^{m-1}\sum_{j=1}^{m} d_{j,i}^2\|u_{k-1,j}^N\|_{\mathbf{L}^2(\Omega)}^2 (t_k^N - t_{k-1}^N) \\
&\quad + c_i^2\|u_{k,i}^N\|_{\mathbf{L}^2(\Omega)}^2 (t_k^N - t_{k-1}^N),
\end{aligned} \tag{5.3.3}$$

where the constants $d_{j,i}$ and c_i are defined from

$$|h_i(y_1, \ldots, y_m) - h_i(x_1, \ldots, x_m)| \leqslant \sum_{j=1}^{m} d_{j,i}|y_j - x_j|,$$

$$|g_i(t) - g_i(s)| \leqslant c_i|t - s|.$$

By summing (5.3.2) and (5.3.3) from $k = 1$ to $k = r$, and noticing that the right-hand sides are bounded for each $r = 1, \ldots, M(N)$, we obtain (5.3.1b) and (5.3.1a) for those i for which $D_i \neq 0$. A separate argument, which we now give, is required for those indices i for which $D_i = 0$. For such indices, the corresponding functional equations hold in a pointwise sense almost everywhere in Ω. By computing the gradient of these pointwise equations, followed by an $\mathbf{L}^2(\Omega)$ inner product of the resultant with $\nabla u_{k,i}^N$, we obtain, after some estimation,

$$\begin{aligned} \tfrac{1}{2}\|\nabla u_{k,i}^N\|_{\mathbf{L}^2(\Omega)}^2 - \tfrac{1}{2}\|\nabla u_{k-1,i}^N\|_{\mathbf{L}^2(\Omega)}^2 &\leqslant \tfrac{1}{2}\sum_{j\neq i} c_{j,i}^2\|\nabla u_{k-1,j}^N\|_{\mathbf{L}^2(\Omega)}^2(t_k^N - t_{k-1}^N) \\ &\quad + \left(c_{i,i} + \frac{m-1}{2}\right)\|\nabla u_{k,i}^N\|_{\mathbf{L}^2(\Omega)}^2(t_k^N - t_{k-1}^N), \end{aligned} \tag{5.3.4}$$

for $i = i, \ldots, i_0$ $(D_i = 0)$, and

$$\begin{aligned} &\tfrac{1}{2}\|\nabla u_{k,i}^N\|_{\mathbf{L}^2(\Omega)}^2 - \tfrac{1}{2}\|\nabla u_{k-1,i}^N\|_{\mathbf{L}^2(\Omega)}^2 \\ &\quad \leqslant \tfrac{1}{2}\sum_{j=1}^{m} d_{j,i}^2\|\nabla u_{k-1,j}^N\|_{\mathbf{L}^2(\Omega)}^2(t_k^N - t_{k-1}^N) + \frac{m}{2}\|\nabla u_{k,i}^N\|_{\mathbf{L}^2(\Omega)}^2(t_k^N - t_{k-1}^N), \end{aligned} \tag{5.3.5}$$

for $i = i_0 + 1, \ldots, m$ $(D_i = 0)$. Summing (5.3.4) from $k = 1$ to $k = r$, we obtain, after transpositions from the right-hand side of (5.3.4) to the left, and vice versa,

$$\begin{aligned} \|\nabla u_{r,i}^N\|_{\mathbf{L}^2(\Omega)}^2 &\leqslant 2\|\nabla u_{0,i}\|_{\mathbf{L}^2(\Omega)}^2 + 2T_0\sum_{j\neq i} c_{j,i}^2\|\nabla u_{0,j}\|_{\mathbf{L}^2(\Omega)}^2 \\ &\quad + 2\sum_{k=1}^{r-1}\sum_{j:D_j\neq 0} c_{j,i}^2\|\nabla u_{k,j}^N\|_{\mathbf{L}^2(\Omega)}^2 \\ &\quad + 2\sum_{k=1}^{r-1}\sum_{j:D_j=0} c_{j,i}^2\|\nabla u_{k,j}^N\|_{\mathbf{L}^2(\Omega)}^2, \end{aligned}$$

provided

$$[2c_{i,i} + (m-1)](t_k^N - t_{k-1}^N) \leqslant \tfrac{1}{2}, \qquad k = 1, \ldots, M(N). \tag{5.3.6}$$

By the previous part of the proof, we see that

$$\|\nabla u_{r,i}^N\|_{\mathbf{L}^2(\Omega)}^2 \leqslant C + 2\sum_{k=1}^{r-1}\sum_{j:D_j=0} c_{j,i}^2\|\nabla u_{k,j}^N\|_{\mathbf{L}^2(\Omega)}^2(t_k^N - t_{k-1}^N). \tag{5.3.7}$$

An estimate similar to (5.3.7) holds, based upon (5.3.5), if $i = i_0 + 1, \ldots, m$. If we sum over i, such that $D_i = 0$, we have, altogether,

$$\sum_{i:D_i=0}\|\nabla u_{r,i}^N\|_{\mathbf{L}^2(\Omega)}^2 \leqslant mC + 2\left(\sum_{i,j=1}^{m} c_{j,i}^2\right)\sum_{k=1}^{r-1}\sum_{i:D_i=0}\|\nabla u_{k,i}^N\|_{\mathbf{L}^2(\Omega)}^2(t_k^N - t_{k-1}^N),$$

and an application of the discrete Gronwall inequality (cf. Proposition 2.2.1) yields the remaining conclusion, since only a finite number of partitions fail to satisfy (5.3.6). ■

Remark 5.3.1. The proof of Proposition 5.3.1 reveals that the dependence of the constant C in (5.3.1) upon f_i (and its components g_i and h_i) is a multivariable polynomial dependence upon ess $\sup|f_i|$ and ess $\sup|(\partial f_i/\partial u_j)|$. In particular, if f_i is perturbed by a parameter-dependent expression in such a way that the perturbation is bounded, with (essentially) bounded partial derivatives, then the estimates of Propositions 2.5.1 and 5.3.1 continue to remain valid for the semidiscrete solutions of the perturbed system. Similarly, if f_i is redefined over part of its domain, which can also be characterized as a perturbation, similar statements hold. Such a redefinition will be necessary to remove the assumption of strict inequality in (2.5.6).

Definition 5.3.1. Suppose $1 \leqslant i \leqslant i_0$, and let $0 < \varepsilon < (b_i - a_i)/2$ be specified. We define the Lipschitz continuous function f_i^ε as follows. If $\eta \in Q_{a_i}$, then, as a function of η_i', set

$$f_i^\varepsilon(\eta_1, \ldots, \eta_{i-1}, \eta_i', \eta_{i+1}, \ldots, \eta_m) = \begin{cases} \ell_i^\varepsilon(\eta_i'), & a_i \leqslant \eta_i' \leqslant a_i + \varepsilon, \\ f_i(\eta_1, \ldots, \eta_{i-1}, \eta_i', \eta_{i+1}, \ldots, \eta_m), & a_i + \varepsilon \leqslant \eta_i' \leqslant \dfrac{a_i + b_i}{2}, \end{cases}$$

where ℓ_i^ε is the linear function, with value $-\varepsilon + f_i(\eta)$ at $\eta_i' = a_i$, and value $f_i(\eta_1, \ldots, \eta_{i-1}, a_i + \varepsilon, \eta_{i+1}, \ldots, \eta_m)$ at $\eta_i' = a_i + \varepsilon$. A similar construction is carried out on $[(a_i + b_i)/2, b_i]$ for $\eta \in Q_{b_i}$, where ℓ_i^ε is determined to have value $f(\eta_1, \ldots, \eta_{i-1}, b_i - \varepsilon, \eta_{i+1}, \ldots, \eta_m)$ at $b_i - \varepsilon$, and $\varepsilon + f_i(\eta)$ at b_i.

Remark 5.3.2. The quantity f_i^ε is a Lipschitz continuous function and the Lipschitz constants are bounded independently of ε. Important for our purposes is the fact that f_i^ε satisfies (2.5.6). Moreover, f_i^ε satisfies the ordering of sign-regions property (1.3.6d), and f_i^ε is uniformly convergent to f_i as $\varepsilon \to 0$ on $\mathbb{R}^m$, if appropriately extended outside Q.

Lemma 5.3.2. Proposition 2.5.3 remains true if strict inequality in (2.5.6) is relaxed to inequality as in (1.3.6b).

Proof: We consider the corresponding solutions $u_{k,i}^{\varepsilon,N}$ of the semidiscrete system (2.5.2), when f_i is replaced by f_i^ε in (2.5.2a). By the estimates obtained in Proposition 5.3.1, together with Remark 5.3.1, it is possible to obtain sequences, satisfying

$$\mathbf{u}_k^{\varepsilon_\nu,N} \rightharpoonup \mathbf{w}_k^N \qquad (\text{weakly in} \quad \mathbf{H}^1(\Omega;\mathbb{R}^m)), \tag{5.3.8a}$$

$$\mathbf{u}_k^{\varepsilon_\nu,N} \to \mathbf{w}_k^N \qquad (\text{in} \quad \mathbf{L}^2(\Omega;\mathbb{R}^m)), \tag{5.3.8b}$$

with limits $\mathbf{w}_k^N$ in $\mathbf{H}^1(\Omega;\mathbb{R}^m) \cap \mathbf{L}^\infty(\Omega;\mathbb{R}^m)$. However, the limits

$$\begin{aligned}(\mathbf{L}^2) \lim_{\varepsilon_\nu \to 0} & f_i^{\varepsilon_\nu}(u_{k-1,1}^{\varepsilon_\nu,N}, \ldots, u_{k-1,i-1}^{\varepsilon_\nu,N}, u_{k,i}^{\varepsilon_\nu,N}, u_{k-1,i+1}^{\varepsilon_\nu,N}, \ldots, u_{k-1,m}^{\varepsilon_\nu,N}) \\ &= f_i(w_{k-1,1}^N, \ldots, w_{k-1,i-1}^N, w_{k,i}^N, w_{k-1,i+1}^N, \ldots, w_{k-1,m}^N),\end{aligned} \tag{5.3.9}$$

obtained via the triangle inequality, the uniform convergence of f_i^ε, (5.3.8b), and the Lebesgue dominated convergence theorem, together with (5.3.8), show that the $\mathbf{w}_k^N$ are also solutions of (2.5.2a). A similar statement is true for (2.5.2b). Hence, $\mathbf{w}_k^N = \mathbf{u}_k^N$, by uniqueness of solutions of these equations. Since, by Proposition 2.5.3, Range$(\mathbf{u}_k^{\varepsilon_\nu,N}) \subset Q$, for each ν in (5.3.8b), it follows that this property is preserved in passing to the limits. ■

We are now ready for the major theorem of this section.

Theorem 5.3.3. Under the hypotheses of Section 1.3, there exists a unique solution of (1.3.11), with Range$(\mathbf{u}) \subset Q$.

Proof: The identity (1.3.11a) follows from the intermediate identity

$$\begin{aligned}&\int_{\mathscr{D}} \left[-u_i \frac{\partial \psi_i}{\partial t} + D_i \nabla u_i \cdot \nabla \psi_i + f_i(u)\psi_i \right] d\mathbf{x}\, dt \\ &\quad + \int_{\Omega \times \{T_0\}} u_i \psi_i \, d\mathbf{x} - \int_{\Omega \times \{0\}} u_{0,i} \psi_i \, d\mathbf{x} + D_i \int_{\partial\Omega \times (0,T_0)} \omega_i u_i \psi_i = 0,\end{aligned} \tag{5.3.10}$$

for all $\psi \in \mathbf{H}^1(\Omega) \otimes \mathbf{C}^\infty([0, T_0])$, upon integration by parts in the first term in (5.3.10), and use of the denseness of $\mathbf{C}^\infty([0, T_0])$ in $\mathbf{L}^2(0, T_0)$. The verification of (5.3.10), however, follows exactly the lines of the proof of Theorem 5.2.1, with the following qualifications. We treat the terms

$$\int_\Omega \nabla u_i \cdot \nabla \psi_i + \int_{\partial\Omega} \omega_i u_i \psi_i$$

as a unit, since this expression represents an equivalent inner product on $\mathbf{H}^1(\Omega)$ (see (1.3.14)). Also, it is evident that only sequences $\{\mathbf{u}^N\}$ and $\{\tilde{\mathbf{u}}^N\}$ need be considered, in the notation of Lemma 5.2.6, satisfying

$$\mathbf{u}^{N_i} \rightharpoonup {}^*\mathbf{u} \qquad (\text{weak* in} \quad \mathbf{L}^\infty(\mathscr{D};\mathbb{R}^m) \text{ and } \mathbf{L}^\infty((0,T_0);\mathbf{H}^1(\Omega;\mathbb{R}^m))), \tag{5.3.11a}$$

$$\mathbf{u}^{N_i} \to \mathbf{u} \qquad (\text{in} \quad \mathbf{L}^2(\mathscr{D};\mathbb{R}^m)), \tag{5.3.11b}$$

$$\tilde{\mathbf{u}}^{N_i} \rightharpoonup \mathbf{u} \qquad (\text{weakly in} \quad \mathbf{H}^1([0,T_0];\mathbf{L}^2(\Omega;\mathbb{R}^m))), \tag{5.3.11c}$$

where we have used the estimates of Propositions 2.5.1 and 5.3.1, and the Rellich compactness principle. Since, by Lemma 5.3.2, Range($\mathbf{u}^{N_i}$) $\subset Q$, this property is inherited by the limit in (5.3.11b). Thus, using (2.5.2) as a starting point, we may proceed, as in the proof of Theorem 5.2.1, to obtain (5.3.10) and, thus, (1.3.11a). Note that membership in (1.3.11b) follows from (5.3.11). This uniqueness was demonstrated in Proposition 1.5.3. ■

Remark 5.3.3. The pointwise estimates of Proposition 2.5.1, made possible by the assumption of $\mathbf{L}^\infty$ initial data, considerably simplify the existence analysis by permitting us to assume that the components f_i of f are Lipschitz continuous. In the analysis of the FitzHugh–Nagumo system, given by the writer in Jerome [25], this assumption was not made and a weaker form of (5.3.1b) was necessitated. This, in turn, required $\mathbf{L}^1$ limit principles for the rapidly growing components and a substitute for the Rellich compactness criterion.

5.4 EXISTENCE FOR THE GENERALIZED FORM OF THE NAVIER–STOKES EQUATIONS FOR INCOMPRESSIBLE FLUIDS

This section will be relatively brief since the key estimate and essential ideas are already in place. Also, we state the result in function space in the format of Remark 1.4.3, rather than in the more general setting suggested by Example 3.2.3.

Definition 5.4.1. Let $\mathbf{V}$ and $\mathbf{V}_s$ be defined as in (1.4.20), (1.4.21), and let $a(\cdot,\cdot,\cdot)$ be a continuous trilinear form on $\mathbf{V} \times \mathbf{V} \times \mathbf{V}_s$, satisfying (1.4.22b) and (1.4.24). The notion of a weak solution, as defined in Definition 1.4.1, is retained. Let $\{\mathscr{P}^N\}$ denote a sequence of partitions of $[0, T_0]$, satisfying (5.2.15), (2.2.41c), and $\|\mathscr{P}^N\| \to 0$, and let $\{\mathbf{u}_k^N\} \subset \mathbf{V}$ denote the recursively generated solutions of (3.3.32), understood as an identity on $\mathbf{V}_s^*$, and guaranteed to exist by Proposition 3.3.4. We assume $\frac{1}{2} < \theta \leqslant 1$. The step function sequence $\{\mathbf{u}^N\}$ and piecewise linear in time sequence $\{\tilde{\mathbf{u}}^N\}$ are defined as in Lemma 5.2.6. We also assume (5.4.2) (see (1.4.14), valid for $n \geqslant 2$).

Theorem 5.4.1. A subsequence of $\{\mathbf{u}^N\}$ (respectively $\{\tilde{\mathbf{u}}^N\}$) is convergent in $\mathbf{L}^2(\mathscr{D})$ to a weak solution of the generalized Navier–Stokes system (1.4.27) and (1.4.28). This assumes $s \geqslant n/2$ for the case (1.4.11).

Proof: The boundedness of $\{\mathbf{u}^N\}$ and $\{\tilde{\mathbf{u}}^N\}$ in $\mathbf{L}^2((0, T_0); \mathbf{V})$ is a direct consequence of Lemma 3.3.5. Moreover, the estimate

$$\sum_{k=1}^{M} \|(t_k^N - t_{k-1}^N)^{-1}(\mathbf{u}_k^N - \mathbf{u}_{k-1}^N)\|_{\mathbf{V}_s^*}^2 (t_k^N - t_{k-1}^N) \leqslant C, \tag{5.4.1}$$

which implies that $\{\tilde{\mathbf{u}}^N\}$ is bounded in $\mathbf{H}^1([0, T_0]; \mathbf{V}_s^*)$, follows by directly estimating the $\|\cdot\|_{\mathbf{V}_s^*}$ norm of (3.3.32), and using the fact that

$$\|a(\mathbf{w}, \mathbf{w}, \cdot)\|_{\mathbf{V}_s^*} \leqslant C\|\mathbf{w}\|_{\mathbf{V}}\|\mathbf{w}\|_{\mathbf{H}}, \tag{5.4.2}$$

in conjunction with (3.3.35). In particular, we conclude that $\{\tilde{\mathbf{u}}^N\}$ is a relatively compact subset of $\mathbf{L}^2((0, T_0); \mathbf{H})$, by Lemma 5.4.2, to follow. Altogether, then, we may select subsequences $\{\mathbf{u}^{N_i}\}$ and $\{\tilde{\mathbf{u}}^{N_i}\}$, and a function $u \in \mathscr{C}$ (see (1.4.28a)), satisfying the following properties:

$$\mathbf{u}^{N_i} \rightharpoonup \mathbf{u} \qquad \text{(weakly in } \mathbf{L}^2((0, T_0); \mathbf{V})), \tag{5.4.3a}$$

$$\tilde{\mathbf{u}}^{N_i} \rightharpoonup \mathbf{u} \qquad \text{(weakly in } \mathbf{L}^2((0, T_0); \mathbf{V}) \text{ and } \mathbf{H}^1([0, T_0]; \mathbf{V}_s^*)), \tag{5.4.3b}$$

$$\mathbf{u}^{N_i} \to \mathbf{u}, \qquad \tilde{\mathbf{u}}^{N_i} \to \mathbf{u} \qquad \text{(in } \mathbf{L}^2(\mathscr{D}; \mathbb{R}^n)). \tag{5.4.3c}$$

Some comments are in order concerning (5.4.3). That the limit in (5.4.3b) is the same for both spaces and coincides with the second limit in (5.4.3c) is standard and has already been discussed. Note the application of relative compactness. That the first limit in (5.4.3c) is a strong limit and coincides with the second limit is a consequence of Lemma 3.3.5. Indeed, by direct estimation, one sees that $\|\mathbf{u}^{N_i} - \tilde{\mathbf{u}}^{N_i}\|_{\mathbf{L}^2(\mathscr{D}; \mathbb{R}^n)} \to 0$ if (3.3.33) holds. Finally, it is then immediate that the limit in (5.4.3a) and the first limit in (5.4.3c) coincide.

Given a test function $\boldsymbol{\phi}$ in the class $\mathbf{C}^\infty([0, T_0]; \mathscr{V})$, where $\mathscr{V}$ is defined by (1.4.20c), we define the averages $\boldsymbol{\phi}_k^N$ as in (5.2.11). In the weak formulation, corresponding to (3.3.32), we select $\boldsymbol{\phi}_k^N$ for the test function, and apply summation by parts as in the proof of Theorem 5.2.1. Analogous to (5.2.27), we obtain

$$\int_0^{t_{M-1}^N} (\mathbf{u}^N, \boldsymbol{\zeta}^N)_{\mathbf{L}^2(\Omega; \mathbb{R}^n)}\, dt + (\mathbf{u}_M^N, \boldsymbol{\phi}_M^N)_{\mathbf{L}^2(\Omega; \mathbb{R}^n)} - (\mathbf{u}_0, \boldsymbol{\phi}_1^N)_{\mathbf{L}^2(\Omega; \mathbb{R}^n)}$$
$$+ \int_0^{T_{M-1}^N} (\mathbf{u}^N, \boldsymbol{\phi}^N)_{\mathbf{V}}\, dt + \int_0^{t_{M-1}^N} a(\mathbf{u}^N, \mathbf{u}^N, \boldsymbol{\phi}^N)\, dt = 0, \tag{5.4.4}$$

where $\boldsymbol{\zeta}^N$ is the vector analog of (5.2.14), and the inner product $(\cdot, \cdot)_{\mathbf{V}}$ is defined as suggested by (3.2.40a). Using Lemma 5.2.5, and the easily verified addendum that

$$\boldsymbol{\phi}^N \rightharpoonup \boldsymbol{\phi} \qquad \text{(strongly in } \mathbf{L}^2((0, T_0); \mathbf{V}_s)), \tag{5.4.5}$$

and using the convergence results (5.4.3), we may take limits in (5.4.4), much as in the proof of Theorem 5.2.1. Note that a relation analogous to (5.2.32) must be employed to conclude that

$$\mathbf{u}_{M(N_i)}^{N_i} \rightharpoonup^* \mathbf{u}(\cdot, T_0) \qquad (\text{weak } * \quad \text{in} \quad \mathbf{V}_s^*). \tag{5.4.6}$$

Of course, (1.4.24b,c) must be used, together with the *a priori* estimates of (3.3.33). The generalization to the test function class $\mathscr{C}_0$ of (1.4.28b) follows the proof of Theorem 5.2.1, and depends upon the characterization of $\mathbf{V}_s$ given by (1.4.21), and $\mathbf{V}$ given by (1.4.20a). ■

We close the section with the compactness property mentioned previously and a statement of the general principle.

Lemma 5.4.2. A set bounded in $\mathbf{L}^2((0, T_0); \mathbf{V})$ and in $\mathbf{H}^1([0, T_0]; \mathbf{V}_s^*)$ is relatively compact in $\mathbf{L}^2((0, T_0); \mathbf{H})$.

This is a special case of the following (Aubin) lemma (see Lions [33] p. 58, Chapter 1 [23] p. 271). The original reference is *C. R. Acad. Sci. Paris* **256**, 5042–5044 (1963).

Lemma 5.4.3. Let $\mathbf{X}_0$, $\mathbf{X}$, and $\mathbf{X}_1$ be three Banach spaces, such that $\mathbf{X}_0$ and $\mathbf{X}_1$ are reflexive, and

$$\mathbf{X}_0 \subset \mathbf{X} \subset \mathbf{X}_1,$$

where the embeddings are continuous and dense, with $\mathbf{X}_0$ compactly embedded in $\mathbf{X}$. Define, for p_0 and p_1 greater than 1, and $T_0 > 0$, the space

$$\mathbf{Y} = \{u : u \in \mathbf{L}^{p_0}((0, T_0); \mathbf{X}_0) \quad \text{and} \quad u_t \in \mathbf{L}^{p_1}((0, T_0); \mathbf{X}_1)\} \tag{5.4.7a}$$

with norm

$$\|u\|_{\mathbf{Y}} = \|u\|_{\mathbf{L}^{p_0}((0, T_0); \mathbf{X}_0)} + \|u_t\|_{\mathbf{L}^{p_1}((0, T_0); \mathbf{X}_1)}. \tag{5.4.7b}$$

Then, $\mathbf{Y}$ is compactly embedded in $\mathbf{L}^{p_0}((0, T_0); \mathbf{X})$.

5.5 BIBLIOGRAPHICAL REMARKS

The notion of a weak solution for the two-phase Stefan problem was introduced by Oleĭnik [35] and Kamenomostskaja [28]. These authors treated the one-dimensional and multidimensional problems, respectively, the latter involving a technique of explicit finite differences. Both of these authors obtained existence results by using the Kirchhoff transformation and the resulting enthalpy formulation. The results of [28] were refined by Friedman [16], who employed a smoothing method suggested by the analysis of [35]. Subsequent results were obtained via the Faedo–Galerkin method by Lions [33], and by the use of maximal monotone operators by Brézis [5]. Uniqueness is standard when the formulation is monotone. Damlamian [13] obtained uniqueness for Lipschitz perturbations by introducing an $\mathbf{L}^1$ contraction theory. This contrasts with the method of lifting, used in Section 1.5.

For the porous-medium equation, existence of continuous weak solutions of the pure initial-value problem in one space variable was demonstrated by Oleĭnik, Kalashnikov, and Yuĭ-Lin [36], and global regularity properties were derived by Kruzhkov [30] and Aronson [2], who also investigated properties of the free boundary; see also Kalashnikov [27]. Models containing convective effects have been investigated by Gilding and Peletier [20]. The multidimensional pure initial-value problem was considered by Sabinina [39], who demonstrated existence of unique weak solutions. Caffarelli and Friedman [7] have established continuity of this weak solution and certain regularity properties of the free boundary. They have also deduced continuity of the solution of the one-phase Stefan problem. Short-time classical solutions of the one-phase Stefan problem have been demonstrated by Hanzawa [22] by use of the Nash–Moser implicit function theorem. Friedman and Kinderlehrer [18] have established regularity of the free boundary for star-shaped regions in the one-phase problem. For constructive approaches to these inequalities, see [26] and the two-volume work of Glowinski, Lions, and Trémolières [21].

For the two-phase Stefan problem, continuity has been established by Caffarelli and Evans [8] and by Di Benedetto [14], who has also successfully studied the porous-medium equation by similar methods, originally suggested by the Di Georgi–Nash–Moser theory. There is a substantial literature on boundary regularity in one space dimension for the two-phase Stefan problem, which we shall not attempt to summarize here. However, the book of Friedman [17] is quite comprehensive in this respect (see Chapter 5). One of the most powerful tools for analyzing regularity of free boundaries, related to variational inequalities, is a $\mathbf{C}^1$-stability result due to Caffarelli [6] (see Friedman also [17] for contextual development).

Brézis [5] observed that the initial/boundary-value problem for the porous-medium equation can be treated in $\mathbf{H}^{-1}$ by the theory of maximal monotone operators. Analysis on $\mathbf{L}^1$ by accretive operators is also possible. A survey of such applications is given by Evans [15]. The theory of accretive operators, the associated generation of nonlinear semigroups, and the underlying differential equation are related by the fully implicit method employed in this section. In fact, this method of proving existence is due to

Rothe [38], and was effectively carried on by the Russian school of mathematicians (see e.g., Ladyženskaja [32]). A summary of the method, as applied to a linear parabolic initial/boundary-value problem is given in the book of Ladyženskaja, Solonnikov, and Ural'ceva (Chapter 2 [10, pp. 241–252]). The historical antecedents and the effectiveness of the method were pointed out to the writer in 1975 by Gunter Meyer†. It was Hille, with his famous exponential formula, lying at the core of the Hille–Yosida theorem (see Section 6.1), who first drew the rigorous connection between the line method and semigroup generation. A nice account of this for accretive operators is given by Crandall and Evans [10] (see also Crandall and Liggett [11]). In particular, it is emphasized in Crandall and Evans [10] that semigroup generation, in conjunction with semigroup differentiability, are equivalent to a solution of the differential equation. The differentiability requirement does present a serious impediment in the application of this theory to specific problems. On the other hand, the convergence of the difference scheme is consistent with certain generalized notions of solution, such as that proposed by Bénilan [3]. It is likely that a rich abstract theory of generalized solutions is latent here in this circle of ideas.

The author developed the methods used for the horizontal-line analysis of this book in a series of papers, beginning with [24]. Rather than use classical maximum principles to deduce stability, we have used methods familiar to numerical analysts. Also, the theorems discussed here are various nonlinear realizations of the famous Lax equivalence principle, which characterizes convergence in terms of consistency and stability. For further discussion, with interesting perspectives, the reader is referred to the paper of Chorin, Hughes, McCracken, and Marsden [9].

We repeat again an earlier statement of Chapter 2 that we have not attempted an $\mathbf{L}^1$ initial data theory, although the methods are flexible enough to handle this. Our point of view, for the most part, has been to require that u_t be a square integrable function, and this has led to the specification of $\mathbf{H}^1$ initial data (see, however, Proposition 5.2.12). One of the very nice features of the horizontal-line method is that the divided difference estimate in time, expressed via spatial norms, leads directly to the property that u_t is square integrable. The more delicate estimate of Proposition 5.1.5, leading to the result that $\partial H(u)/\partial t$ is essentially bounded on $(0, T_0)$, with range in $\mathbf{M}(\bar{\Omega})$, uses a semidiscrete version of an idea due to Kruzhkov [31]. Kruzhkov's idea was used by the writer and Rose (Chapter 2 [9]), where a different proof of this result is given. This estimate is fundamental for the error estimates of Chapter 4.

We believe that the semidiscrete version of invariant regions, obtained in Chapter 2, and used as the basis for the corresponding parabolic principle

† Private communication.

here, is a new result. However, the parabolic principle, as stated here, is certainly less general in certain respects than that contained in the references of Chapter 2, e.g., Chapter 2 [4]†. The reader may compare the difference in approach presented here with that of Rauch and Smoller (Chapter 2 [14]) and Rauch [37]. Periodicity properties of solutions of parabolic systems have been investigated by Amann [1].

The results we have presented in this chapter concerning weak solutions of the Navier–Stokes system do not differ appreciably from those presented by Temam (Chapter 1 [23]), though some superficial generalizations have been made. Actually, we have presented these results as background for the results of Chapter 7 concerning classical solutions. As preparation, we shall give a brief literature survey concerning this problem.

Following the pioneering work of Leray‡ early writers, including Hopf [23], Kiselev and Ladyžhenskaja [29], Serrin [40, 41], and Ladyžhenskaja (Chapter 1 [11]), introduced notions of global weak solutions and local classical solutions for this problem. Specifically, Hopf considered solenoidal (distribution-wise) $\mathbf{L}^2$ initial velocities and Kiselev and Ladyžhenskaja considered $\mathbf{H}^2$ initial solenoidal velocities. Whereas Hopf's weak solution is global in time, but possibly nonunique, the latter authors obtain a local time result within an existence–uniqueness class. It should be noted that the difficult problem here is for $n \geqslant 3$; the case $n = 2$ is well-understood (see Lions [33]). Several authors generalized the results of Kiselev and Ladyžhenskaja [29], obtained for $n = 3$, to $n > 3$, e.g., Serrin [41] for $n = 4$ and $\mathbf{H}^2$ solenoidal velocities, and Fujita and Kato [19] for initial datum in an interpolation space roughly corresponding to $\mathbf{H}^{1/2}$. We have quoted these results for situations neglecting external effects, such as gravitation, or assuming these effects are derivable from a potential. The result for global weak solutions, given by Temam (Chapter 1 [23]), assumes only $\mathbf{L}^2$ solenoidal initial velocities. The proof of Section 5.4 does not differ, in essential respects, from that of Temam (Chapter 1 [23]), and could readily be modified to relax the assumption $\mathbf{u}_0 \in \mathbf{V}$ to $\mathbf{u}_0 \in \mathbf{H}$.

Concerning uniqueness, it is known that $\mathbf{L}^p((0, T_0); \mathbf{L}^r(\Omega))$ (abbreviated $\mathbf{L}^{p,r}$) is a uniqueness class, if

$$\frac{2}{p} + \frac{n}{r} = 1, \qquad n \leqslant r \leqslant \infty$$

(see Serrin [41]; also Lions and Prodi [34]).

Classical regularity results were obtained by Serrin [40]. In the case $\Omega = \mathbb{R}^n$, these results guarantee that $\mathbf{u}$ is $\mathbf{C}^\infty$ in all variables, if $\mathbf{u} \in L^{p,r}$ for

$$\frac{2}{p} + \frac{n}{r} < 1.$$

† It is, however, a multidimensional principle.

‡ *J. Math. Pures et Appl.* **XII** 1–82, (1933); *ibid.*, **XIII** 331–418, (1934).

This implies that the Kiselev–Ladyžhenskaja solution is $\mathbf{C}^\infty$ in all variables for $\Omega = \mathbb{R}^3$.

A fundamental point of view has recently been adopted by Bona and Di Benedetto [4], who have studied local properties of solutions without reference to initial and boundary data. These authors have studied $\mathbf{C}^\infty$ regularity in $\mathbf{x}$ for almost all t, and have used the equations themselves as the regularizing tool. These investigations were carried out in the case $n = 2$.

We now make some final remarks. The reader will recall the fundamental role played by the lifting operators in Section 5.1 (see especially Proposition 5.1.1). We have made some historical comments concerning these operators in Chapter 2. Recent use of these operators to define suitable renormings has been made by Showalter [42], in applying Hilbert space methods to partial differential equations. We also call attention to very recent work of Alt and Luckhaus† on quasi-linear elliptic and parabolic equations, including degenerate equations, such as the porous-medium equation and the equation describing the two-phase Stefan problem. Mixing of elliptic and parabolic regions is discussed for these general problems. Complicating the qualitative behavior of solutions of equations, such as the porous-medium equation, is the finite propagation speed associated with such equations, due to the fact that $H(\cdot)$ is not Lipschitz continuous. Thus, compact support of initial data is not instantaneously globally diffused as in truly parabolic problems. The reader is referred to this paper and to [2] for further details.

REFERENCES

[1] H. Amann, Periodic solutions of semilinear parabolic equations, *in* "Nonlinear Analysis" (L. Cesari, R. Kannan, and H. Weinberger, eds.), pp. 1–29. Academic Press, New York, 1978.

[2] D. G. Aronson, Regularity properties of flows through porous media. *SIAM J. Appl. Math.* **17**, 461–467 (1969); A counterexample, *Ibid.* **19**, 299–307 (1970); The interface, *Arch. Rational Mech. Anal.* **37**, 1–10 (1970).

[3] P. Bénilan, Solutions intégrales d'équations d'évolution dans un espace de Banach, *C. R. Acad. Sci. Paris* **Sér A–B 274**, A47–A50 (1972).

[4] J. Bona and E. Di Benedetto, On the interior regularity of weak solutions of the Navier–Stokes equations in dimension 2, manuscript.

[5] H. Brézis, On some degenerate nonlinear parabolic equations, *in* "Nonlinear Functional Analysis" (F. Browder, ed.) Part I, pp. 28–38 (*Proc. Symp. Pure Math.* **18**). American Mathematical Society, Providence, Rhode Island, 1970.

[6] L. A. Caffarelli, Compactness methods in free boundary problems, *Comm. Partial Differential Equations* **5**, 427–448 (1980).

† Unpublished manuscript.

[7] L. A. Caffarelli and A. Friedman, Continuity of the temperature in the Stefan problem, *Indiana Univ. Math. J.* **28**, 53–70 (1979); Continuity of the density of a gas flow in a porous medium, *Trans. Amer. Math. Soc.* **252**, 99–113 (1979); Regularity of the free boundary of a gas flow in an n-dimensional porous medium, *Indiana Univ. Math. J.* **29**, 361–391 (1980).

[8] L. A. Caffarelli and L. C. Evans, Continuity of the temperature in the two-phase Stefan problem, *Arch. Rational Mech. Anal.* (to appear).

[9] A. J. Chorin, T. J. R. Hughes, M. F. McCracken, and J. E. Marsden, Product formulas and numerical algorithms, *Comm. Pure Appl. Math.* **31**, 205–256 (1978).

[10] M. G. Crandall and L. C. Evans, On the relation of the operator $\partial/\partial s + \partial/\partial\tau$ to evolution governed by accretive operators, *Israel J. Math.* **21**, 261–278 (1975).

[11] M. G. Crandall and T. M. Liggett, Generation of semigroups of nonlinear transformations on general Banach spaces, *Amer. J. Math.* **93**, 265–293 (1972).

[12] M. G. Crandall and A. Pazy, Semigroups of nonlinear contractions and dissipative sets, *J. Functional Anal.* **3**, 376–418 (1969).

[13] A. Damlamian, Some results on the multi-phase Stefan problem, *Comm. Partial Differential Equations* **2**, 1017–1044 (1977).

[14] E. Di Benedetto, Continuity of weak solutions to certain singular parabolic equations, *Ann. Mat. Pura Appl.* **130**, 131–177 (1982); Continuity of weak solutions to a general porous media equation, *Indiana Univ. Math. J.*, 1983, to appear.

[15] L. C. Evans, Application of nonlinear semigroup theory to certain partial differential equations, *in* "Nonlinear Evolution Equations" (M. G. Crandall, ed.), pp. 163–188. Academic Press, New York, 1978.

[16] A. Friedman, The Stefan problem in several space variables, *Trans. Amer. Math. Soc.* **133**, 51–87 (1968); One dimensional Stefan problems with nonmonotone free boundary, *Ibid.* **133**, 89–114 (1968).

[17] A. Friedman, "Variational Principles and Free Boundary Problems." Wiley, New York, 1982.

[18] A. Friedman and D. Kinderlehrer, A one phase Stefan problem, *Indiana Univ. Math. J.* **24**, 1005–1035 (1975).

[19] H. Fujita and T. Kato, On the Navier–Stokes initial value problem I, *Arch. Rational Mech. Anal.* **16**, 269–315 (1964).

[20] B. H. Gilding and L. A. Peletier, The Cauchy problem for an equation in the theory of infiltration, *Arch. Rational Mech. Anal.* **61**, 126–152 (1976).

[21] R. Glowinski, J. Lions and R. Trémolières, "Analyse Numérique Des Inéquations Variationnelles," Vols. I and II. Dunod, Paris, 1976.

[22] E-I. Hanzawa, Classical solutions of the Stefan problem, *Tôhoku Math. J.* **33**, 297–335 (1981).

[23] E. Hopf, Über die Anfangswertaufgabe für die hydrodynamischen Grundleichungen, *Math. Nachr.* **4**, 213–231 (1951).

[24] J. W. Jerome, Nonlinear equations of evolution and a generalized Stefan problem, *J. Differential Equations* **26**, 240–261 (1977).

[25] J. W. Jerome, Convergence of successive iterative semidiscretizations for FitzHugh–Nagumo reaction diffusion systems, *SIAM J. Numer. Anal.* **17**, 192–206 (1980).

[26] J. W. Jerome, Uniform convergence of the horizontal line method for solutions and free boundaries in Stefan evolution inequalities, *Math. Meth. Appl. Sci.* **2**, 149–167 (1980).

[27] A. S. Kalashnikov, Formation of singularities in the solutions of the equation of nonstationary filtration, *Ž. Vyčist. Mat. i. Mat. Fiz.* **7**, 440–444 (1967) (Russian).

[28] S. L. Kamenomostskaja, On the Stefan problem, *Mat. Sb.* **53**, 489–514 (1961) (Russian).

[29] A. Kiselev and O. Ladyžhenskaja, On existence and uniqueness of the solution of the nonstationary problem for a viscous incompressible fluid, *Izv. Akad. Nauk SSSR* **21**, 655–680 (1957) (Russian).

[30] S. N. Kruzhkov, Results on the character of the regularity of solutions of parabolic equations and some of their applications, *Mat. Zametki* **6**, 97–108 (1969) (Russian).

[31] S. N. Kruzhkov, First order quasilinear equations in several independent variables, *Math. USSR Sb.* **10**, 217–243 (1970).

[32] O. A. Ladyženskaja, The solution in the large of the first boundary-value problem for quasilinear parabolic equations, *Trudy Moskov. Mat. Obšč.* **7**, 149–177 (1958) (Russian).

[33] J. L. Lions, "Quelques Méthodes de Résolution des Problèmes aux Limites non Linéaires." Dunod, Paris, 1969.

[34] J. L. Lions and G. Prodi, Un théorème d'existence et d'unicité dans les équations de Navier–Stokes en dimension 2, *C. R. Acad. Sci. Paris* **248**, 3519–3521 (1959).

[35] O. A. Oleĭnik, A method of solution of the general Stefan problem, *Soviet Math. Dokl.* **1**, 1350–1354 (1961).

[36] O. A. Oleĭnik, A. S. Kalashnikov and C. Yuĭ-Lin, The Cauchy problem and boundary problems for equations of the type of nonstationary filtration, *Izv. Akad. Nauk SSSR Ser. Math.* **22**, 667–704 (1958) (Russian).

[37] J. Rauch, Global existence for the FitzHugh–Nagumo equations, *Comm. Partial Differential Equations* **1**, 609–621 (1976).

[38] E. H. Rothe, Zweidimensionale parabolische Randwertaufgaben als Grenzfall eindimensionaler Randwertaufgaben, *Math. Ann.* **102**, 650–670 (1930); Über die Wärmeleitungsgleichung mit nicht-konstante Koeffizienten im räumlichen Falle, Erste Mitteilung, *ibid.* **104**, 340–354 (1931); *Ibid.*, Zweite Mitteilung, 355–362.

[39] E. S. Sabinina, On the Cauchy problem for the equation of nonstationary gas filtration in several space variables, *Dokl. Akad. Nauk SSSR* **136**, 1034–1037 (1961) (Russian).

[40] J. Serrin, On the interior regularity of weak solutions of Navier–Stokes equations, *Arch. Rational Mech. Anal.* **9**, 187–195 (1962).

[41] J. Serrin, The initial-value problem for the Navier–Stokes equations, *in* "Nonlinear Problems" (R. Langer, ed.), pp. 69–98. Univ. of Wisconsin Press, Madison, Wisconsin, 1963.

[42] R. E. Showalter, Nonlinear degenerate evolution equations and partial differential equations of mixed type, *SIAM J. Math. Anal.* **6**, 25–42 (1975); "Hilbert Space Methods for Partial Differential Equations." Pitman, London, 1977.

LOCAL SMOOTH SOLUTIONS

II

6 LINEAR EVOLUTION OPERATORS

6.0 INTRODUCTION

In this chapter, we begin with a given family $\{-A(t)\}$ of closed generators of strongly continuous semigroups on a Banach space **X**, and construct the associated evolution operators $\{U(t, s)\}$, under an assumption of quasi-stability on $\{A(t)\}$, certain continuity assumptions, as well as the similarity of this family to the perturbation family $\{A(t) + B(t)\}$, for an appropriately measurable family $\{B(t)\}$ of bounded operators. The summarizing theorem containing the major result is Theorem 6.3.7. It is sufficiently strong to cover the nonlinear applications of the next chapter where $\{A(t)\}$ is typically stable on **X** and also stable on an appropriate dense smooth subspace **Y**. In order to avoid a needlessly technical development, we carry out the details of the construction in Section 6.2 in the case where $\{A(t)\}$ is stable on **X** and **Y**. The modifications required for quasi-stability are technical considerations of integration theory, and, for these, the reader is referred to the appropriate literature. In this chapter, **X** and **Y** are not assumed reflexive.

The properties developed in Section 6.2 for the operators $\{U(t, s)\}$ are not sufficiently strong to classify them as evolution operators. The remaining property is the invariant action of $U(t, s)$ on a specified smooth space $\mathbf{Y} \subset \mathbf{D}_{A(t)}$. This is achieved in Section 6.3 by a similarity relation between $U(t, s)$

and W(t, s) where the latter are defined as bounded operators on **X** via a Volterra integral equation. The convolution structure associated with solutions of such integral equations is compatible with the quasi-stability assumption on $\{\mathrm{A}(t)\}$ and the measurability of $\{\mathrm{B}(t)\}$.

For the reader's convenience, we have presented, without proof, in Section 6.1, a brief development of strongly continuous semigroups, particularly those properties required in the sequel. In the final section we prove the validity of the formal representation of solutions of abstract linear Cauchy problems, and use the classical example of a linear symmetric hyperbolic system to illustrate the concepts of the linear theory. This application, however, requires only the stability of $\{A(t)\}$, not the more general notion of quasi-stability. Finally, it is possible to weaken the norm continuity assumption of this chapter to strong continuity. The reader is referred to the bibliographical comments.

6.1 SEMIGROUP PRELIMINARIES

We shall summarize the basic facts concerning semigroups of linear operators in this section. For proofs and further amplification, the reader is invited to consult the book of Butzer and Berens [3] and the classic treatise of Hille and Phillips [10]. The semigroup notion may be thought of as an appropriate generalization of the separation-of-variables solution

$$\mathrm{T}(t)u_0 = u(t) = \sum_k a_k e^{-\lambda_k t} u_k \tag{6.1.1}$$

of the equation $u_t = -\mathrm{A}u$, where A is an appropriate positive-definite, self-adjoint Hilbert space operator, with eigenvalues $\{\lambda_k\}$ and normalized eigenfunctions $\{u_k\}$, and $a_k = (u_0, u_k)$. The family $\{\mathrm{T}(t)\}_{t\geqslant 0}$ is called the semigroup generated by $-\mathrm{A}$. The generalization applies equally to operators A, not possessing compact inverses, where (6.1.1) is replaced by an appropriate convolution. In general, the semigroup satisfies the properties

$$\mathrm{T}(t+s) = \mathrm{T}(t)\mathrm{T}(s), \tag{6.1.2a}$$

$$\mathrm{T}(0) = \mathrm{I}, \tag{6.1.2b}$$

$$\lim_{t\to 0+} \mathrm{T}(t)f = f, \qquad f \in \mathbf{X}, \tag{6.1.2c}$$

$$\frac{d}{dt}(\mathrm{T}(t)f) = -\mathrm{A}\mathrm{T}(t)f, \qquad f \in \mathbf{D}_{\mathrm{A}} \subset \mathbf{X}, \tag{6.1.2d}$$

where $\{\mathrm{T}(t)\}_{t\geqslant 0}$ is defined on the normed linear space **X**.

Definition 6.1.1. Designate by **B**[**X**] the Banach space of continuous linear transformations of a given Banach space **X** into itself. Then,

$$\mathrm{T}(t):[0, \infty) \to \mathbf{B}[\mathbf{X}]$$

is called a semigroup of operators (in **B**[**X**]) if (6.1.2a,b) hold. The family $\{\mathrm{T}(t)\}_{t \geqslant 0}$ is said to be strongly continuous (of class (C_0)) if (6.1.2c) holds.

Remark 6.1.1. All semigroups considered will be of class (C_0).

Proposition 6.1.1.†

(1) $\|\mathrm{T}(t)\|$ is bounded on every finite subinterval of $[0, \infty)$.

(2) For each $f \in \mathbf{X}$, the function

$$\mathrm{T}(t)f:[0, \infty) \to \mathbf{X}$$

is (strongly) continuous.

(3) If

$$\omega_0 = \inf\left\{\left(\frac{1}{t}\right)\ln\|\mathrm{T}(t)\| : t > 0\right\}, \tag{6.1.3a}$$

then

$$\omega_0 = \lim_{t \to \infty}\left(\left(\frac{1}{t}\right)\ln\|\mathrm{T}(t)\|\right) < \infty. \tag{6.1.3b}$$

(4) For each $\omega > \omega_0$, there exists a constant M_ω, such that

$$\|\mathrm{T}(t)\| \leqslant M_\omega e^{\omega t}, \qquad t \geqslant 0. \tag{6.1.4}$$

Definition 6.1.2. The infinitesimal generator U of the semigroup $\{\mathrm{T}(t)\}_{t \geqslant 0}$ is defined by

$$\mathrm{U}f = \lim_{\tau \to 0+} \mathrm{U}_\tau f, \qquad \mathrm{U}_\tau = \frac{1}{\tau}[\mathrm{T}(\tau) - \mathrm{I}], \tag{6.1.5}$$

whenever the limit exists. The domain of U, $\mathbf{D}_\mathrm{U}$, is the set of elements f for which this limit exists.

† Propositions 6.1.1, 6.1.2, 6.1.4 and Theorems 6.1.6, 6.1.7 are adapted from Butzer and Berens [3].

Proposition 6.1.2.
(1) The set $\mathbf{D}_U$ is a linear manifold and U is a linear operator.
(2) The family T(t) leaves $\mathbf{D}_U$ invariant for $t \geqslant 0$, and, for $f \in \mathbf{D}_U$ and $t \geqslant 0$,

$$\frac{d}{dt}(\mathrm{T}(t)f) = \mathrm{UT}(t)f = \mathrm{T}(t)\mathrm{U}f, \tag{6.1.6a}$$

$$\mathrm{T}(t)f - f = \int_0^t \mathrm{T}(u)\mathrm{U}f\,du. \tag{6.1.6b}$$

(3) The manifold $\mathbf{D}_U$ is dense in **X**, and U is a closed linear operator.

Definition 6.1.3. Let **X** be a Banach space, and let U be a linear operator with domain and range in $\mathbf{X}: \mathbf{D}_U \subset \mathbf{X}$, $\mathbf{R}_U \subset \mathbf{X}$. Set

$$\mathrm{U}_\lambda = \lambda \mathrm{I} - \mathrm{U} = \lambda - \mathrm{U}.$$

The spectrum of U is defined as the set

$$\sigma(\mathrm{U}) = \{\lambda \in \mathbb{C} : \mathrm{U}_\lambda \text{ does not have a continuous, densely defined inverse}\}.$$

The resolvent set $\rho(\mathrm{U})$ of U is the complement of $\sigma(\mathrm{U})$ in $\mathbb{C}$. Thus, $\lambda \in \rho(\mathrm{U})$ if and only if U_λ^{-1} exists as a bounded operator with dense domain. Set

$$\mathrm{R}(\lambda, \mathrm{U}) = \mathrm{U}_\lambda^{-1}, \qquad \lambda \in \rho(\mathrm{U}). \tag{6.1.7}$$

The operator $\mathrm{R}(\cdot, \mathrm{U})$ is called the resolvent of U.

Proposition 6.1.3. If U is closed, then $\mathbf{R}_{\mathrm{U}\lambda} = \mathbf{X}$ if $\lambda \in \rho(\mathrm{U})$ and, thus, $\mathrm{R}(\lambda, \mathrm{U})$ transforms **X** one-to-one onto $\mathbf{D}_U$, i.e.,

$$(\lambda \mathrm{I} - \mathrm{U})\mathrm{R}(\lambda, \mathrm{U})f = f, \qquad f \in \mathbf{X}, \tag{6.1.8a}$$

$$\mathrm{R}(\lambda, \mathrm{U})(\lambda \mathrm{I} - \mathrm{U})g = g, \qquad g \in \mathbf{D}_U. \tag{6.1.8b}$$

Proposition 6.1.4. Let $\{\mathrm{T}(t)\}_{t \geqslant 0}$ be a semigroup of class (C_0) in $\mathbf{B}[\mathbf{X}]$, with infinitesimal generator U. If ω_0 is given by (6.1.3), and $\mathrm{Re}\,\lambda > \omega_0$, then, $\lambda \in \rho(\mathrm{U})$, and

$$\mathrm{R}(\lambda, \mathrm{U})f = \int_0^\infty e^{-\lambda t}\mathrm{T}(t)f\,dt, \qquad f \in \mathbf{X}. \tag{6.1.9}$$

Furthermore, for each $f \in \mathbf{X}$,

$$\lim_{|\lambda| \to \infty} \lambda \mathrm{R}(\lambda, \mathrm{U})f = f \qquad \left(|\arg \lambda| \leqslant \alpha_0 < \frac{\pi}{2}\right). \tag{6.1.10}$$

Remark 6.1.2. The resolvent is an analytic function on $\rho(\mathrm{U})$:

$$\mathrm{R}(\lambda, \mathrm{U}) = \sum_{k=0}^{\infty} (\lambda_0 - \lambda)^k (\lambda_0 - \mathrm{U})^{-k-1}, \tag{6.1.11}$$

$|\lambda_0 - \lambda| < \|\mathrm{R}(\lambda_0, \mathrm{U})\|^{-1}$. Thus, for such λ,

$$\|\mathrm{R}(\lambda, \mathrm{U})\| \leqslant \sum_{k=0}^{\infty} |\lambda_0 - \lambda|^k \|\mathrm{R}(\lambda_0, \mathrm{U})\|^{k+1}. \tag{6.1.12}$$

It follows from (6.1.11) that

$$[\mathrm{R}(\lambda, \mathrm{U})]^{n+1} f = \frac{1}{n!}(-1)^n \left(\frac{d}{d\lambda}\right)^n \mathrm{R}(\lambda, \mathrm{U}) f, \tag{6.1.13}$$

and, if (6.1.13) is applied to (6.1.9), there results

$$[\mathrm{R}(\lambda, \mathrm{U})]^{n+1} f = \frac{1}{n!}(-1)^n \int_0^\infty (-t)^n e^{-\lambda t} \mathrm{T}(t) f \, dt.$$

In particular, if

$$\|\mathrm{T}(t)\| \leqslant M e^{\omega t}, \tag{6.1.14a}$$

then

$$\|[\mathrm{R}(\lambda, \mathrm{U})]^r\| \leqslant \frac{M}{(\sigma - \omega)^r}, \qquad \sigma = \operatorname{Re} \lambda > \omega, \quad r \geqslant 1. \tag{6.1.14b}$$

Proposition 6.1.5. Let U be a closed linear operator with dense domain and range in a Banach space **X**. Suppose there exist real numbers M and ω, such that (6.1.14b) holds. For $\lambda > \omega$, set

$$\mathrm{B}_\lambda = \lambda^2 \mathrm{R}(\lambda, \mathrm{U}) - \lambda \mathrm{I} = \lambda \mathrm{U} \mathrm{R}(\lambda, \mathrm{U}), \tag{6.1.15a}$$

and

$$\mathrm{S}_\lambda(t) = \exp(t \mathrm{B}_\lambda) = e^{-t\lambda} \sum_{k=0}^{\infty} \frac{(\lambda^2 t)^k}{k!} [\mathrm{R}(\lambda, \mathrm{U})]^k. \tag{6.1.15b}$$

Then,

$$\lim_{\lambda \to \infty} \mathrm{B}_\lambda f = \mathrm{U} f, \qquad f \in \mathbf{D}_\mathrm{U}.$$

Moreover, the limit

$$\lim_{\lambda \to \infty} \mathrm{S}_\lambda(t) f = \mathrm{T}(t) f, \qquad f \in \mathbf{X}, \tag{6.1.16}$$

exists uniformly for t in any finite interval $[0, b]$. The family $\{\mathrm{T}(t)\}_{t \geqslant 0}$ defines a (C_0) semigroup, satisfying

$$\|\mathrm{T}(t)\| \leqslant Me^{\omega t}, \qquad t \geqslant 0,$$

with infinitesimal generator U.

Remark 6.1.3. The estimate (6.1.14b) directly implies

$$\begin{aligned}\|\mathbf{S}_\lambda(t)\| &\leqslant e^{-\lambda t} \sum_{k=0}^{\infty} \frac{(\lambda^2 t)^k}{k!} \frac{M}{(\lambda - \omega)^k}. \\ &= M \exp\left[t\lambda\omega(\lambda - \omega)^{-1}\right],\end{aligned} \tag{6.1.17}$$

and (6.1.14a) follows, if $\lambda \to \infty$ in (6.1.17). Thus, (6.1.14a) and (6.1.14b) are seen to be equivalent, via the Laplace transform (6.1.9) and the exponential formula (6.1.16).

Theorem 6.1.6. A necessary and sufficient condition for a closed linear operator U, with domain dense in a Banach space **X** and range in **X**, to generate a semigroup $\{\mathrm{T}(t): 0 \leqslant t < \infty\}$ of class (C_0) is that there exist real numbers M and ω, such that $\lambda \in \rho(\mathrm{U})$ for every real $\lambda > \omega$, and, moreover,

$$\|[\mathrm{R}(\lambda, \mathrm{U})]^r\| \leqslant M(\lambda - \omega)^{-r}, \qquad r \geqslant 1. \tag{6.1.18}$$

In this case, $\|\mathrm{T}(t)\| \leqslant Me^{\omega t}$, $t \geqslant 0$.

Corollary 6.1.7. (Hille–Yosida). If U is a closed linear operator with domain dense in a Banach space **X** and range in **X**, and if $\mathrm{R}(\lambda, \mathrm{U})$ exists for all real λ larger than some real number ω, and satisfies the inequality

$$\|\mathrm{R}(\lambda, \mathrm{U})\| \leqslant (\lambda - \omega)^{-1}, \qquad \lambda > \omega, \tag{6.1.19a}$$

then U is the infinitesimal generator of a semigroup of class (C_0), such that

$$\|\mathrm{T}(t)\| \leqslant e^{\omega t}, \qquad t \geqslant 0. \tag{6.1.19b}$$

Proposition 6.1.8. (Trotter–Kato). The strong limit

$$\mathrm{T}(t)f = \lim_{n \to \infty} (\mathrm{I} - t\mathrm{U}/n)^{-n} f, \qquad t \geqslant 0, \tag{6.1.20}$$

holds for every $f \in \mathbf{X}$, if U is the infinitesimal generator of the (C_0) semigroup $\mathrm{T}(t)$.

Remark 6.1.4. One can prove that, if $t \in [0, T_0]$,

$$\|[\mathrm{T}(t) - (\mathrm{I} - t\mathrm{U}/n)^{-n}]f\| \leqslant M_{T_0} n^{-1} t^2 \|\mathrm{U}^2 f\|, \tag{6.1.21}$$

for $f \in \mathbf{D}_{U^2}$ (see Yosida [14] p. 269). Thus, (6.1.20) follows from (6.1.21), by the uniform boundedness of

$$(I - tU/n)^{-n} = [nR(n, tU)]^n,$$

and the denseness of $\mathbf{D}_{U^2}$ in $\mathbf{X}$.

Definition 6.1.4. The set $\{-U:U$ is closed and satisfies (6.1.18)$\}$ will be denoted by $G(\mathbf{X}, M, \omega)$. Write

$$G(\mathbf{X}, M) = \bigcup_{-\infty<\omega<\infty} G(\mathbf{X}, M, \omega), \qquad G(\mathbf{X}) = \bigcup_{M>0} G(\mathbf{X}, M).$$

Remark 6.1.5. There is an equivalent formulation of the hypothesis of the Hille–Yosida theorem, if $\mathbf{X}$ is a (real) Hilbert space. In particular, if

$$(Af, f)_{\mathbf{X}} \geqslant -\omega\|f\|_{\mathbf{X}}^2, \tag{6.1.22a}$$

and

$$\lambda \in \rho(A) \qquad \text{for all} \quad \lambda < -\omega, \tag{6.1.22b}$$

then $A \in G(\mathbf{X}, 1, \omega)$ (see Bellini-Morante [2] p. 142).

6.2 THE LINEAR EVOLUTION EQUATION AND EVOLUTION OPERATORS

We shall be motivated by the desire to represent, in terms of evolution operators, the solution of the abstract Cauchy problem for linear equations

$$\frac{du}{dt} + A(t)u = F(t), \qquad 0 \leqslant t \leqslant T_0, \tag{6.2.1}$$

in a Banach space $\mathbf{X}$. Equation (6.2.1) is of hyperbolic type, in the sense that the linear operators $-A(t)$ are the infinitesimal generators of (C_0) semigroups on $\mathbf{X}$. The subspace $\mathbf{D}_{A(t)}$ will not be required to be constant in t, but rather to contain a dense linear manifold $\mathbf{Y}$ for all t. The associated semigroups will be required to satisfy a certain stability or quasi-stability condition.

Definition 6.2.1. Let $\mathbf{X}$ be a Banach space, and suppose that a family $A(t) \in G(\mathbf{X})$ of linear operators is given on $0 \leqslant t \leqslant T_0$. The family $\{A(t)\}$ is said to be stable if there are (stability) constants M and ω, such that

$$\left\| \prod_{j=1}^{k} [A(t_j) + \lambda]^{-1} \right\| \leqslant M(\lambda - \omega)^{-k}, \qquad \lambda > \omega, \tag{6.2.2a}$$

for any finite family $\{t_j\}_{j=1}^k$, with $0 \leqslant t_1 \leqslant \cdots \leqslant t_k \leqslant T_0, k = 1, 2, \ldots$. Moreover, $\prod$ is time-ordered, i.e., $[\mathrm{A}(t_j) + \lambda]^{-1}$ is to the left of $[\mathrm{A}(t_{j'}) + \lambda]^{-1}$ if $j > j'$. More generally, $\{\mathrm{A}(t)\}$ is said to be quasi-stable if there exists a constant M and a real-valued, upper Lebesgue integrable function ω, defined almost everywhere on $[0, T_0]$, such that

$$\left\| \prod_{j=1}^{k} (\mathrm{A}(t_j) + \lambda)^{-1} \right\| \leqslant M \prod_{j=1}^{k} (\lambda - \omega(t_j))^{-1}, \qquad \lambda > \omega(t_j), \tag{6.2.2b}$$

for any finite increasing family $\{t_j\}$ taken almost everywhere on $[0, T_0]$. In this case, (M, ω) is a stability index for $\{\mathrm{A}(t)\}$, and the domain of definition of $\mathrm{A}(\cdot)$ may omit a set of measure zero on $[0, T_0]$. In the notation $G(\mathbf{X}, M, \omega)$, introduced in Definition 6.1.4, we permit only constant values of ω.

Remark 6.2.1. Quasi-stability is preserved, with only a change in M, under equivalence of norms. An upper Lebesgue integrable function ψ is, of course, dominated, pointwise, by an integrable function ϕ, and the value, $\int^* \psi$, of the upper integral is the infimum $\int \phi$. For the nonlinear theory of the next chapter, we shall not require the result of Theorem 6.2.5 in the full generality of quasi-stability, thus we shall give the complete details of the construction of the evolution operators, described therein, only in the stable case, referring the reader to the literature for the required technical modifications. For the other results of this chapter, we shall prepare the statements, for the most part, in the more general format, but give the details of proof only in the stable case for ease of reading. Here, the required changes are essentially "cosmetic."

Proposition 6.2.1.† The stability (respectively, quasi-stability) condition (6.2.2) is equivalent to each of the following conditions, in which $\{t_j\}_{j=1}^k$ varies over all admissible finite families, as in Definition 6.2.1:

$$\left\| \prod_{j=1}^{k} \exp\{-s_j \mathrm{A}(t_j)\} \right\| \leqslant M \exp\{\omega(s_1 + \cdots + s_k)\} \quad \left(\text{resp. } M \exp \sum_j s_j \omega(t_j)\right), \tag{6.2.3}$$

for $s_j \geqslant 0$;

$$\left\| \prod_{j=1}^{k} [\mathrm{A}(t_j) + \lambda_j]^{-1} \right\| \leqslant M \prod_{j=1}^{k} (\lambda_j - \omega)^{-1} \quad \left(\text{resp. } M \prod_{j=1}^{k} (\lambda_j - \omega(t_j))^{-1}\right), \tag{6.2.4}$$

† Propositions 6.2.1–6.2.4, Theorem 6.2.5, and Lemmas 6.2.6, 6.2.7 are adapted from Kato [11, 12].

for $\lambda_j > \omega$ and $\lambda_j > \omega(t_j)$, respectively. Here, we have written e^{-tA} for the (C_0) semigroup generated by $-A$.

Proof: We prove, in the case of stability,

$$(6.2.3) \Rightarrow (6.2.4) \Rightarrow (6.2.2) \Rightarrow (6.2.3).$$

By (6.1.9), we have

$$\begin{aligned}\left\|\prod_{j=1}^{k} [A(t_j) + \lambda_j]^{-1} f\right\| &= \left\|\prod_{j=1}^{k} \int_0^\infty \exp(-\lambda_j s_j) \exp\{-s_j A(t_j)\} f \, ds_j\right\| \\ &\leqslant \int_0^\infty \cdots \int_0^\infty \exp\left\{-\sum_{j=1}^{k} \lambda_j s_j\right\} \left\|\prod_{j=1}^{k} \exp\{-s_j A(t_j)\}\right\| ds_1 \cdots ds_k \|f\| \\ &\leqslant M \int_0^\infty \cdots \int_0^\infty \exp\left\{-\sum_{j=1}^{k} (\lambda_j - \omega) s_j\right\} ds_1 \cdots ds_k \|f\|,\end{aligned}$$

where we have used (6.2.3). The evaluation of the preceding integral is

$$M \prod_{j=1}^{k} (\lambda_j - \omega)^{-1} \|f\|,$$

so that we have proved (6.2.3) $\Rightarrow$ (6.2.4). The implication (6.2.4) $\Rightarrow$ (6.2.2) is immediate upon taking $\lambda_j = \lambda, j = 1, \ldots, k$.

Now, by (strong) continuity, we note that (6.2.2) $\Rightarrow$ (6.2.3) certainly holds, if it can be established whenever s_j is a rational number. By selecting non-negative integers m_j, and a rational number s, satisfying

$$s_j = m_j s, \qquad j = 1, \ldots, k,$$

for given rationals s_j, we see that it suffices to prove the implication for $s_j = m_j s$. Finally, by absorbing the repetition of s into the equivalent repetition of t_j, we see that it suffices to verify (6.2.2) $\Rightarrow$ (6.2.3) for $s_j = s$. Thus, by (6.1.20) and (6.2.2),

$$\begin{aligned}\left\|\prod_{j=1}^{k} \exp\{-sA(t_j)\} f\right\| &= \left\|\prod_{j=1}^{k} \lim_{m\to\infty} [I + sA(t_j)/m]^{-m} f\right\| \\ &\leqslant \lim_{m\to\infty} \left\{\left(\frac{s}{m}\right)^{-km} \left\|\prod_{j=1}^{k} \left[\frac{m}{s} + A(t_j)\right]^{-m} f\right\|\right\} \\ &\leqslant \lim_{m\to\infty} \left\{\left(\frac{s}{m}\right)^{-km} M \left(\frac{m}{s} - \omega\right)^{-km} \|f\|\right\} \\ &= M \lim_{m\to\infty} \{(1 - \omega s/m)^{-km} \|f\|\} \\ &= M e^{\omega k s} \|f\|,\end{aligned}$$

so that (6.2.3) holds. ■

Remark 6.2.2. It is difficult, in general, to decide whether a given family $\{A(t)\}$ is quasi-stable. A criterion is given by the following.

Proposition 6.2.2. For each $t \in [0, T_0]$, let $\|\cdot\|_t$ be a new norm in $\mathbf{X}$, let $\mathbf{X}_t = (\mathbf{X}, \|\cdot\|_t)$, and suppose there is a real number c, such that

$$\|f\|_t/\|f\|_s \leqslant e^{c|t-s|}, \qquad 0 \neq f \in \mathbf{X}, \quad s, t \in [0, T_0]. \tag{6.2.5}$$

Suppose $A(t) \in G(\mathbf{X}_t, 1, \omega(t))$, where ω is an upper Lebesgue integrable function on $[0, T_0]$, for almost all t. Then $\{A(t)\}$ is quasi-stable, with stability index $(\exp(2cT_0), \omega)$, with respect to $\|\cdot\|_t$ for almost every $t \in [0, T_0]$.

Proof: We consider only the case where ω is constant, assuming the stable case. Applying (6.2.5), with $t = T_0$ and $s = t_k$, gives

$$\begin{aligned}\left\|\prod_{j=1}^{k} [A(t_j) + \lambda]^{-1} f\right\|_{T_0} &\leqslant e^{c(T_0 - t_k)} \left\|\prod_{j=1}^{k} [A(t_j) + \lambda]^{-1} f\right\|_{t_k} \\ &\leqslant e^{c(T_0 - t_k)} (\lambda - \omega)^{-1} \left\|\prod_{j=1}^{k-1} [A(t_j) + \lambda]^{-1} f\right\|_{t_k},\end{aligned}$$

where we have used $A(t_k) \in G(\mathbf{X}_{t_k}, 1, \omega)$. Inductive application of (6.2.5) and the hypothesis give

$$\begin{aligned}\left\|\prod_{j=1}^{k} [A(t_j) + \lambda]^{-1} f\right\|_{T_0} &\leqslant (\lambda - \omega)^{-k} e^{c(T_0 - t_k)} \cdots e^{c(t_2 - t_1)} e^{ct_1} \|f\|_0 \\ &= (\lambda - \omega)^{-k} e^{cT_0} \|f\|_0 \\ &\leqslant (\lambda - \omega)^{-k} e^{2cT_0} \|f\|_{T_0},\end{aligned}$$

where we have used (6.2.5) to obtain the last inequality. The proof for arbitrary $t \in [0, T_0]$ is similar. ■

Definition 6.2.2. Let $\mathbf{Y}$ be a Banach space densely and continuously embedded in a Banach space $\mathbf{X}$. If Q is a linear operator on $\mathbf{X}$, then the part of Q in $\mathbf{Y}$, $\tilde{Q}$, is the restriction of Q to

$$\mathbf{D}_{\tilde{Q}} = \{f \in \mathbf{Y} \cap \mathbf{D}_Q : Qf \in \mathbf{Y}\}.$$

Let $A \in G(\mathbf{X}, M, \omega)$. The space $\mathbf{Y}$ is said to be A-admissible if $\{e^{-tA}\}$ leaves $\mathbf{Y}$ invariant, and forms a semigroup of class (C_0) on $\mathbf{Y}$.

Proposition 6.2.3. For $A \in G(\mathbf{X})$, $\mathbf{Y}$ is A-admissible if and only if, for sufficiently large λ,

$$(A + \lambda)^{-1}\mathbf{Y} \subset \mathbf{Y}, \qquad \|(A + \lambda)^{-n}\|_{\mathbf{Y}} \leqslant \tilde{M}(\lambda - \tilde{\omega})^{-n}, \qquad n = 1, 2, \ldots, \tag{6.2.6a}$$

for constants $\tilde{M}$ and $\tilde{\omega}$, and

$$(\mathrm{A} + \lambda)^{-1}\mathbf{Y} \qquad \text{is dense in} \quad \mathbf{Y}. \tag{6.2.6b}$$

In this case, the part $-\tilde{\mathrm{A}}$ of $-\mathrm{A}$ in $\mathbf{Y}$ generates the part of $\{e^{-t\mathrm{A}}\}$ in $\mathbf{Y}$, the part of $(\mathrm{A} + \lambda)^{-1}$ in $\mathbf{Y}$ is just $(\tilde{\mathrm{A}} + \lambda)^{-1}$, and

$$\|e^{-t\mathrm{A}}\|_{\mathbf{Y}} \leqslant \tilde{M}e^{\tilde{\omega}t}. \tag{6.2.6c}$$

Proof: Suppose $\mathbf{Y}$ is A-admissible and let $-\tilde{\mathrm{A}}$ be the generator of the part of $\{e^{-t\mathrm{A}}\}$ in $\mathbf{Y}$, so that $\tilde{\mathrm{A}} \in G(\mathbf{Y}, \tilde{M}, \tilde{\omega})$. Let $g \in \mathbf{Y}$. Since

$$e^{-t\mathrm{A}}g = e^{-t\tilde{\mathrm{A}}}g, \qquad t \geqslant 0,$$

the Laplace transform (6.1.9) shows that

$$(\mathrm{A} + \lambda)^{-1}g = (\tilde{\mathrm{A}} + \lambda)^{-1}g \in \mathbf{Y}, \qquad \lambda > \max(\omega, \tilde{\omega}).$$

Thus, $(\mathrm{A} + \lambda)^{-1}\mathbf{Y} \subset \mathbf{Y}$, and $(\tilde{\mathrm{A}} + \lambda)^{-1} \in \mathbf{B}[\mathbf{Y}]$ is exactly the part of $(\mathrm{A} + \lambda)^{-1}$ in $\mathbf{Y}$. In particular, (6.2.6a) follows from $\tilde{\mathrm{A}} \in G(\mathbf{Y}, \tilde{M}, \tilde{\omega})$. Note that (6.2.6b) follows from Propositions 6.1.2, 6.1.3, and

$$\mathbf{D}_{\tilde{\mathrm{A}}} = (\tilde{\mathrm{A}} + \lambda)^{-1}\mathbf{Y} = (\mathrm{A} + \lambda)^{-1}\mathbf{Y}.$$

Suppose, conversely, that (6.2.6a,b) hold, and let $\tilde{\mathrm{A}}$ be the part of A in $\mathbf{Y}$. For $g \in \mathbf{Y}$, let

$$f = (\mathrm{A} + \lambda)^{-1}g, \qquad \lambda \quad \text{sufficiently large.}$$

Then, $f \in \mathbf{Y}$ by hypothesis, and, by Proposition 6.1.3, $f \in \mathbf{D}_{\mathrm{A}}$, $\mathrm{A}f = g - \lambda f \in \mathbf{Y}$. Hence, $f \in \mathbf{D}_{\tilde{\mathrm{A}}}$, with $\tilde{\mathrm{A}}f = \mathrm{A}f$, and

$$g = (\mathrm{A} + \lambda)f = (\tilde{\mathrm{A}} + \lambda)f.$$

Thus, $(\tilde{\mathrm{A}} + \lambda)^{-1}$ is defined on $\mathbf{Y}$ and coincides with the part of $(\mathrm{A} + \lambda)^{-1}$ on $\mathbf{Y}$. Thus, by (6.2.6a),

$$\|(\tilde{\mathrm{A}} + \lambda)^{-n}\|_{\mathbf{Y}} \leqslant \tilde{M}(\lambda - \tilde{\omega})^{-n}, \qquad n = 1, 2, \ldots.$$

Since $\tilde{\mathrm{A}}$ is densely defined by (6.2.6b), we have $\tilde{\mathrm{A}} \in G(\mathbf{Y}, \tilde{M}, \tilde{\omega})$ by Theorem 6.1.6. The limit characterization (6.1.20) shows that $e^{-t\mathrm{A}}g = e^{-t\tilde{\mathrm{A}}}g \in \mathbf{Y}$ for $g \in \mathbf{Y}$. In particular, $\mathbf{Y}$ is A-admissible and (6.2.6c) holds. ■

Proposition 6.2.4. Let S be an isomorphism of $\mathbf{Y}$ onto $\mathbf{X}$, where $\mathbf{X}$ and $\mathbf{Y}$ are described in Definition 6.2.2. For $\mathrm{A} \in G(\mathbf{X}, M, \omega)$, $\mathbf{Y}$ is A-admissible if and only if $\mathrm{A}_1 = \mathrm{SAS}^{-1}$ is in $G(\mathbf{X})$. In this case, $\mathrm{S}e^{-t\mathrm{A}}\mathrm{S}^{-1} = e^{-t\mathrm{A}_1}$, $t \geqslant 0$. Here, $\mathbf{D}_{\mathrm{A}_1} = \{f \in \mathbf{X} : \mathrm{S}^{-1}f \in \mathbf{D}_{\mathrm{A}}, \mathrm{AS}^{-1}f \in \mathbf{Y}\}$.

Proof: Since $A_1 + \lambda = S(A + \lambda)S^{-1}$, it follows that

$$(A_1 + \lambda)^{-1} = S(A + \lambda)^{-1}S^{-1}, \qquad \lambda > \omega. \tag{6.2.7}$$

Now suppose that **Y** is A-admissible. By (6.2.7) and the previous proposition, $(A_1 + \lambda)^{-1} \in \mathbf{B}[\mathbf{X}]$, $\lambda > \omega$, so that A_1 is closed, since $(A_1 + \lambda)^{-1}$ is similar to $(\tilde{A} + \lambda)^{-1} \in \mathbf{B}[\mathbf{Y}]$, with $\tilde{A} \in G(\mathbf{Y})$. Moreover, $\mathbf{D}_{A_1}$ is dense in **X**, so that, by direct estimation of (6.2.7) and Theorem 6.1.6, it follows that $A_1 \in G(\mathbf{X})$. The converse is similar. The relation $e^{-tA_1} = Se^{-tA}S^{-1}$ follows from (6.1.20). ∎

Theorem 6.2.5. Let **X** and **Y** be Banach spaces, such that **Y** is densely and continuously embedded in **X**. Let $A(t) \in G(\mathbf{X})$, $0 \leqslant t \leqslant T_0$, and assume the following.

(1) The family $\{A(t)\}$ is quasi-stable on **X**, say with stability index (M, ω).
(2) The space **Y** is $A(t)$-admissible for each t. If $\tilde{A}(t) \in G(\mathbf{Y})$ is the part of $A(t)$ in **Y**, $\{\tilde{A}(t)\}$ is quasi-stable on **Y**, say with stability index $(\tilde{M}, \tilde{\omega})$.
(3) The space $\mathbf{Y} \subset \mathbf{D}_{A(t)}$, and the map $A(\cdot): [0, T_0] \to \mathbf{B}(\mathbf{Y}, \mathbf{X})$, defined by $t \mapsto A(t)$, is continuous.

Under these conditions, there exists a unique family of operators $U(t, s) \in \mathbf{B}[\mathbf{X}]$, defined for $0 \leqslant s \leqslant t \leqslant T_0$, with the following properties.

(a) The operator function $U(t, s)$ is strongly continuous (**X**) jointly in t and s, with $U(s, s) = I$, and $\|U(t, s)\|_{\mathbf{X}} \leqslant M \exp\{\int_{(s,t)}^{*} |\omega| \, d\tau\}$.
(b) $U(t, r) = U(t, s)U(s, r), \qquad r \leqslant s \leqslant t.$
(c) $[D_t^+ U(t, s)g]_{t=s} = -A(s)g, \qquad g \in \mathbf{Y}, \qquad 0 \leqslant s < T_0.$
(d) $(d/ds)U(t, s)g = U(t, s)A(s)g, \qquad g \in \mathbf{Y}, \qquad 0 \leqslant s \leqslant t \leqslant T_0.$

Here, D^+ denotes right derivative in the strong sense and d/ds the corresponding two-sided derivative in the strong sense. When $s = t$, the latter is simply a left derivative.

Remark 6.2.3. The family $\{U(t, s)\}$ will be referred to as the evolution operators for the family $\{A(t)\}$.

Proof: We give the details only in the case where $\{A(t)\}$ and $\{\tilde{A}(t)\}$ are stable. We define a sequence $\{A_n(\cdot)\}$ of step-function approximations to $A(\cdot)$ by

$$A_n(t) = A(T_0[nt/T_0]/n), \qquad 0 \leqslant t \leqslant T_0,$$

where $[s]$ denotes the greatest integer less than or equal to s. Since $A(\cdot)$ is norm continuous $(\mathbf{Y}, \mathbf{X})$, we have

$$\|A_n(t) - A(t)\|_{\mathbf{Y},\mathbf{X}} \to 0, \qquad n \to \infty \tag{6.2.8}$$

uniformly for $t \in [0, T_0]$. Moreover, $\{A_n(t)\}$ and $\{\tilde{A}_n(t)\}$ are stable, with constants M, ω and $\tilde{M}$, $\tilde{\omega}$, respectively, independent of n.

We define $\{U_n(t,s)\}$ as follows. If $s \leqslant t$, and s and t belong to the closure of an interval, in which $A_n(\cdot) \equiv \text{constant} = A$, then set $U_n(t,s) = e^{-(t-s)A}$. For other values of $s \leqslant t$, $U_n(t,s)$ is determined by conditions (a) and (b). Explicitly,

$$U_n(t,r) = U_n(t,s)U_n(s,r), \qquad 0 \leqslant r \leqslant s \leqslant t.$$

For example, if

$$\left(\frac{i-2}{n}\right)T_0 \leqslant r < \left(\frac{i-1}{n}\right)T_0 < t \leqslant \left(\frac{i}{n}\right)T_0, \qquad 1 < i \leqslant n,$$

then set $s_* = ((i-1)/n)T_0$, and define

$$U_n(t,r) = \exp\left\{(s_* - t)A\left[\left(\frac{i-1}{n}\right)T_0\right]\right\}\exp\left\{(r - s_*)A\left[\left(\frac{i-2}{n}\right)T_0\right]\right\}.$$

Now, $U_n(t,s)$ clearly satisfies (c) and (d), if $s \neq (j/n)T_0$, $1 \leqslant j \leqslant n-1$, and

$$\frac{d}{dt}U_n(t,s)g = -A_n(t)U_n(t,s)g, \qquad g \in \mathbf{Y}, \tag{6.2.9}$$

if $t \neq (j/n)T_0$; it is necessary to note here that $U_n(t,s)\mathbf{Y} \subset \mathbf{Y}$ by (2).

By the stability of $\{A(t)\}$, $\{\tilde{A}(t)\}$, and (6.2.3), we have

$$\|U_n(t,s)\|_{\mathbf{X}} \leqslant M\exp\{\omega(t-s)\} \qquad \|U_n(t,s)\|_{\mathbf{Y}} \leqslant \tilde{M}\exp\{\tilde{\omega}(t-s)\} \tag{6.2.10}$$

We shall show that

$$\lim_{n\to\infty} U_n(t,s) := U(t,s) \tag{6.2.11}$$

exists strongly $(\mathbf{X})$ uniformly in t, s, $t \geqslant s$. It will then be clear that $U(t,s)$ satisfies (a) and (b) by inheriting the corresponding properties of $\{U_n(t,s)\}$. By the first inequality in (6.2.10), it suffices to show that $\lim_{n\to\infty} U_n(t,r)g$ exists in $\mathbf{X}$, for each $g \in \mathbf{Y}$, uniformly in t, r. Now, differentiating $U_n(t,s)U_m(s,r)g$ with respect to s, by using (d) and (6.2.9), and then integrating in s, we obtain the identity

$$-\mathrm{U}_n(t,s)\mathrm{U}_m(s,r)g\Big|_{s=r}^{s=t} = \mathrm{U}_n(t,r)g - \mathrm{U}_m(t,r)g$$
$$= -\int_r^t \mathrm{U}_n(t,s)[\mathrm{A}_n(s) - \mathrm{A}_m(s)]\mathrm{U}_m(s,r)g\,ds, \quad (6.2.12a)$$

which immediately gives the estimate

$$\|\mathrm{U}_n(t,r)g - \mathrm{U}_m(t,r)g\|_{\mathbf{X}} \leqslant M\tilde{M}e^{\gamma(t-r)}\|g\|_{\mathbf{Y}} \int_r^t \|\mathrm{A}_n(s) - \mathrm{A}_m(s)\|_{\mathbf{Y},\mathbf{X}}\,ds, \quad (6.2.12b)$$

where $\gamma = \max(\omega, \tilde{\omega})$. By (6.2.8) and the triangle inequality, we have $\|\mathrm{A}_n(s) - \mathrm{A}_m(s)\|_{\mathbf{Y},\mathbf{X}} \to 0$, $n, m \to \infty$ uniformly in s, so that $\{\mathrm{U}_n(t,r)g\}$ is Cauchy and, hence, strongly convergent.

We are now ready to prove (c) and (d). Fix $0 \leqslant r < T_0$, and set $\mathrm{A}'(s) \equiv \mathrm{A}(r)$ for $s \in [0, T_0]$. We may identify this family with that described in Lemma 6.2.6, to follow. In particular, the right member of (6.2.14) is $o(t - r)$ as $t \downarrow r$. On the left, we have

$$\mathrm{U}'(t,r)g = e^{-(t-r)\mathrm{A}(r)}g,$$

which has the right derivative $-\mathrm{A}(r)g$ at $t = r$. Hence, division of (6.2.14) by $(t - r)$ yields (c) upon letting $t \downarrow r$. By fixing $0 \leqslant t \leqslant T_0$, and setting $\mathrm{A}'(s) \equiv \mathrm{A}(t)$ for $s \in [0, T_0]$, we obtain for the left-hand derivative

$$[\mathrm{D}_s^- \mathrm{U}(t,s)g]_{s=t} = \mathrm{A}(t)g. \quad (6.2.13)$$

Property (d) is now proved by reductions to (c) and (6.2.13). Thus, if $s < t$, we have

$$\mathrm{D}_s^+ \mathrm{U}(t,s)g = \lim_{h \downarrow 0} \{h^{-1}[\mathrm{U}(t, s+h)g - \mathrm{U}(t,s)g]\}$$
$$= \lim_{h \downarrow 0} \{\mathrm{U}(t, s+h)h^{-1}[g - \mathrm{U}(s+h, s)g]\}$$
$$= \mathrm{U}(t,s)\mathrm{A}(s)g,$$

by (c) and the strong continuity ($\mathbf{X}$) of $\mathrm{U}(t,s)$. Again, for $s \leqslant t$,

$$\mathrm{D}_s^- \mathrm{U}(t,s)g = \lim_{h \downarrow 0} \{h^{-1}[\mathrm{U}(t,s)g - \mathrm{U}(t, s-h)g]\}$$
$$= \lim_{h \downarrow 0} \{\mathrm{U}(t,s)h^{-1}[g - \mathrm{U}(s, s-h)g]\}$$
$$= \mathrm{U}(t,s)\mathrm{A}(s)g,$$

by (6.2.13). This proves (d).

To verify uniqueness, suppose $\{\mathrm{V}(t,s)\}$ satisfies (a) and (d). Then, differentiation of $\mathrm{V}(t,s)\mathrm{U}_n(s,r)g$ in the variable s, by use of (d), and (6.2.9) gives, after integration,

$$V(t,r)g - U_n(t,r)g = -\int_r^t V(t,s)[A(s) - A_n(s)]U_n(s,r)g\,ds,$$

for $g \in \mathbf{Y}$. Using (a), (6.2.8), and (6.2.10), we find, as above, that

$$V(t,r)g = \lim_{n\to\infty} U_n(t,r)g = U(t,r)g,$$

hence, $V(t,r) = U(t,r)$. ■

We cite now the lemma referred to in the proof of Theorem 6.2.5. We omit the proof, which follows the verification of (6.2.12), with an additional limiting process.

Lemma 6.2.6. Let $\{A'(t)\}$ be another family, satisfying the assumptions of Theorem 6.2.5, with stability constants M', ω', $\tilde{M}'$, $\tilde{\omega}'$. Let $\{U'(t,s)\}$ be constructed from $\{A'(t)\}$, as in the construction (6.2.11). Then,

$$\|U'(t,r)g - U(t,r)g\|_{\mathbf{X}} \leqslant M'\tilde{M}e^{\gamma(t-r)}\|g\|_{\mathbf{Y}} \int_r^t \|A'(s) - A(s)\|_{\mathbf{Y},\mathbf{X}}\,ds, \quad (6.2.14)$$

where $\gamma = \max(\omega', \tilde{\omega})$.

Remark 6.2.4. In the more general case of quasi-stability, considerable care must be exercised in the construction of both the partitions and the step-function sequence, itself. Fundamental to these constructions is the approximation of measurable functions, with values in a Banach space, by Riemann step functions. These step functions, and the associated partitions, provide the basis for the definition of $A_n(t)$. The reader is referred to the paper of Kato [12], especially the Appendix, for details. We note here that the additional modifications involve the replacement of (6.2.10) by

$$\|U_n(t,s)\|_{\mathbf{X}} \leqslant M \exp\left\{\int_{(s,t)}^{*} |\omega|\,d\tau\right\}, \qquad \|U_n(t,s)\|_{\mathbf{Y}} \leqslant \tilde{M} \exp\left\{\int_{(s,t)}^{*} |\tilde{\omega}|\,d\tau\right\}, \quad (6.2.15)$$

and the replacement of Lemma 6.2.6 by the more special application required to prove the differentiability.

The following proposition gives a sufficient condition for hypothesis (2) of Theorem 6.2.5.

Proposition 6.2.7. Hypothesis (2) of Theorem 6.2.5 is implied by: (2′) There is a family $\{S(t)\}$ of isomorphisms of $\mathbf{Y}$ onto $\mathbf{X}$, such that

$$S(t)A(t)S(t)^{-1} = A_1(t) \in G(\mathbf{X}), \qquad 0 \leqslant t \leqslant T_0, \quad (6.2.16)$$

and that $\{A_1(t)\}$ is a quasi-stable family on $\mathbf{X}$, say with index (M_1, ω_1). Furthermore, there is a constant c, such that $\|S(t)\|_{\mathbf{Y},\mathbf{X}} \leqslant c$, $\|S(t)^{-1}\|_{\mathbf{X},\mathbf{Y}} \leqslant c$, and the map $t \mapsto S(t)$ is of bounded variation in $\mathbf{B}(\mathbf{Y}, \mathbf{X})$ norm. If $\{A_1(t)\}$ is stable, then a stronger form of hypothesis (2) holds in which $\{\tilde{A}(t)\}$ is stable on $\mathbf{Y}$.

Proof: We give the details only in the case of stability. From Proposition 6.2.4, we conclude that $\mathbf{Y}$ is $A(t)$-admissible. Let $\tilde{A}(t) \in G(\mathbf{Y})$ be the part of $A(t)$ in $\mathbf{Y}$. To show that $\{\tilde{A}(t)\}$ is stable, we write, by (6.2.7) and Proposition 6.2.3,

$$(\tilde{A}(t) + \lambda)^{-1} = S(t)^{-1}(A_1(t) + \lambda)^{-1}S(t),$$

so that

$$\prod_{j=1}^{k} [\tilde{A}(t_j) + \lambda]^{-1} = \prod_{j=1}^{k} S(t_j)^{-1}[A_1(t_j) + \lambda]^{-1}S(t_j). \tag{6.2.17}$$

Setting

$$P_j = [S(t_j) - S(t_{j-1})]S(t_{j-1})^{-1}, \tag{6.2.18}$$

so that

$$S(t_j)S(t_{j-1})^{-1} = I + P_j,$$

we have, for (6.2.17),

$$S(t_k)^{-1}[(A_1(t_k) + \lambda)^{-1}(I + P_k) \cdots (I + P_2)(A_1(t_1) + \lambda)^{-1}]\, S(t_1). \tag{6.2.19}$$

To estimate (6.2.19), we expand $[\cdot]$ into a polynomial in the P_j, with the general term

$$\prod_{j=j_1+1}^{k} (A_1(t_j) + \lambda)^{-1}P_{j_1+1} \prod_{j=j_2+1}^{j_1} (A_1(t_j) + \lambda)^{-1}P_{j_2+1} \cdots P_{j_r+1}$$
$$\times \prod_{j=1}^{j_r} (A_1(t_j) + \lambda)^{-1},$$

where $k > j_1 > j_2 > \cdots > j_r > 0$. If (6.2.2a) is used to estimate the product resolvent blocks, and the estimator polynomial is then recombined, we obtain

$$\left\| \prod_{j=1}^{k} (\tilde{A}(t_j) + \lambda)^{-1} \right\|_{\mathbf{Y}} \leqslant c^2 M_1(\lambda - \omega_1)^{-k}(1 + M_1\|P_k\|_{\mathbf{X}}) \cdots (1 + M_1\|P_2\|_{\mathbf{X}}). \tag{6.2.20}$$

Since

$$\|P_j\|_{\mathbf{X}} \leqslant c\|S(t_j) - S(t_{j-1})\|_{\mathbf{Y},\mathbf{X}},$$

it follows that $\sum_{j=2}^{k} \|\mathrm{P}_j\|_{\mathbf{X}} \leqslant cV$, where V denotes the total variation of $\mathrm{S}(\cdot)$. The right-hand side of (6.2.20) is, thus, bounded from above by $c^2 M_1(\lambda - \omega_1)^{-k} e^{cM_1 V}$. ■

Remark 6.2.5. If (2′) holds, then $\{\tilde{\mathrm{A}}(t)\}$ is quasi-stable on **Y** with index $(M_1 c^2 e^{cM_1 V}, \omega_1)$.

6.3 PERTURBATIONS OF GENERATORS AND REGULARITY OF EVOLUTION OPERATORS

We begin with a standard type perturbation result.

Proposition 6.3.1.

(1) If $\mathrm{A} \in G(\mathbf{X}, M, \omega)$ and $\mathrm{B} \in \mathbf{B}[\mathbf{X}]$, then $\mathrm{A} + \mathrm{B} \in G(\mathbf{X}, M, \omega + \|\mathrm{B}\| M)$.

(2) Let $\{\mathrm{A}(t)\}$ be quasi-stable on $[0, T_0]$, with index (M, ω), and let $\mathrm{B}(t) \in \mathbf{B}[\mathbf{X}]$ be a family defined almost everywhere on $[0, T_0]$, such that $\|\mathrm{B}(\cdot)\|$ is upper Lebesgue integrable. Then $\{\mathrm{A}(t) + \mathrm{B}(t)\}$ forms a quasi-stable family on $[0, T_0]$, with index $(M, \omega + M\|\mathrm{B}(\cdot)\|)$. The expression $\{\mathrm{A}(t) + \mathrm{B}(t)\}$ is stable, with index $(M, \omega + M\beta)$, if $\{\mathrm{A}(t)\}$ is stable and $\|\mathrm{B}(t)\| \leqslant \beta$ for all $t \in [0, T_0]$.

Proof: We follow [10, Theorem 13.2.1] with minor modifications. Suppose the hypotheses of (1) hold. Then, the resolvent identity (when $\mathrm{R}(\lambda, -\mathrm{A} - \mathrm{B})$ is defined),

$$\mathrm{R}(\lambda, -\mathrm{A}-\mathrm{B}) - \mathrm{R}(\lambda, -\mathrm{A}) = \mathrm{R}(\lambda, -\mathrm{A}-\mathrm{B})(-\mathrm{B})\mathrm{R}(\lambda, -\mathrm{A}),$$

leads to

$$\mathrm{R}(\lambda, -\mathrm{A}-\mathrm{B}) = \sum_{j=0}^{\infty} \mathrm{R}(\lambda, -\mathrm{A})[-\mathrm{B}\mathrm{R}(\lambda, -\mathrm{A})]^j, \qquad (6.3.1)$$

if

$$\|\mathrm{B}\mathrm{R}(\lambda, -\mathrm{A})\| < 1.$$

In particular, the latter holds, if

$$\lambda > \omega_1 = \omega + M\|\mathrm{B}\|, \qquad (6.3.2)$$

since, in this case,

$$\|\mathrm{BR}(\lambda, -\mathrm{A})\| \leqslant \frac{\|B\|M}{\lambda - \omega} < 1.$$

Now (1) and (2) are verified by the estimates

$$\begin{aligned} \|[\mathrm{R}(\lambda, -\mathrm{A}-\mathrm{B})]^k\| &= \left\| \left\{ \sum_{j=0}^{\infty} \mathrm{R}(\lambda, -\mathrm{A})[-\mathrm{BR}(\lambda, -\mathrm{A})]^j \right\}^k \right\| \\ &\leqslant M(\lambda - \omega_1)^{-k}, \end{aligned} \tag{6.3.3a}$$

and

$$\begin{aligned} \left\| \prod_{j=1}^{k} \mathrm{R}(\lambda, -\mathrm{A}(t_j) - \mathrm{B}(t_j)) \right\| &= \left\| \prod_{j=1}^{k} \sum_{n=0}^{\infty} \mathrm{R}(\lambda, -\mathrm{A}(t_j))[-\mathrm{B}(t_j)\mathrm{R}(\lambda, -\mathrm{A}(t_j))]^j \right\| \\ &\leqslant M \prod_{j=1}^{k} (\lambda - \omega(t_j))^{-1}, \end{aligned} \tag{6.3.3b}$$

respectively. We shall derive the first estimate, since the second is proved in a parallel manner.

A typical term containing ℓ of the B's in the expansion within $\|\cdot\|$ in (6.3.3a) is of the form

$$[\mathrm{R}(\lambda, -\mathrm{A})]^{r_1}(-\mathrm{B})[\mathrm{R}(\lambda, -\mathrm{A})]^{r_2}(-\mathrm{B}) \cdots [\mathrm{R}(\lambda, -\mathrm{A})]^{r_\ell}(-\mathrm{B})[\mathrm{R}(\lambda, -\mathrm{A})]^{r_{\ell+1}},$$

where $\sum_{i=1}^{\ell+1} r_i = k + \ell$, and $r_i > 0$. Such a term is estimated from above, in norm, by

$$\begin{aligned} &M(\lambda - \omega)^{-r_1}\|\mathrm{B}\|M(\lambda - \omega)^{-r_2}\|\mathrm{B}\| \cdots M(\lambda - \omega)^{-r_\ell}\|\mathrm{B}\|M(\lambda - \omega)^{-r_{\ell+1}} \\ &\quad = M^{\ell+1}\|\mathrm{B}\|^{\ell}(\lambda - \omega)^{-(k+\ell)}. \end{aligned}$$

Now, the number of terms containing ℓ of the B's is the binomial coefficient C_ℓ^k in $(1 - x)^{-k} = \Sigma_{\ell=0}^{\infty} C_\ell^k x^\ell$. Thus,

$$\begin{aligned} \|[R(\lambda, -\mathrm{A}-\mathrm{B})]^k\| &\leqslant \sum_{\ell=0}^{\infty} C_\ell^k M^{\ell+1}\|\mathrm{B}\|^{\ell}(\lambda - \omega)^{-(k+\ell)} \\ &= M(\lambda - \omega)^{-k}[1 - M\|\mathrm{B}\|(\lambda - \omega)^{-1}]^{-k} \\ &= M[(\lambda - \omega) - M\|\mathrm{B}\|]^{-k} \\ &= M(\lambda - \omega_1)^{-k}. \quad \blacksquare \end{aligned}$$

Remark 6.3.1. The second topic in this section deals with the regularity of the evolution operators, or, more precisely, the invariance of **Y** under U(t, s), and the differentiability of the latter with respect to the first argument. To achieve this, we shall introduce an operator convolution structure, and examine the solution of certain Volterra integral equations. We prepare for this with a definition and technical lemma.

Definition 6.3.1. Set $J=[0, T_0]$, and $\Delta=\{(t,s):0\leqslant s\leqslant t\leqslant T_0\}$. For $f:J\to\mathbf{X}$ almost everywhere, write

$$\|f\|_{\mathbf{L}^\infty(J;\,\mathbf{X})} = \operatorname{ess\,sup}\{\|f(t)\|_{\mathbf{X}} : t \in J\},$$

$$\|f\|_{\mathbf{L}^1_*(J;\,\mathbf{X})} = \int_J^* \|f(t)\|_{\mathbf{X}}\, dt.$$

These are sometimes abbreviated to $\|f\|_{\infty,\mathbf{X}}$ and $\|f\|_{1,\mathbf{X}}$. Similar notation holds for functions on Δ to $\mathbf{X}$. For an operator-valued function $\mathrm{F}: J \to \mathbf{B}(\mathbf{X}, \mathbf{Y})$ almost everywhere or $\mathrm{F}:\Delta \to \mathbf{B}(\mathbf{X}, \mathbf{Y})$ almost everywhere we use the obvious symbols, sometimes abbreviated to $\|\mathrm{F}\|_{\infty,\mathbf{X},\mathbf{Y}}$ and $\|\mathrm{F}\|_{1,\mathbf{X},\mathbf{Y}}$, respectively. We even suppress $\mathbf{X}, \mathbf{Y}$ without ambiguity on occasion.

Lemma 6.3.2. Suppose that $\mathbf{X}', \mathbf{Y}', \mathbf{X}''$, and $\mathbf{Y}''$ are given Banach spaces. Let $\mathrm{G}':\Delta \to \mathbf{B}(\mathbf{X}', \mathbf{Y}')$ and $\mathrm{G}'':\Delta \to \mathbf{B}(\mathbf{X}'', \mathbf{Y}'')$ be strongly continuous. Let $\mathrm{F}:J \to \mathbf{B}(\mathbf{Y}', \mathbf{X}'')$ almost everywhere be strongly measurable with $\|\mathrm{F}\|_1 < \infty$. Then, there is $\mathrm{G}:\Delta \to \mathbf{B}(\mathbf{X}', \mathbf{Y}'')$, denoted by $\mathrm{G} = \mathrm{G}''\mathrm{F}\mathrm{G}'$, such that

$$\mathrm{G}(t, r)f = \int_r^t \mathrm{G}''(t, s)\mathrm{F}(s)\mathrm{G}'(s, r)f\, ds, \tag{6.3.4}$$

$(t, r) \in \Delta$, for each $f \in X'$. The function G is strongly continuous on Δ to $\mathbf{B}(\mathbf{X}', \mathbf{Y}'')$, and

$$\|\mathrm{G}f\|_\infty \leqslant \|\mathrm{G}''\|_\infty \|\mathrm{F}\|_1 \|\mathrm{G}'f\|_\infty, \tag{6.3.5a}$$

$$\|\mathrm{G}\|_\infty \leqslant \|\mathrm{G}''\|_\infty \|\mathrm{F}\|_1 \|\mathrm{G}'\|_\infty. \tag{6.3.5b}$$

Proof: We omit the proof of this important technical result and refer the reader to Kato [12, p. 652].

Corollary 6.3.3. Let $\mathrm{D}:J \to \mathbf{B}[\mathbf{X}]$ almost everywhere be strongly measurable, with $\|\mathrm{D}\|_1 < \infty$, and let $\{\mathrm{U}(t, s)\}$ be the evolution operators of Theorem 6.2.5. Then, for each integer $p \geqslant 0$, the p–fold convolution estimate

$$\|(\mathrm{UD})^p\mathrm{U}\|_{\infty,\mathbf{X}} \leqslant Me^{\|\omega\|_1} M^p \|\mathrm{D}\|_1^p / p! \tag{6.3.6}$$

holds on Δ. Thus, the Neumann series

$$\mathrm{Z} = \sum_{p=0}^{\infty} (-\mathrm{UD})^p\mathrm{U} \tag{6.3.7}$$

converges uniformly on Δ in $\mathbf{B}[\mathbf{X}]$ norm; Z satisfies the integral equation

$$\mathrm{Z}(t,r)f = \mathrm{U}(t,r)f - \int_r^t \mathrm{Z}(t,s)\mathrm{D}(s)\mathrm{U}(s,r)f\,ds, \tag{6.3.8}$$

for $f \in \mathbf{X}$ and $(t,r) \in \Delta$.

Proof: To verify (6.3.6), let $\phi: J \to [0,\infty)$ be Lebesgue integrable, with $M\|D(\cdot)\|_{\mathbf{X}} \leqslant \phi$. Then, by (6.3.5), if $t = t_{p+1}$ and $\mathbf{X}' = \mathbf{X}'' = \mathbf{Y}' = \mathbf{Y}'' = \mathbf{X}$,

$$\begin{aligned}\|[(\mathrm{UD})^p\mathrm{U}](t,r)\| &\leqslant Me^{\|\omega\|_1}\int_0^{t_{p+1}}\phi(t_p)\cdots\int_0^{t_2}\phi(t_1)dt_1\cdots dt_p\\ &\leqslant Me^{\|\omega\|_1}\int_0^{T_0}\frac{1}{p!}\left(\frac{d}{dt_p}\right)[\Phi(t_p)]^p\,dt_p,\end{aligned}$$

where

$$\Phi(u) = \int_0^u \phi(s)\,ds.$$

Note that $e^{\|\omega\|_1}$ is the *cumulative* exponential part of the evolution operator estimates. Relations (6.3.6) and, thus, (6.3.7) are now immediate, upon taking an infimum over ϕ and a supremum over (t,r). The solution of (6.3.8) may be characterized by

$$\mathrm{Z}(t,r)f = (\mathrm{I} + {*}(\cdot))^{-1}\mathrm{U}(t,r)f, \tag{6.3.9a}$$

where the operation $*$ is defined by

$${*}\mathrm{G}'(t,r)f = \int_r^t \mathrm{G}'(t,s)\mathrm{D}(s)\mathrm{U}(s,r)f. \tag{6.3.9b}$$

The expansion of (6.3.9) is just (6.3.7), which shows the equivalence of (6.3.7) and (6.3.8). ■

We return now to the notational framework of the previous section. The use of the Volterra equation follows in the key identities (6.3.12) and (6.3.13). We require one further technical result which affords a transition between the operators $\{\mathrm{S}(t)\}$, introduced in Proposition 6.2.7, and the hypotheses of the previous corollary.

Lemma 6.3.4. Let $\mathrm{S}: J \to \mathbf{B}(\mathbf{Y},\mathbf{X})$ be an indefinite strong integral of a strongly measurable function $\dot{\mathrm{S}}: J \to \mathbf{B}(\mathbf{Y},\mathbf{X})$, with $\|\dot{\mathrm{S}}(\cdot)\|_{\mathbf{Y},\mathbf{X}}$ upper Lebesgue integrable on J. Let $\mathrm{S}^{-1}(\cdot)$ exist and, hence, be bounded on J. Then $\mathrm{S}(\cdot)$ and $\mathrm{S}^{-1}(\cdot)$ are absolutely continuous in operator norm on J, and $(d/ds)\mathrm{S}^{-1}(\cdot)$ exists strongly almost everywhere on J, with

$$(d/ds)[\mathrm{S}(s)^{-1}f] = -\mathrm{S}(s)^{-1}\dot{\mathrm{S}}(s)\mathrm{S}(s)^{-1}f, \qquad f \in \mathbf{X}. \tag{6.3.10}$$

Proof: The absolute continuity of $\mathrm{S}(\cdot)$ in operator norm is immediate from

$$\|\mathrm{S}(t) - \mathrm{S}(s)\| \leqslant \int_{(s,t)}^{*} \|\dot{\mathrm{S}}(r)\|\, dr, \qquad 0 \leqslant s \leqslant t \leqslant T_0.$$

The boundedness of $\mathrm{S}^{-1}(\cdot)$ on J is due to the fact that the image of the closed unit ball in $\mathbf{Y}$, under $\mathrm{S}(t)$, contains a fixed ball in $\mathbf{X}$, independent of t, due to the continuity of $\mathrm{S}(\cdot)$ on the compact set J. The absolute continuity of $\mathrm{S}^{-1}(\cdot)$ follows from

$$\mathrm{S}(t)^{-1} - \mathrm{S}(s)^{-1} = \mathrm{S}(t)^{-1}[\mathrm{S}(s) - \mathrm{S}(t)]\mathrm{S}(s)^{-1},$$

the absolute continuity of $\mathrm{S}(\cdot)$, and the boundedness of $\mathrm{S}^{-1}(\cdot)$. The strong differentiability of $\mathrm{S}^{-1}(\cdot)$ and (6.3.10) follow from

$$\frac{\mathrm{S}(t)^{-1} - \mathrm{S}(s)^{-1}}{t - s} = \mathrm{S}(t)^{-1}\left(\frac{-\mathrm{S}(t) + \mathrm{S}(s)}{t - s}\right)\mathrm{S}(s)^{-1},$$

the strong differentiability of $\mathrm{S}(\cdot)$, and the uniform boundedness of $\mathrm{S}^{-1}(\cdot)$. ■

Theorem 6.3.5. Suppose that hypotheses (1) and (3) of Theorem 6.2.5 are satisfied together with the following hypothesis.

(2″) There is a family $\{\mathrm{S}(t)\}$ of isomorphisms of $\mathbf{Y}$ onto $\mathbf{X}$, such that

$$\mathrm{S}(t)\mathrm{A}(t)\mathrm{S}(t)^{-1} = \mathrm{A}(t) + \mathrm{B}(t), \qquad \mathrm{B}(t) \in \mathbf{B}[\mathbf{X}], \tag{6.3.11}$$

almost everywhere on $[0, T_0]$, where $\mathrm{B}(\cdot)$ is strongly measurable, with $\|\mathbf{B}(\cdot)\|_{\mathbf{X}}$ upper Lebesgue integrable on $[0, T_0]$. Furthermore, there is a strongly measurable function $\dot{\mathrm{S}}:[0, T_0] \to \mathbf{B}[\mathbf{Y}, \mathbf{X}]$, with $\|\dot{\mathrm{S}}(\cdot)\|_{\mathbf{Y},\mathbf{X}}$ upper integrable on $[0, T_0]$, such that S is an indefinite strong integral of $\dot{\mathrm{S}}$.

Then (2) of Theorem 6.2.5 holds, hence, the evolution operators $\{\mathrm{U}(t,s)\}$ exist. Moreover, if $\{\mathrm{W}(t,s)\} \subset \mathbf{B}[\mathbf{X}]$ are uniquely defined by

$$\mathrm{W}(t,r)f = \mathrm{U}(t,r)f - \int_r^t \mathrm{W}(t,s)[\mathrm{B}(s) - \mathrm{C}(s)]\mathrm{U}(s,r)f\, ds, \tag{6.3.12}$$

for $f \in \mathbf{X}$ and $0 \leqslant r \leqslant t \leqslant T_0$, where $\mathrm{C}(s) = \dot{\mathrm{S}}(s)\mathrm{S}(s)^{-1}$, then

$$\mathrm{U}(t,s) = \mathrm{S}(t)^{-1}\mathrm{W}(t,s)\mathrm{S}(s). \tag{6.3.13}$$

In particular, the following additional conditions hold.

(e) $\mathrm{U}(t,s)\mathbf{Y} \subset \mathbf{Y}, \qquad 0 \leqslant s \leqslant t \leqslant T_0.$
(f) The operator function $\mathrm{U}(t,s)$ is strongly continuous $(\mathbf{Y})$, jointly in s and t.

(g) For each fixed $g \in \mathbf{Y}$, and $0 \leqslant s < T_0$,

$$\frac{d}{dt}\mathrm{U}(t,s)g = -\mathrm{A}(t)\mathrm{U}(t,s)g, \tag{6.3.14}$$

for $s \leqslant t \leqslant T_0$, and this derivative is continuous ($\mathbf{X}$).

Proof: The implication (2″) ⇒ (2) is a consequence of Propositions 6.2.7 and 6.3.1, and Lemma 6.3.4. The existence of a unique solution $\mathrm{W}(t,s) \in \mathbf{B}[\mathbf{X}]$ of (6.3.12), strongly continuous jointly in s, t, is guaranteed by Corollary 6.3.3. Note that $\mathrm{D} = \mathrm{B} - \mathrm{C}$ is strongly measurable, $\|\mathrm{D}\|_1 < \infty$, by (2″) and Lemma 6.3.4.

The verification of (6.3.13) is the heart of the theorem, since (e) and (f) follow directly; (g) is then seen to follow, via the right and left difference quotient relationships

$$\frac{\mathrm{U}(t+h,s)g - \mathrm{U}(t,s)g}{h} = \frac{(\mathrm{U}(t+h,t) - \mathrm{I})\mathrm{U}(t,s)g}{h},$$

$$\frac{\mathrm{U}(t-h,s)g - \mathrm{U}(t,s)g}{-h} = \frac{(\mathrm{U}(t,t-h) - \mathrm{I})\mathrm{U}(t-h,s)g}{h},$$

and (a)–(f).

We now verify (6.3.13).[†] Define $\mathrm{Q}(t,s) = \mathrm{U}(t,s)\mathrm{S}(s)^{-1}$. It is enough to show

$$\mathrm{Q}(t,s) = \mathrm{S}(t)^{-1}\mathrm{W}(t,s). \tag{6.3.15}$$

It follows from (6.3.10) and (d) that, for each $f \in \mathbf{X}$, and almost all s

$$\begin{aligned}(d/ds)\mathrm{Q}(t,s)f &= \mathrm{U}(t,s)\mathrm{A}(s)\mathrm{S}(s)^{-1}f - \mathrm{U}(t,s)\mathrm{S}(s)^{-1}\dot{\mathrm{S}}(s)\mathrm{S}(s)^{-1}f \\ &= \mathrm{U}(t,s)\mathrm{A}(s)\mathrm{S}(s)^{-1}f - \mathrm{Q}(t,s)\mathrm{C}(s)f.\end{aligned}$$

Since $\mathrm{A}(s)\mathrm{S}(s)^{-1}\mathbf{Y} \subset \mathbf{Y}$ by (2″), we then have

$$\mathrm{A}(s)\mathrm{S}(s)^{-1}g = \mathrm{S}(s)^{-1}\mathrm{S}(s)\mathrm{A}(s)\mathrm{S}(s)^{-1}g = \mathrm{S}(s)^{-1}[\mathrm{A}(s) + \mathrm{B}(s)]g,$$

for each $g \in \mathbf{Y}$. Thus, for $g \in \mathbf{Y}$ and $0 \leqslant s \leqslant t \leqslant T_0$, we have, for almost all s on this interval,

$$(d/ds)\mathrm{Q}(t,s)g = \mathrm{Q}(t,s)[\mathrm{A}(s) + \mathrm{B}(s) - \mathrm{C}(s)]g. \tag{6.3.16}$$

If $\{\mathrm{U}_n(t,s)\}$ are the approximating evolution operators constructed in the proof of Theorem 6.2.5, or the modified operators constructed in [12],

[†] Adapted from Dorroh [6]. This paper represents a significant simplification of the earlier development of the theory.

then, by (6.2.9) and (6.3.16),

$$(d/ds)\mathrm{Q}(t,s)\mathrm{U}_n(s,r)g = \mathrm{Q}(t,s)[\mathrm{A}(s) + \mathrm{B}(s) - \mathrm{C}(s) - \mathrm{A}_n(s)]\mathrm{U}_n(s,r)g, \tag{6.3.17}$$

for each $g \in \mathbf{Y}$ and for almost all s. Integrating (6.3.17) from r to t, and utilizing the absolute continuity of the integrand, we obtain

$$\mathrm{S}(t)^{-1}\mathrm{U}_n(t,r)g - \mathrm{Q}(t,r)g = \int_r^t \mathrm{Q}(t,s)[\mathrm{A}(s) + \mathrm{B}(s) - \mathrm{C}(s) - \mathrm{A}_n(s)]\mathrm{U}_n(s,r)g\,ds,$$

for $g \in \mathbf{Y}$. Using (6.2.8) and (6.2.10) (or (6.2.15)), and letting $n \to \infty$, we obtain

$$\mathrm{Q}(t,r)g = \mathrm{S}(t)^{-1}\mathrm{U}(t,r)g - \int_r^t \mathrm{Q}(t,s)[\mathrm{B}(s) - \mathrm{C}(s)]\mathrm{U}(s,r)g\,ds, \tag{6.3.18}$$

for each $g \in \mathbf{Y}$; by continuity, (6.3.18) holds on all of $\mathbf{X}$. Thus, $\mathrm{Q}(t,s) = \mathrm{U}(t,s)\mathrm{S}(s)^{-1}$ may be characterized as the unique solution of (6.3.18). However, by (6.3.12),

$$\mathrm{S}(t)^{-1}\mathrm{W}(t,r)f = \mathrm{S}(t)^{-1}\mathrm{U}(t,r)f - \int_r^t \mathrm{S}(t)^{-1}\mathrm{W}(t,s)[\mathrm{B}(s) - \mathrm{C}(s)]\mathrm{U}(s,r)f\,ds, \tag{6.3.19}$$

where we have interchanged $\mathrm{S}(t)^{-1}$ and $\int_r^t(\cdot)f\,ds$ by continuity. Comparison of (6.3.18) and (6.3.19) gives (6.3.15). ∎

Corollary 6.3.6. The estimate on $\mathbf{Y}$,

$$\|\mathrm{U}(t,s)\|_{\infty,\mathbf{Y}} \leqslant \|\mathrm{S}\|_{\infty,\mathbf{Y},\mathbf{X}} M \exp\{\|\omega\|_1 + M\|\mathrm{B} - \mathrm{C}\|_{1,\mathbf{X}}\}\|\mathrm{S}^{-1}\|_{\infty,\mathbf{X},\mathbf{Y}}, \tag{6.3.20}$$

holds for the operators satisfying the hypotheses of Theorem 6.3.5.

Proof: This follows directly from (6.3.6), (6.3.12), and (6.3.13), with $\mathrm{D} = \mathrm{B} - \mathrm{C}$. ∎

We summarize the discussion of these two sections with the following theorem.

Theorem 6.3.7. Assume that hypotheses (1) and (3) of Theorem 6.2.5, and (2″) of Theorem 6.3.5 hold. Then, there exists a unique family $\{\mathrm{U}(t,s)\}$, defined on $\Delta: 0 \leqslant s \leqslant t \leqslant T_0$, with the following properties.

(a′) The operator function U is strongly continuous on Δ to $\mathbf{B}[\mathbf{X}]$ with $\mathrm{U}(s,s) = \mathrm{I}$.

(b′) $U(t,s)U(s,r) = I$.
(c′) $U(t,s)\mathbf{Y} \subset \mathbf{Y}$, and U is strongly continuous on Δ to $\mathbf{B}[\mathbf{Y}]$.
(d′) $dU(t,s)/dt = -A(t)U(t,s)$, $dU(t,s)/ds = U(t,s)A(s)$, which exist in the strong sense in $\mathbf{B}[\mathbf{Y},\mathbf{X}]$, and are strongly continuous on Δ to $\mathbf{B}[\mathbf{Y},\mathbf{X}]$.

6.4 THE INHOMOGENEOUS PROBLEM AND AN APPLICATION TO LINEAR SYMMETRIC HYPERBOLIC SYSTEMS

Let $F: J \to \mathbf{X}$ be given. We consider the linear Cauchy problem

$$\frac{du}{dt} + A(t)u(t) = F(t), \qquad 0 \leqslant t \leqslant T_0, \tag{6.4.1a}$$

$$u(0) = u_0, \tag{6.4.1b}$$

in the Banach space $\mathbf{X}$. The solution is formally given by

$$u(t) = U(t,0)u_0 + \int_0^t U(t,s)F(s)\,ds, \qquad 0 \leqslant t \leqslant T_0, \tag{6.4.2}$$

where $\{U(t,s)\}$ satisfy Theorem 6.3.7.

Proposition 6.4.1.† Let u be given by (6.4.2).

(1) If $u_0 \in \mathbf{X}$ and $F \in \mathbf{L}^1(J;\mathbf{X})$, then $u \in \mathbf{C}(J;\mathbf{X})$ and

$$\|u\|_{\infty,\mathbf{X}} \leqslant \|U\|_{\infty,\mathbf{X}}(\|u_0\|_{\mathbf{X}} + \|F\|_{1,\mathbf{X}}). \tag{6.4.3}$$

(2) If $u_0 \in \mathbf{Y}$ and $F \in \mathbf{L}^1(J;\mathbf{Y})$, then $u \in \mathbf{C}(J;\mathbf{Y})$ and

$$\|u\|_{\infty,\mathbf{Y}} \leqslant \|U\|_{\infty,\mathbf{Y}}(\|u_0\|_{\mathbf{Y}} + \|F\|_{1,\mathbf{Y}}). \tag{6.4.4}$$

(3) If $u_0 \in \mathbf{Y}$ and $F \in \mathbf{C}(J;\mathbf{X}) \cap \mathbf{L}^1(J;\mathbf{Y})$, then $u \in \mathbf{C}(J;\mathbf{Y}) \cap \mathbf{C}^1(J;\mathbf{X})$, and u satisfies (6.4.1) and the estimate

$$\|du/dt\|_{\infty,\mathbf{X}} \leqslant \|F\|_{\infty,\mathbf{X}} + \|A\|_{\infty,\mathbf{Y},\mathbf{X}}\|U\|_{\infty,\mathbf{Y}}(\|u_0\|_{\mathbf{Y}} + \|F\|_{1,\mathbf{Y}}) \tag{6.4.5}$$

holds.

Proof: Results (1) and (2) follow routinely from the properties of $\{U(t,s)\}$; (6.4.5) follows from (6.4.4) and (6.4.1). To prove the latter, we set up the

† Reproduced from Kato [12]; used with permission.

difference quotient

$$\frac{u(t)-u(r)}{t-r}=\frac{\mathrm{U}(t,0)u_0-\mathrm{U}(r,0)u_0}{t-r}$$
$$+\frac{\int_0^t \mathrm{U}(t,s)F(s)\,ds-\int_0^r \mathrm{U}(t,s)F(s)\,ds}{t-r}$$
$$+\int_0^r\left(\frac{\mathrm{U}(t,s)-\mathrm{U}(r,s)}{t-r}\right)F(s)\,ds. \tag{6.4.6}$$

The second term on the right-hand side of (6.4.6) tends to the limit $\mathrm{U}(r,r)F(r)=F(r)$ as $t\to r$, by the property $F\in\mathbf{C}(J;\mathbf{X})$, and the third term tends to

$$\int_0^r -\mathrm{A}(r)\mathrm{U}(r,s)F(s)\,ds=-\mathrm{A}(r)\int_0^r \mathrm{U}(r,s)F(s)\,ds,$$

by the property $F\in\mathbf{L}^1(J;\mathbf{Y})$. The first term, of course, tends to $-\mathrm{A}(r)\mathrm{U}(r,0)u_0$. This verifies (6.4.1). The regularity of u follows from (6.4.1) and (6.4.2), and the properties of $\{\mathrm{U}(t,s)\}$ and $\{\mathrm{A}(t)\}$. ■

We now give an application to symmetric hyperbolic systems.

Definition 6.4.1. Consider the linear Cauchy problem,

$$\frac{\partial\mathbf{u}}{\partial t}+\sum_{j=1}^{n}a_j(\mathbf{x},t)\frac{\partial\mathbf{u}}{\partial x_j}+b(\mathbf{x},t)\mathbf{u}=F(\mathbf{x},t), \tag{6.4.7a}$$

$$\mathbf{u}(\mathbf{x},0)=\mathbf{u}_0(\mathbf{x}), \tag{6.4.7b}$$

$\mathbf{x}\in\mathbb{R}^n$, $0<t\leqslant T_0$. Here, $\mathbf{u}=(u_1,\ldots,u_m)$, a_j, and b are $m\times m$ matrix functions, and the a_j are assumed to be Hermitian symmetric. Regarding the regularity of the a_j and b, we assume:

The functions a_j and b are in $\mathbf{C}(J;\mathbf{C}_b^1(\mathbb{R}^n))$ as mappings of J into $\mathbf{C}_b^1(\mathbb{R}^n)$.

Here, $\mathbf{C}_b^1(\mathbb{R}^n)$ is the set of all $m\times m$ matrix-valued functions a, such that a and $\partial a/\partial x_j$, $1\leqslant j\leqslant n$, are in the space $\mathbf{C}_b(\mathbb{R}^n)$ of continuous, bounded functions on $\mathbb{R}^n$. The supremum norm for $\mathbf{C}_b(\mathbb{R}^n)$ and $\mathbf{C}_b^1(\mathbb{R}^n)$ is taken on the Euclidean norm of a. In the remainder of this section, we make the identifications $\mathbf{X}=\mathbf{L}^2(\mathbb{R}^n;\mathbb{R}^m)$ and $\mathbf{Y}=\mathbf{H}^1(\mathbb{R}^n;\mathbb{R}^m)$.

The following lemma, proved in Friedrichs [8], permits the recasting of (6.4.7) into (6.4.1).

Lemma 6.4.2. Let $\mathrm{A}_0(t)$ be defined by

$$(\mathrm{A}_0(t)\phi)(\mathbf{x},t)=\sum_{j=1}^{m}a_j(\mathbf{x},t)\frac{\partial\phi}{\partial x_j}+b(\mathbf{x},t)\phi, \tag{6.4.8}$$

for $\boldsymbol{\phi} \in D_{\mathrm{A}_0} = \mathbf{C}_0^\infty(\mathbb{R}^n)$. Then, the linear operator $\mathrm{A}_w(t)$, defined by the formal adjoint, or weak, relation,

$$(\mathbf{u}, \mathrm{A}_0^*(t)\boldsymbol{\phi})_{\mathbf{L}^2(\mathbb{R}^n)} = (\mathbf{v}, \boldsymbol{\phi})_{\mathbf{L}^2(\mathbb{R}^n)} \qquad \text{for all} \quad \boldsymbol{\phi} \in \mathbf{C}_0^\infty(\mathbb{R}^n), \tag{6.4.9}$$

where $\mathbf{v} = \mathrm{A}_w(t)\mathbf{u}$, $\mathbf{u} \in \mathbf{D}_{\mathrm{A}_w(t)}$, is identical to the closure of $\mathrm{A}_0(t)$ in $\mathbf{L}^2(\mathbb{R}^n, \mathbb{R}^m)$, sometimes called the strong extension $\mathrm{A}_s(t)$ of $\mathrm{A}_0(t)$.

Definition 6.4.2. We denote by $\mathrm{A}(t)$ the common operator $\mathrm{A}_w(t) = \mathrm{A}_s(t)$. The formal adjoint, as usual, is defined by

$$(\mathrm{A}_0^*(t)\boldsymbol{\phi})(\mathbf{x}, t) = \sum_{j=1}^{n} -\frac{\partial}{\partial x_j}[a_j(\mathbf{x}, t)\boldsymbol{\phi}(\mathbf{x}, t)] + b^*(\mathbf{x}, t)\boldsymbol{\phi}(x, t).$$

Remark 6.4.1. The space $\mathbf{H}^1(\mathbb{R}^n; \mathbb{R}^m)$ is continuously embedded in $\mathbf{D}_{\mathrm{A}(t)}$, $t \in J$.

Lemma 6.4.3. The energy inequality

$$(\mathrm{A}(t)\mathbf{u}, \mathbf{u})_{\mathbf{L}^2(\mathbb{R}^n)} \geqslant -\omega_t(\mathbf{u}, \mathbf{u})_{\mathbf{L}^2(\mathbb{R}^n)} \tag{6.4.10}$$

holds for all $u \in \mathbf{D}_{\mathrm{A}(t)}$, where

$$\omega_t = \tfrac{1}{2}\sum_{j=1}^{n} \|a_j(\cdot, t)\|_{\mathbf{C}_b^1(\mathbb{R}^n)} + \|b(\cdot, t)\|_{\mathbf{C}_b(\mathbb{R}^n)}. \tag{6.4.11}$$

Proof: It suffices to prove (6.4.10) for $\mathbf{u} = \boldsymbol{\phi} \in \mathbf{D}_{\mathrm{A}_0}$, since $\mathrm{A}(t)$ is closed. Writing

$$A_0(t)\boldsymbol{\phi} = \sum_{j=1}^{n}\left(a_j\frac{\partial\boldsymbol{\phi}}{\partial x_j} + \frac{\partial}{\partial x_j}(a_j\boldsymbol{\phi})\right) + \left(b\boldsymbol{\phi} - \sum_{j=1}^{n}\frac{\partial}{\partial x_j}(a_j\boldsymbol{\phi})\right),$$

$$A_0^*(t)\boldsymbol{\phi} = -\sum_{j=1}^{n}\left(a_j\frac{\partial\boldsymbol{\phi}}{\partial x_j} + \frac{\partial}{\partial x_j}(a_j\boldsymbol{\phi})\right) + \left(b^*\boldsymbol{\phi} + \sum_{j=1}^{n} a_j\frac{\partial\boldsymbol{\phi}}{\partial x_j}\right),$$

and

$$(\mathrm{A}_0(t)\boldsymbol{\phi}, \boldsymbol{\phi})_{\mathbf{L}^2(\mathbb{R}^n)} = \tfrac{1}{2}[(\mathrm{A}_0(t)\boldsymbol{\phi}, \boldsymbol{\phi})_{\mathbf{L}^2(\mathbb{R}^n)} + (\mathrm{A}_0^*(t)\boldsymbol{\phi}, \boldsymbol{\phi})_{\mathbf{L}^2(\mathbb{R}^n)}],$$

we obtain

$$(\mathrm{A}_0(t)\boldsymbol{\phi}, \boldsymbol{\phi})_{\mathbf{L}^2(\mathbb{R}^n)} = \left(\tfrac{1}{2}\left(b + b^* - \sum_{j=1}^{n}\frac{\partial a_j}{\partial x_j}\right)\boldsymbol{\phi}, \boldsymbol{\phi}\right)_{\mathbf{L}^2(\mathbb{R}^n)}. \tag{6.4.12}$$

Inequality (6.4.10) follows immediately from (6.4.12). ■

Lemma 6.4.4. The relation

$$\{\mu : \operatorname{Re} \mu < -\omega_t\} \subset \rho(A(t)) \tag{6.4.13}$$

holds for each $t \in J$.

Proof: The operator

$$A_\lambda(t) = A(t) + (\omega_t + \lambda)I \tag{6.4.14}$$

is clearly injective for $\lambda > 0$ by (6.4.10), and has dense range by (6.4.9). The (uniform) continuity of $[A_\lambda(t)]^{-1}$ on the range of $A_\lambda(t)$ follows from (6.4.10). In particular, $0 \in \rho(A_\lambda(t))$. The conclusion (6.4.13) is now standard (see Agmon [1] Theorem 12.8). ■

Remark 6.4.2. Lemmas 6.4.3 and 6.4.4, taken together, imply $A(t) \in G(\mathbf{X}, 1, \omega_t)$, via the Hille–Yosida theorem, as expressed in Remark 6.1.5. In particular, $\{A(t)\}$ is stable, with stability constants $M = 1$ and $\omega = \sup\{\omega_t : t \in J\}$.

Definition 6.4.3. We now define

$$S(t) = S = (I - \Delta)^{1/2}, \tag{6.4.15}$$

which is an isomorphism of $\mathbf{H}^1(\mathbb{R}^n; \mathbb{R}^m)$ onto $\mathbf{L}^2(\mathbb{R}^n; \mathbb{R}^m)$. Here, we may use the definition, which makes S an isometry,

$$\mathbf{H}^1(\mathbb{R}^n; \mathbb{R}^m) = \{\mathbf{v} \in \mathbf{L}^2(\mathbb{R}^n; \mathbb{R}^m) : \int_{\mathbb{R}^n} (1 + |\xi|^2)|\hat{\mathbf{v}}(\xi)|^2 \, d\xi = \|\mathbf{v}\|^2_{\mathbf{H}^1(\mathbb{R}^n)} < \infty\},$$

where ^ denotes the Fourier transform, defined by

$$F\mathbf{g}(\xi) = \hat{\mathbf{g}}(\xi) = (2\pi)^{-n/2} \int_{\mathbb{R}^n} e^{-i\xi \cdot \mathbf{x}} \mathbf{g}(\mathbf{x}) \, d\mathbf{x},$$

for $\mathbf{g}$ in the Schwartz class $\mathscr{S}$ of rapidly decreasing functions at infinity. With this notation, as usual, if S is understood to induce an action on each vector component,

$$S = (I - \Delta)^{1/2} = F^{-1}(1 + |\xi|^2)^{1/2} F,$$
$$S^{-1} = (I - \Delta)^{-1/2} = F^{-1}(1 + |\xi|^2)^{-1/2} F.$$

Remark 6.4.3. The derivative $\partial/\partial x_j$ commutes with S on $\mathscr{S}$, and $\|(\partial/\partial x_j)S^{-1}\|_{\mathbf{L}^2(\mathbb{R}^n)} \leqslant 1$.

Theorem 6.4.5. The operator S of (6.4.15) satisfies (2″) of Theorem 6.3.5. In particular, the evolution operators $\{U(t,s)\}$ exist for (6.4.7), satisfying (a′)–(d′) of Theorem 6.3.7.

Proof: The only condition to be verified is that

$$SA(t)S^{-1} = A(t) + B(t), \tag{6.4.16}$$

where $B(t) \in \mathbf{B}[\mathbf{X}]$ is, say, strongly continuous ($\mathbf{X}$) as a mapping from J into $\mathbf{B}[\mathbf{X}]$. In fact, we shall show that the mapping is norm continuous. One computes, for $\mathbf{u} \in \mathcal{S}$,†

$$\begin{aligned} SA(t)S^{-1}u &= \sum_{j=1}^{n} Sa_j\left(\frac{\partial}{\partial x_j}\right)S^{-1}\mathbf{u} + SbS^{-1}\mathbf{u} \\ &= \sum_{j=1}^{n} a_j S\left(\frac{\partial}{\partial x_j}\right)S^{-1}\mathbf{u} + \sum_{j=1}^{n}[S, a_j]\left(\frac{\partial}{\partial x_j}\right)S^{-1}\mathbf{u} + b\mathbf{u} + [S, b]S^{-1}\mathbf{u} \\ &= A(t)u + \sum_{j=1}^{n}[S, a_j]\left(\frac{\partial}{\partial x_j}\right)S^{-1}\mathbf{u} + [S, b]S^{-1}\mathbf{u}. \end{aligned}$$

According to a result of Calderón [4], $[S, a(\cdot)]$ can be extended to a bounded operator from $\mathbf{L}^2(\mathbb{R}^n;\mathbb{R}^m)$ to $\mathbf{L}^2(\mathbb{R}^n;\mathbb{R}^m)$, if $a \in \mathbf{C}_b^1(\mathbb{R}^n)$, with bound $\leqslant c\|a\|_{\mathbf{C}_b^1(\mathbb{R}^n)}$. Since

$$[S, a_j(\cdot, t)] - [S, a_j(\cdot, t')] = [S, a_j(\cdot, t) - a_j(\cdot, t')],$$

it follows from Remark 6.4.3 that

$$\sum_{j=1}^{n}[S, a_j]\left(\frac{\partial}{\partial x_j}\right)S^{-1} \in \mathbf{C}([0, T_0]; \mathbf{B}[\mathbf{X}])$$

with a similar statement for $[S, b]S^{-1}$. Thus,

$$SA(t)S^{-1}\mathbf{u} = A(t)\mathbf{u} + B(t)\mathbf{u}, \qquad \mathbf{u} \in \mathcal{S}, \tag{6.4.17}$$

where $B(\cdot) \in \mathbf{C}(J; \mathbf{B}[\mathbf{X}])$. It remains to prove that (6.4.17) holds on $\mathbf{D}_{A(t)}$. Thus, let $\mathbf{v} \in \mathbf{D}_{A(t)}$, and let

$$\mathbf{u}_k \to \mathbf{v}, \qquad A(t)\mathbf{u}_k \to A(t)\mathbf{v} \qquad \text{in} \quad \mathbf{L}^2(\mathbb{R}^n;\mathbb{R}^m),$$

with $\{\mathbf{u}_k\} \subset \mathcal{S}$. Then, $B(t)\mathbf{u}_k \to B(t)\mathbf{v}$, $S^{-1}\mathbf{u}_k \to S^{-1}\mathbf{v}$, and

$$A(t)S^{-1}\mathbf{u}_k = S^{-1}(A(t) + B(t))\mathbf{u}_k \to S^{-1}(A(t) + B(t))\mathbf{v}.$$

Since $A(t)$ is closed, it follows that $A(t)S^{-1}\mathbf{v}$ exists and equals $S^{-1}(A(t) + B(t))\mathbf{v}$. Thus,

$$SA(t)S^{-1} \supset A(t) + B(t).$$

† The symbol $[a_j]$ is used to emphasize operator action.

It follows that

$$S(A(t) + \lambda)^{-1}S^{-1} \supset (A(t) + B(t) + \lambda)^{-1}, \qquad (6.4.18)$$

for sufficiently large λ. Since the right-hand side of (6.4.18) has domain $\mathbf{L}^2(\mathbb{R}^n; \mathbb{R}^m)$, it follows that (6.4.18) represents an equality on $\mathbf{L}^2(\mathbb{R}^n; \mathbb{R}^m)$. Taking inverses leads to (6.4.16). ■

Remark 6.4.4. A comprehensive analysis of linear symmetric hyperbolic systems was given by Friedrichs [9] for the pure Cauchy problem, as well as the mixed initial/boundary-value problem.

6.5 BIBLIOGRAPHICAL REMARKS

We have followed the development of Kato in his two fundamental papers [11, 12], and have incorporated the logical simplification, due to Dorroh [6], concerning the invariant action of $U(t, s)$ on the smooth space $\mathbf{Y}$. The requirement that $A(\cdot)$ be continuous from $[0, T_0]$ to $B(\mathbf{Y}, \mathbf{X})$ may be appropriately relaxed (see Kobayashi [13]) to strong continuity.

Recent alternative treatments have been given by Dorroh and Graff [7], and by Dollard and Friedman [5], and the latter exposition, based on product integration, serves as a nice introduction to the present development. It should be recognized that we have presented this particular development not only for its generality, but also because it is sufficiently robust to serve as a basis for the nonlinear theory developed in the next chapter.

REFERENCES

[1] S. Agmon, "Elliptic Boundary Value Problems." Van Nostrand-Reinhold, New York, 1965.

[2] A. Bellini-Morante, "Applied Semigroups and Evolution Equations." Oxford Univ. Press (Clarendon), London and New York, 1979.

[3] P. L. Butzer and H. Berens, "Semigroups of Operators and Approximation." Springer–Verlag, Berlin and New York, 1967.

[4] A. P. Calderón, Commutators of singular integral operators, *Proc. Nat. Acad. Sci.* **53** 1092–1099 (1965).

[5] J. Dollard and C. Friedman, On strong product integration, *J. Functional Anal.* **28** 309–354 (1978).

[6] J. R. Dorroh, A simplified proof of a theorem of Kato on linear evolution equations, *J. Math. Soc. Japan* **27**, 474–478 (1975).

[7] J. R. Dorroh and R. A. Graff, Integral equations in Banach spaces; a general approach to the linear Cauchy problem and applications to the nonlinear problem, *J. Integral Equations* **1**, 309–359 (1979).

[8] K. Friedrichs, The identity of weak and strong extensions of differential operators, *Trans. Amer. Math. Soc.* **55**, 132–151 (1944).

[9] K. Friedrichs, Symmetric hyperbolic linear differential equations, *Comm. Pure Appl. Math.* **7**, 345–392 (1954); Symmetric positive linear differential equations, *ibid.* **11**, 333–418 (1958).

[10] E. Hille and R. S. Phillips, "Functional Analysis and Semigroups." American Mathematical Society Col. Publ. 31, Providence, Rhode Island, 1957.

[11] T. Kato, Linear equations of "hyperbolic" type, *J. Fac. Sc. Univ. Tokyo* **17**, 241–258 (1970).

[12] T. Kato, Linear evolution equations of "hyperbolic" type II, *J. Math. Soc. Japan* **25**, 648–666 (1973).

[13] K. Kobayashi, On a theorem for linear evolution equations of hyperbolic type, *J. Math Soc. Japan* **31**, 647–654 (1979).

[14] K. Yosida, "Functional Analysis," 5th. ed. Springer–Verlag, Berlin and New York, 1978.

7 QUASI-LINEAR EQUATIONS OF EVOLUTION

7.0 INTRODUCTION

An existence, uniqueness, and well-posedness theory, which is local in time, is presented in Section 7.2 for quasi-linear equations of evolution in separable reflexive Banach spaces (see Theorem 7.2.4); the preliminaries for this theory are presented in the first section. The robustness of the theory is demonstrated in Section 7.3, where it is shown that a significant class of quasi-linear second-order hyperbolic systems is covered by the abstract theory (see Theorem 7.3.6). A nontrivial application is presented in Section 7.4, where the vacuum field equations of general relativity are discussed. The standard reduction afforded by harmonic coordinates is utilized. In the final section, an invariant time interval is determined on which the solutions $\mathbf{u}^{\nu}$ of the Navier–Stokes system for an incompressible viscous fluid tend to the solution of the Euler equations for an ideal fluid as $\nu \to 0$. This result (see Theorem 7.5.4) is discussed in the format of much more general results, including a stability analysis for the horizontal-line method. The basic ideas could be used, if desired, to present an alternative approach to that of Section 7.2. The advantage here is that the evolution operators, and the attendant measure theoretic questions concerning their construction, are totally bypassed. In addition, an implicit estimate for the length of the time interval is presented.

7.1 PERTURBATION OF THE LINEAR PROBLEM AND NONLINEAR PRELIMINARIES

Suppose we have another equation of the same type as (6.4.1), say,

$$\frac{du'}{dt} + \mathrm{A}'(t)u' = F'(t), \qquad 0 \leqslant t \leqslant T_0, \quad u'(0) = u'_0, \tag{7.1.1}$$

in the (same) Banach space $\mathbf{X}$. For $\{\mathrm{A}'(t)\}$, we make the same basic assumptions (1), (2″) and (3) (cf. Theorem 6.3.7), and use corresponding notation $\mathrm{S}'(t)$, $\mathrm{B}'(t)$, etc. The space $\mathbf{Y} \subset \mathbf{X}$ is assumed common to the two systems. For this section, we assume only that $\mathbf{X}$ and $\mathbf{Y}$ are Banach spaces, with $\mathbf{Y}$ densely and continuously embedded in $\mathbf{X}$.

Proposition 7.1.1.† Let $u_0 \in \mathbf{Y}$, $F \in \mathbf{L}^1(J;\mathbf{Y})$, and $u'_0 \in \mathbf{X}$, $F' \in \mathbf{L}^1(J;\mathbf{X})$. Let u and u' be given by (6.4.2), written $u = \mathrm{U}(\delta \otimes u_0 \oplus F)$, and $u' = \mathrm{U}'(\delta \otimes u'_0 \oplus F')$. Then,

$$\|u' - u\|_{\infty,\mathbf{X}} \leqslant K'[\|u'_0 - u_0\|_{\mathbf{X}} + \|F' - F\|_{1,\mathbf{X}} + \|(\mathrm{A}' - \mathrm{A})u\|_{1,\mathbf{X}}], \tag{7.1.2}$$

where K' depends only on A' and $\mathrm{A}' - \mathrm{A}$ is regarded as a multiplication operator from $\mathbf{C}(J;\mathbf{Y})$ to $\mathbf{L}^1(J;\mathbf{X})$. The quantity K', in fact, may be taken equal to $\|\mathrm{U}'\|_{\infty,\mathbf{X}}$.

Proof: By differentiating the expression $\mathrm{U}'(t,s)\mathrm{U}(s,r)g$ with respect to s, and integrating from r to t, we obtain the identity, for $g \in \mathbf{Y}$,

$$\mathrm{U}'(t,r)g - \mathrm{U}(t,r)g = -\int_r^t \mathrm{U}'(t,s)[\mathrm{A}'(s) - \mathrm{A}(s)]\mathrm{U}(s,r)g\,ds. \tag{7.1.3}$$

Setting $r = 0$ and $g = u_0$ in (7.1.3) gives

$$(\mathrm{U}' - \mathrm{U})(\delta \otimes u_0) = -\mathrm{U}'[\mathrm{A}' - \mathrm{A}]\mathrm{U}(\delta \otimes u_0),$$

where the notation of Lemma 6.3.2 has been used on the right-hand side. Setting $g = F(r)$, and integrating (7.1.3), with respect to r, gives

$$(\mathrm{U}' - \mathrm{U})F = -[\mathrm{U}'[\mathrm{A}' - \mathrm{A}]\mathrm{U}]F.$$

Hence, addition of these two equations gives

$$\begin{aligned}(\mathrm{U}' - \mathrm{U})(\delta \otimes u_0 \oplus F) &= -\mathrm{U}'[\mathrm{A}' - \mathrm{A}]\mathrm{U}(\delta \otimes u_0 \oplus F)\\ &= -\mathrm{U}'[\mathrm{A}' - \mathrm{A}]u.\end{aligned}$$

† Reproduced from Kato (Chapter 6 [12]), with permission.

On the other hand,

$$\begin{aligned} u' - u &= \mathrm{U}'(\delta \otimes u_0') - \mathrm{U}(\delta \otimes u_0) + \mathrm{U}'\mathrm{F}' - \mathrm{U}F \\ &= \mathrm{U}'[\delta \otimes (u_0' - u_0) \oplus (F' - F)] + (\mathrm{U}' - \mathrm{U})(\delta \otimes u_0 \oplus F) \\ &= U'[\delta \otimes (u_0' - u_0) \oplus (F' - F)] - \mathrm{U}'[\mathrm{A}' - \mathrm{A}]u. \end{aligned}$$

Thus,

$$\|u' - u\|_{\infty,\mathbf{X}} \leqslant \|\mathrm{U}'\|_{\infty,\mathbf{X}}[\|u_0' - u_0\|_{\mathbf{X}} + \|F' - F\|_{1,\mathbf{X}} + \|(\mathrm{A}' - \mathrm{A})u\|_{1,\mathbf{X}}],$$

so that (7.1.2) holds, with $K' = \|\mathrm{U}'\|_{\infty,\mathbf{X}}$. ■

Definition 7.1.1. Denote by $\mathbf{E} = \mathbf{E}_L$ the metric space of functions $v\colon J \to \mathbf{X}$, such that

$$\|v(t') - v(t)\|_{\mathbf{X}} \leqslant L|t' - t|, \qquad t, t' \in J, \tag{7.1.4}$$

for some fixed positive constant L, with metric $d(u, v) = \|u - v\|_{\infty,\mathbf{X}}$. Let $\{\mathrm{A}(t, w)\}$, $\{\mathrm{B}(t, w)\}$, and $\{\mathrm{S}(t, w)\}$ be given, for $0 \leqslant t \leqslant T_0$, and w in some subset $\mathbf{W}$ of $\mathbf{X}$ at our disposal, satisfying, generically, the following relations:

(A)′ $\|\mathrm{A}(t, w') - \mathrm{A}(t, w)\| \leqslant c_{\mathbf{A}}\|w' - w\|_{\mathbf{X}}$,

for $\mathrm{A}(t, w) \in \mathbf{B}(\mathbf{Y}, \mathbf{X})$, with $t \mapsto \mathrm{A}(t, w) \in \mathbf{B}(\mathbf{Y}, \mathbf{X})$ continuous in norm for each w;

(S)′ $\|\mathrm{S}(t', w') - \mathrm{S}(t, w)\|_{\mathbf{Y},\mathbf{X}} \leqslant c_{\mathbf{S}}(|t' - t| + \|w' - w\|_{\mathbf{X}})$,

for $\mathrm{S}(t, w)$ an isomorphism of $\mathbf{Y}$ onto $\mathbf{X}$;

(B)′ $\mathrm{S}(t, w)\mathrm{A}(t, w)\mathrm{S}(t, w)^{-1} = \mathrm{A}(t, w) + \mathrm{B}(t, w)$,

for $t \mapsto \mathrm{B}(t, w) \in \mathbf{B}[\mathbf{X}]$ bounded, independently of w.

Denote by $\mathrm{A}^v(t)$, $\mathrm{B}^v(t)$, and $\mathrm{S}^v(t)$ the expressions $\mathrm{A}(t, v(t))$, $\mathrm{B}(t, v(t))$, and $\mathrm{S}(t, v(t))$, respectively, for v in a prescribed subset $\mathbf{E}_0$ of $\mathbf{E}$ with $\mathbf{Range}(v) \subset \mathbf{W}$ for $v \in \mathbf{E}_0$. Assume, finally,

$(\mathrm{S}^{-1})'$ $t \mapsto \mathrm{S}^v(t)^{-1} \in \mathbf{B}(\mathbf{X}, \mathbf{Y})$ is norm continuous.

Proposition 7.1.2. The mappings $t \mapsto \mathrm{A}^v(t) \in \mathbf{B}(\mathbf{Y}, \mathbf{X})$ and $t \mapsto \mathrm{S}^v(t) \in \mathbf{B}(\mathbf{Y}, \mathbf{X})$ are continuous in norm.

Proof: The first statement is immediate from

$$\begin{aligned} \|\mathrm{A}^v(t') - \mathrm{A}^v(t)\|_{\mathbf{Y},\mathbf{X}} &\leqslant \|\mathrm{A}(t', v(t')) - \mathrm{A}(t', v(t))\|_{\mathbf{Y},\mathbf{X}} \\ &\quad + \|\mathrm{A}(t', v(t)) - \mathrm{A}(t, v(t))\|_{\mathbf{Y},\mathbf{X}} \\ &\leqslant c_{\mathbf{A}}\|v(t') - v(t)\|_{\mathbf{X}} + \|\mathrm{A}(t', v(t)) - \mathrm{A}(t, v(t))\|_{\mathbf{Y},\mathbf{X}}, \end{aligned}$$

by virtue of (7.1.4) and (A)′. The second is similar. ■

The following lemma is a preparation for the strong measurability of the mapping $t \mapsto \mathrm{B}^v(t)$.

Lemma 7.1.3.†

(1) Any closed, convex and bounded subset $\mathscr{K}$ of $\mathbf{Y}$ is closed in $\mathbf{X}$ if $\mathbf{Y}$ is reflexive.

(2) If a function $G: J \to \mathbf{Y}$ is bounded in $\mathbf{Y}$–norm and continuous in $\mathbf{X}$–norm, then G is weakly continuous as a $\mathbf{Y}$–valued function if $\mathbf{Y}$ is reflexive. In particular, G is strongly measurable ($\mathbf{Y}$) in this case.

Proof:

(1) The subset $\mathscr{K}$ is weakly compact in $\mathbf{Y}$ since $\mathbf{Y}$ is assumed reflexive. Since the inclusion map of $\mathbf{Y}$ into $\mathbf{X}$ is continuous and, hence, weakly continuous, we conclude that $\mathscr{K}$ is weakly compact in $\mathbf{X}$. It follows that $\mathscr{K}$ is closed in $\mathbf{X}$.

(2) Let ℓ be a linear, continuous (real-valued) functional on $\mathbf{Y}$. We must show that

$$m = \ell G: J \to \mathbb{R}^1$$

is continuous. It is enough to show that, if $t_n \to t_0$ in J, every subsequence of $\{m(t_n)\}$ has a subsequence convergent to $m(t_0)$. Since G is $\mathbf{Y}$–bounded, every subsequence of $\{m(t_n)\}$ has a subsequence convergent, say, to m_*; however, since G is $\mathbf{X}$–continuous and $\mathbf{Y}$ is reflexive, $m_* = m(t_0)$. The final assertion is a consequence of the Dunford–Pettis theorem (see Chapter 6 [14], p. 131), since $G(J) \subset \mathbf{X}$ is compact and, hence, separable in $\mathbf{X}$ and $\mathbf{Y}$. ■

Proposition 7.1.4. The map $t \mapsto \mathrm{B}^v(t) \in \mathbf{B}[\mathbf{X}]$ is weakly continuous on $\mathbf{X}$ if $\mathbf{Y}$ is reflexive. If, in addition $\mathbf{X}$ is separable, the mapping is strongly measurable ($\mathbf{X}$) and, in fact, strongly integrable.

Proof: By (S)′ and $(\mathrm{S}^{-1})'$, the map $t \mapsto \mathrm{B}^v(t)f \in \mathbf{X}$ is weakly continuous on $\mathbf{X}$ if and only if the map $t \mapsto \mathrm{S}^v(t)^{-1}\mathrm{B}^v(t)f \in \mathbf{Y}$ is weakly continuous on $\mathbf{Y}$. Identify the latter map with the map G of the previous lemma. If $g \in \mathbf{Y}$, then, by (B)′,

$$\mathrm{S}^v(t)^{-1}\mathrm{B}^v(t)g = \mathrm{A}^v(t)\mathrm{S}^v(t)^{-1}g - \mathrm{S}^v(t)^{-1}\mathrm{A}^v(t)g. \tag{7.1.5}$$

Since $t \mapsto \mathrm{S}^v(t)^{-1}g \in \mathbf{Y}$ is continuous, it follows, from (7.1.5) and from Proposition 7.1.2, that $t \mapsto \mathrm{S}^v(t)^{-1}\mathrm{B}^v(t)g$ is $\mathbf{X}$–continuous for $g \in \mathbf{Y}$. Since

$$\|\mathrm{S}^v(t)^{-1}\mathrm{B}^v(t)\|_{\mathbf{X},\mathbf{Y}} \leqslant C_v, \tag{7.1.6}$$

† Adapted from Kato [17].

and since **Y** is dense in **X**, it follows that $t \mapsto \mathrm{S}^v(t)^{-1}\mathrm{B}^v(t)f$ is **X**–continuous for $f \in \mathbf{X}$. The weak continuity is now immediate from (7.1.6) and Lemma 7.1.3. The strong measurability follows from the Dunford–Pettis theorem and the **X** integrability from the Bochner theorem (see Chapter 6 [14] p. 133). Note that $\|\mathrm{B}^v(\cdot)f\|_{\mathbf{X}}$ is lower semicontinuous, hence integrable. ■

Proposition 7.1.5. Let $\mathbf{X}'$ be a reflexive Banach space, and let $F:[0, T_0] \to \mathbf{X}'$ be absolutely continuous. Then F is strongly differentiable almost everywhere, and the fundamental theorem of calculus holds:

$$F(t) - F(0) = \int_0^t \left(\frac{dF}{ds}\right) ds \tag{7.1.7}$$

(see Komura [20]).

Proposition 7.1.6. Let $\{F(t, w)\}$ be given, for $0 \leqslant t \leqslant T_0$ and w in some subset **W** of **X**, satisfying

(F)′ $\|F(t, w') - F(t, w)\|_{\mathbf{X}} \leqslant c_F \|w' - w\|_{\mathbf{X}}$,

and $t \mapsto F(t, w) \in \mathbf{Y}$ is uniformly continuous in the **X**-norm and bounded in the **Y**–norm.

Then, the mapping $t \mapsto F^v(t)$ is continuous in the **X**–norm and, if **Y** is reflexive, weakly continuous (hence, strongly measurable if **Y** is also separable) in the **Y**–norm. In this case, $F^v \in \mathbf{L}^1(J; \mathbf{Y})$.

Proof: Continuity in the **X**–norm follows the proof of Proposition 7.1.2. The weak continuity follows from Lemma 7.1.3, the strong measurability from the Dunford–Pettis theorem and the (**Y**) integrability from the theorem of Bochner, via the lower semicontinuity of $\|F^v(\cdot)\|_{\mathbf{Y}}$. ■

7.2 THE QUASI-LINEAR CAUCHY PROBLEM IN BANACH SPACE

In this section, we consider the abstract Cauchy problem

$$\frac{du}{dt} + \mathrm{A}(t, u)u = F(t, u), \qquad 0 \leqslant t \leqslant T_0, \tag{7.2.1a}$$

$$u(0) = u_0, \tag{7.2.1b}$$

where the unknown u takes values in a Banach space, and $A(t, u)$ is a linear, possibly unbounded, operator depending on t and u.

We shall start from four real Banach spaces

$$\mathbf{Y} \subset \mathbf{X} \subset \mathbf{V} \subset \mathbf{Z}, \tag{7.2.2}$$

with each of the spaces reflexive and separable, and the inclusions continuous and dense. We shall split the roles of $\mathbf{X}$, as developed in the linear theory, and assign them to the three spaces $\mathbf{X}$, $\mathbf{V}$, and $\mathbf{Z}$, so that $\dot{u}(t) \in \mathbf{X}$, $\dot{u}(\cdot)$ is strongly continuous on $\mathbf{V}$ and $-A(t, u)$ generates a (C_0) semigroup on $\mathbf{Z}$; this semigroup is assumed to be quasi-contractive (see Proposition 7.2.1), with respect to an equivalent norm $N(t, u)$ on $\mathbf{Z}$. The dependence on u is made precise in the sequel. For the moment, $u(t) \in \mathbf{E}_0 \subset \mathbf{E}$, where $\mathbf{E}$ is defined in Definition 7.1.1.

Definition 7.2.1. Let $\mathbf{N}(\mathbf{Z}) = \{\|\cdot\|_\mu\}$ denote the set of all norms in $\mathbf{Z}$ equivalent to the given norm, with metric

$$d(\|\cdot\|_\mu, \|\cdot\|_\nu) = \ln \max\left\{ \sup_{0 \neq z \in \mathbf{Z}} \|z\|_\mu / \|z\|_\nu,\ \sup_{0 \neq z \in \mathbf{Z}} \|z\|_\nu / \|z\|_\mu \right\}. \tag{7.2.3}$$

Remark 7.2.1. The triangle inequality follows from the following estimates:

$$\begin{aligned}
d(\|\cdot\|_\mu, \|\cdot\|_\nu) &= \ln \max\left\{ \sup_{0 \neq z \in \mathbf{Z}} (\|z\|_\mu / \|z\|_\nu),\ \sup_{0 \neq z \in \mathbf{Z}} (\|z\|_\nu / \|z\|_\mu) \right\} \\
&= \ln \max\left\{ \sup_{0 \neq z \in \mathbf{Z}} \left[\left(\frac{\|z\|_\mu}{\|z\|_\tau} \right) \left(\frac{\|z\|_\tau}{\|z\|_\nu} \right) \right],\ \sup_{0 \neq z \in \mathbf{Z}} \left[\left(\frac{\|z\|_\nu}{\|z\|_\tau} \right) \left(\frac{\|z\|_\tau}{\|z\|_\mu} \right) \right] \right\} \\
&\leqslant \ln \left[\max\left\{ \sup_{0 \neq z \in \mathbf{Z}} \left(\frac{\|z\|_\mu}{\|z\|_\tau} \right),\ \sup_{0 \neq z \in \mathbf{Z}} \left(\frac{\|z\|_\tau}{\|z\|_\mu} \right) \right\} \right. \\
&\qquad \left. \times \max\left\{ \sup_{0 \neq z \in \mathbf{Z}} \left(\frac{\|z\|_\tau}{\|z\|_\nu} \right),\ \sup_{0 \neq z \in \mathbf{Z}} \left(\frac{\|z\|_\nu}{\|z\|_\tau} \right) \right\} \right] \\
&= d(\|\cdot\|_\mu, \|\cdot\|_\tau) + d(\|\cdot\|_\tau, \|\cdot\|_\nu).
\end{aligned}$$

Remark 7.2.2. Note that $\|\cdot\|_\mu / \|\cdot\|_\nu \leqslant \exp\{d(\|\cdot\|_\mu, \|\cdot\|_\nu)\}$.

Proposition 7.2.1. Suppose $\{N(t, w)\} \subset \mathbf{N}(\mathbf{Z})$ is given for $0 \leqslant t \leqslant T_0$ and w in some subset $\mathbf{W}$ of $\mathbf{Y}$, satisfying

$$d(N^v(t'), N^v(t)) \leqslant c|t' - t|,\ d(N^v(t), \|\cdot\|_{\mathbf{Z}}) \leqslant c_1, \tag{7.2.4}$$

for $0 \leqslant t', t \leqslant T_0$ and all $v \in \mathbf{E}_0$. Suppose that, for all t and w, $\mathrm{A}(t, w) \in G(\mathbf{Z}_{N(t,w)}, 1, \omega)$, where $\mathbf{Z}_{N(t,w)}$ denotes the space $\mathbf{Z}$ normed by the equivalent norm $N(t, w)$. Then, $\{\mathrm{A}^v(t)\} \subset G(\mathbf{Z})$ is stable on any subinterval J' of $[0, T_0]$, with stability constants ω and

$$M = \exp(2\{c_1 + c|J'|\}).$$

In particular, (7.2.4) holds, with $c_1 = \lambda_N$, $c = (1 + L)\mu_N$, if $\{N(t, w)\}$ satisfies

(N) $N(t, w) \in \mathbf{N}(\mathbf{Z})$, with

$$d(N(t, w), \|\cdot\|_{\mathbf{Z}}) \leqslant \lambda_N,$$
$$d(N(t', w'), N(t, w)) \leqslant \mu_N(|t' - t| + \|w' - w\|_{\mathbf{X}}).$$

Proof: By the proof of Proposition 6.2.2 and Remark 7.2.2, we have, for $J' = [0, T_0']$,

$$\begin{aligned}\left\|\prod_{j=1}^{k} [A^v(t_j) + \lambda]^{-1} f\right\|_{\mathbf{Z}} &\leqslant e^{c_1}\left\|\prod_{j=1}^{k} [A^v(t_j) + \lambda]^{-1} f\right\|_{T_0'}\\ &\leqslant (\lambda - \omega)^{-k} e^{c_1} e^{2cT_0'}\|f\|_{T'}\\ &\leqslant (\lambda - \omega)^{-k} e^{2c_1} e^{2cT_0'}\|f\|_{\mathbf{Z}},\end{aligned}$$

so that $\{\mathrm{A}^v(t)\} \subset G(\mathbf{Z})$ is stable as stated. The second statement is immediate. ∎

Definition 7.2.2. Let $\mathbf{W}$ be an open subset of $\mathbf{Y}$, let $y_0 \in \mathbf{W}$, and let $R = \operatorname{dist}(y_0, \mathbf{Y}\backslash\mathbf{W})$. We introduce families $\{\mathrm{A}(t, w)\}$, $\{\mathrm{S}(t, w)\}$, $\{\mathrm{B}(t, w)\}$, $\{F(t, w)\}$, and $\{N(t, w)\}$, such that, for $t, t' \in [0, T_0]$ and $w, w' \in \mathbf{W}$, the following conditions hold:

(N) See (N) of Proposition 7.2.1;

(S) $\mathrm{S}(t, w)$ is an isomorphism of $\mathbf{Y}$ onto $\mathbf{Z}$, with

$$\|\mathrm{S}(t, w)\|_{\mathbf{Y},\mathbf{Z}} \leqslant \lambda_{\mathrm{S}}, \qquad \|\mathrm{S}(t, w)^{-1}\|_{\mathbf{Z},\mathbf{Y}} \leqslant \lambda_{\mathrm{S}}',$$
$$\|\mathrm{S}(t', w') - \mathrm{S}(t, w)\|_{\mathbf{Y},\mathbf{Z}} \leqslant \mu_{\mathrm{S}}(|t' - t| + \|w' - w\|_{\mathbf{X}});$$

(A1) $\mathrm{A}(t, w) \in G(\mathbf{Z}_{N(t,w)}, 1, \omega)$;

(A2) $\mathrm{S}(t, w)\mathrm{A}(t, w)\mathrm{S}(t, w)^{-1} = \mathrm{A}(t, w) + \mathrm{B}(t, w)$, where

$$\mathrm{B}(t, w) \in \mathbf{B}[\mathbf{Z}], \qquad \|\mathrm{B}(t, w)\|_{\mathbf{Z}} \leqslant \lambda_{\mathrm{B}};$$

(A3) $\mathrm{A}(t, w) \in \mathbf{B}(\mathbf{Y}, \mathbf{X})$, with $\|\mathrm{A}(t, w)\|_{\mathbf{Y},\mathbf{X}} \leqslant \lambda_{\mathrm{A}}$, and

$$\|\mathrm{A}(t, w') - \mathrm{A}(t, w)\|_{\mathbf{Y},\mathbf{V}} \leqslant \mu_A\|w' - w\|_{\mathbf{V}},$$

with $t \mapsto \mathrm{A}(t, w) \in \mathbf{B}(\mathbf{Y}, \mathbf{V})$ continuous in norm for each w;

(A4)[†] $\mathrm{A}(t, w)y_0 \in \mathbf{Y}$, $\|\mathrm{A}(t, w)y_0\|_{\mathbf{Y}} \leqslant \lambda_0$;
(F) $F(t, w) \in \mathbf{Y}$, $\|F(t, w)\|_{\mathbf{Y}} \leqslant \lambda_F$, with

$$\|F(t, w') - F(t, w)\|_{\mathbf{V}} \leqslant \mu_F\|w' - w\|_{\mathbf{V}},$$

and with $t \mapsto F(t, w) \in \mathbf{V}$ strongly continuous for each w.

Definition 7.2.3. For $0 < T_0' \leqslant T_0$, $L' > 0$ to be specified below, let $\mathbf{E}_0$ denote the (complete) metric space of functions $v:[0, T_0'] \to \mathbf{Y}$, such that

$$\|v(t) - y_0\|_{\mathbf{Y}} \leqslant \tfrac{3}{4}R, \tag{7.2.5a}$$

$$\|v(t') - v(t)\|_{\mathbf{X}} \leqslant L'|t' - t|, \tag{7.2.5b}$$

with metric given by

$$d(v, w) = \sup_{0 \leqslant t \leqslant T_0'} \|v(t) - w(t)\|_{\mathbf{Z}}. \tag{7.2.5c}$$

Remark 7.2.3. The set $\mathbf{E}_0$ is a complete metric space. Indeed, the set $\mathbf{E}'$ satisfying (7.2.5b,c) is routinely complete. Since $\mathbf{W}_0 = \{w \in \mathbf{Y}: \|w - y_0\| \leqslant (\tfrac{3}{4})R\}$ is closed in $\mathbf{Z}$ (cf. Lemma 7.1.3), it follows that

$$\mathbf{E}_0 = \bigcap_{0 \leqslant t \leqslant T_0'} \{v \in \mathbf{E}': v(t) \in \mathbf{W}_0\}$$

is closed in $\mathbf{E}'$ and, hence, complete.

Remark 7.2.4. The hypotheses (N)–(F) imply that Propositions 7.1.2, 7.1.4, 7.1.5, and 7.1.6 are valid for $t \in J'$, $v \in \mathbf{E}_0$, and $\mathbf{X}$ replaced by $\mathbf{V}$, respectively, $\mathbf{Z}$ for statements concerning $\{\mathrm{A}^v(\cdot)\}$ and $\{F^v(\cdot)\}$, respectively, $\{\mathrm{S}^v(\cdot)\}$ and $\{\mathrm{B}^v(\cdot)\}$. Moreover, $\mathrm{S}^v(\cdot)$ is Lipschitz continuous for $v \in \mathbf{E}_0$, i.e.,

$$\|\mathrm{S}^v(t') - \mathrm{S}^v(t)\|_{\mathbf{Y},\mathbf{Z}} \leqslant \mu_S(1 + L')|t' - t|.$$

In particular by Proposition 7.1.5, $\mathrm{S}^v(\cdot)$ is a strong indefinite integral of a strongly integrable function $\dot{\mathrm{S}}^v(\cdot)$, such that

$$\|\dot{\mathrm{S}}^v(t)\|_{\mathbf{Y},\mathbf{Z}} \leqslant \mu_S(1 + L') \qquad \text{almost everywhere on } [0, T_0']. \tag{7.2.6}$$

Moreover, $\mathrm{B}^v(\cdot)$ is strongly measurable ($\mathbf{Z}$) by Proposition 7.1.4, and $F^v(\cdot)$ is integrable ($\mathbf{Y}$) by Proposition 7.1.6. Since $\|\dot{\mathrm{S}}^v(\cdot)\|_{\mathbf{Y},\mathbf{Z}}$ and $\|\mathrm{B}^v(\cdot)\|_{\mathbf{Z}}$ are upper integrable, it follows that (2″) of Theorem 6.3.5 holds for $\{\mathrm{A}^v(t)\}$, $\{\mathrm{B}^v(t)\}$, and $\{\mathrm{S}^v(t)\}$. Thus, by Proposition 7.2.1 and Theorem 6.3.7, the evolution operators $\{\mathrm{U}^v(t, s)\} \subset \mathrm{B}[\mathbf{Z}] \cap \mathrm{B}[\mathbf{Y}]$ exist. It is of interest that $\{\mathrm{A}^v(t)\}$ is stable both on $\mathbf{Z}$ and (when properly restricted) on $\mathbf{Y}$, by Propositions 6.2.7 and 6.3.1, so that the construction of the evolution operators given in Chapter 6 is sufficient.

[†] This hypothesis can be shown to be redundant (see Kato [19]).

Remark 7.2.5. By Theorem 6.2.5(a), Proposition 7.2.1, and Corollary 6.3.6, we note that, for each $v \in \mathbf{E}_0$,

$$\|\mathrm{U}^v(\cdot,\cdot)\|_{\infty,\mathbf{Z}} \leqslant \exp\{2\lambda_N + \gamma_1 T'_0\}, \qquad \gamma_1 = 2\mu_N(1+L') + \omega, \tag{7.2.7a}$$

$$\|\mathrm{U}^v(\cdot,\cdot)\|_{\infty,\mathbf{Y}} \leqslant \lambda_{\mathrm{S}}\lambda'_{\mathrm{S}} \exp\{2\lambda_N + \gamma_2 T'_0\} \tag{7.2.7b}$$

where $\gamma_2 = \gamma_1 + e^{2\lambda_N + 2\mu_N(1+L')T'_0}[\lambda_{\mathrm{B}} + \lambda'_{\mathrm{S}}\mu_{\mathrm{S}}(1+L')]$.

Note that we have used (6.3.20), with

$$\begin{aligned}\|\mathrm{B}^v - \mathrm{C}^v\|_{1,\mathbf{Z}} &= \|\mathrm{B}^v - \dot{\mathrm{S}}^v(\cdot)\mathrm{S}^v(\cdot)^{-1}\|_{1,\mathbf{Z}} \\ &\leqslant [\lambda_{\mathrm{B}} + \lambda'_{\mathrm{S}}\mu_{\mathrm{S}}(1+L')]T'_0.\end{aligned}$$

Lemma 7.2.2. Suppose the initial datum $u_0 \in \mathbf{Y}$ satisfies

$$\|u_0 - y_0\|_{\mathbf{Y}} \leqslant e^{-2\lambda_N} R/(2\lambda_{\mathrm{S}}\lambda'_{\mathrm{S}}) = \rho' \; (\leqslant R/2). \tag{7.2.8}$$

Then, there exist constants L' and T'_0, such that the mapping $v \mapsto u = \Phi(v)$, given by

$$u(t) = \Phi(v)(t) = \mathrm{U}^v(t,0)u_0 + \int_0^t \mathrm{U}^v(t,s)F^v(s)\,ds, \tag{7.2.9}$$

$0 \leqslant t \leqslant T'_0$, is a mapping of $\mathbf{E}_0$ into itself. The function u satisfies (6.4.1), with $\mathrm{A} = \mathrm{A}^v$ and $F = F^v$, and $t \mapsto (du/dt)(t) \in \mathbf{V}$ is continuous.

Proof: We first represent $u(t) - y_0$ by

$$u(t) - y_0 = \mathrm{U}^v(t,0)(u_0 - y_0) + \int_0^t \mathrm{U}^v(t,s)[F^v(s) - \mathrm{A}^v(s)y_0]\,ds.$$

This gives

$$\|u(t) - y_0\|_{\mathbf{Y}} \leqslant \lambda_{\mathrm{S}}\lambda'_{\mathrm{S}} e^{2\lambda_N + \gamma_2 T'_0}[\|u_0 - y_0\|_{\mathbf{Y}} + (\lambda_F + \lambda_0)T'_0]. \tag{7.2.10}$$

By Proposition 6.4.1, (F), and Remark 7.2.4,

$$\frac{du}{dt} = F^v(t) - \mathrm{A}^v(t)u(t),$$

and the right-hand side of this equation is continuous into $\mathbf{V}$ by (A3) and (F). Direct estimation of this differential equation gives

$$\left\|\frac{du}{dt}\right\|_{\mathbf{X}} \leqslant c\lambda_F + \lambda_{\mathrm{A}}\left(\|y_0\|_{\mathbf{Y}} + \sup_{0 \leqslant t \leqslant T'_0} \|u(t) - y_0\|_{\mathbf{Y}}\right), \tag{7.2.11}$$

where c is chosen so that $\|\cdot\|_{\mathbf{X}} \leqslant c\|\cdot\|_{\mathbf{Y}}$.

Now choose L', such that

$$\tfrac{1}{2}L' = c\lambda_F + \lambda_{\mathrm{A}}(\|y_0\|_{\mathbf{Y}} + \tfrac{1}{2}R). \tag{7.2.12}$$

By (7.2.8), (7.2.11), and (7.2.12), we see that $\|(du/dt)(0)\|_{\mathbf{X}} \leqslant L'/2$, so that, by (7.2.10) and (7.2.11),

$$\|u(t) - y_0\|_{\mathbf{Y}} \leqslant (\tfrac{3}{4})R, \qquad \left\|\frac{du}{dt}\right\|_{\mathbf{X}} \leqslant L',$$

for T'_0 sufficiently small, and independent of v. In particular, $v \in \mathbf{E}_0$. ■

Lemma 7.2.3. If u_0 satisfies (7.2.8), and L' is chosen by (7.2.12), there exists T'_0 sufficiently small, and independent of u_0, such that the mapping Φ is a (strict) contraction of $\mathbf{E}_0$ into itself. The contraction factor γ does not depend on u_0.

Proof: By (7.2.9) and Proposition 7.1.1 (cf. (7.1.2)), with $\mathbf{X}$ replaced by $\mathbf{Z}$, we have, using (7.2.7a), (F), and (A3),

$$\begin{aligned} d(\Phi w, \Phi v) &= \|\Phi w - \Phi v\|_{\infty,\mathbf{Z}} \\ &\leqslant T'_0 \exp\{2\lambda_N + \gamma_1 T'_0\} \cdot [\mu_F + \mu_A(\|y_0\|_{\mathbf{Y}} + R)]\, d(w, v), \end{aligned} \tag{7.2.13}$$

where $\|y_0\|_{\mathbf{Y}} + R$ is an estimate for $\|u(t)\|_{\infty,\mathbf{Y}}$. The result is immediate from (7.2.13). ■

Theorem 7.2.4.† Let the hypotheses of Definition 7.2.2 be satisfied. Then, there are positive constants ρ' and $T'_0 \leqslant T_0$, such that, if $u_0 \in \mathbf{Y}$, with $\|u_0 - y_0\| \leqslant \rho'$, (7.2.1) has a unique solution u on $[0, T'_0] = J'$, with

$$u \in \mathbf{C}(J'; \mathbf{W}), \tag{7.2.14a}$$

$$\frac{du}{dt} \in \mathbf{C}(J'; \mathbf{V}) \cap \mathbf{L}^{\infty}(J'; \mathbf{X}). \tag{7.2.14b}$$

When u_0 varies in $\mathbf{Y}$, subject to $\|u_0 - y_0\|_{\mathbf{Y}} \leqslant \rho'$, the map $u_0 \mapsto u(t)$ is Lipschitz continuous in the $\mathbf{Z}$–norm, uniformly in $t \in [0, T'_0]$.

Proof: The unique fixed point u of Φ, guaranteed by Lemma 7.2.3, and the contraction mapping principle, satisfies

$$u(t) = \mathrm{U}^u(t, 0)u_0 + \int_0^t \mathrm{U}^u(t, s)F(s, u(s))\, ds. \tag{7.2.15}$$

By the final statement of Lemma 7.2.2, we conclude that u satisfies (7.2.1) and $(du/dt) \in \mathbf{C}(J'; \mathbf{V})$. Since $\|(du/dt)\|_{\infty,\mathbf{X}} \leqslant L'$, $(du/dt) \in \mathbf{L}^{\infty}(J; \mathbf{X})$. Of course, $u \in \mathbf{C}(J'; \mathbf{Y})$ follows from Proposition 6.4.1. The uniqueness of solutions of (7.2.1) is equivalent to the uniqueness of fixed points of Φ.

† This theorem and its proof are adapted from Hughes, Kato, and Marsden [13].

To prove the Lipschitz continuity, we let u_0' be another initial datum, defining a map Φ' and a solution u'. Then,

$$d(u,u') = d(\Phi u, \Phi' u') \leqslant d(\Phi u, \Phi' u) + d(\Phi' u, \Phi' u') \leqslant \sup_t \|U^u(t,0)\|_{\mathbf{Z}} \|u_0 - u_0'\|_{\mathbf{Z}} + \gamma d(u,u'),$$

by (7.2.9) and the contractive property of Φ'; note that γ does not depend on the mapping Φ. The result follows. ■

Remark 7.2.6. Set $\kappa = e^{-2\lambda_N}/(2\lambda_s\lambda_s')$. By (7.2.8), we may select ρ' to be exactly equal to κR. Now, suppose u_0 is given, and y_0 is undetermined. If there is a dense subset of $\mathbf{W}$ satisfying (A4), then select an admissible y_0 as follows. Let $R_{u_0} = \operatorname{dist}(u_0, \mathbf{Y}\backslash\mathbf{W})$, and choose y_0 to satisfy

$$\|y_0 - u_0\|_{\mathbf{Y}} < \left(\frac{\kappa}{1+\kappa}\right) R_{u_0}.$$

By the triangle inequality,

$$R = \operatorname{dist}(y_0, \mathbf{Y}\backslash\mathbf{W}) \geqslant \left(\frac{1}{1+\kappa}\right) R_{u_0}.$$

Thus, $R_{u_0} \leqslant (1+\kappa)R$, and

$$\|y_0 - u_0\|_{\mathbf{Y}} < \kappa R = \rho',$$

as required.

7.3 QUASI-LINEAR SECOND-ORDER HYPERBOLIC SYSTEMS

We shall consider quasi-linear second-order hyperbolic systems of the form

$$a_{00}\frac{\partial^2 \psi}{\partial t^2} = \sum_{i,j=1}^{n} a_{ij}\frac{\partial^2 \psi}{\partial x_i \partial x_j} + \sum_{i=1}^{n} (a_{0i} + a_{i0})\frac{\partial^2 \psi}{\partial t\, \partial x_i} + \mathbf{b}, \tag{7.3.1}$$

where the unknown $\psi = (\psi_1, \ldots, \psi_m)$ is an m-vector valued function of $t \in [0, T_0]$ and of $\mathbf{x} = (x_1, \ldots, x_n) \in \mathbb{R}^n$, where $\{a_{ij}: i,j = 0, 1, \ldots, n\}$ is a collection of $(m \times m)$-matrix valued functions of the (suppressed) arguments $t, \mathbf{x}, \psi, \partial\psi/\partial t, \nabla\psi$, and $\mathbf{b}$ is a function of these same arguments. The formal definition of solution of (7.3.1) is given in the following definition and remark.

Definition 7.3.1. Let $\Omega \subset \mathbb{R}^m \times \mathbb{R}^m \times \mathbb{R}^{mn}$ be an open set containing the origin which is contractible to the origin, and let $a_{ij}: i, j = 0, 1, \ldots, n$, and $\mathbf{b}$ be defined on $[0, T_0] \times \mathbb{R}^n \times \Omega$. These variables will be denoted by $(t, \mathbf{x}, \mathbf{p}) \in [0, T_0] \times \mathbb{R}^n \times \Omega$. Let $\mathbf{C}_b^k(\mathbb{R}^n \times \Omega; \mathbb{R}^N)$ denote the class of N–component functions of class $\mathbf{C}^k$ in $\mathbf{x}$ and $\mathbf{p}$, whose $\mathbf{x}$-derivatives up to order k are bounded.

Regarding the functions a_{ij} and $\mathbf{b}$, we make the following hypotheses (s is specified below).

(a1) $a_{ij} \in \mathbf{Lip}([0, T_0]; \mathbf{C}_b^{s+1}(\mathbb{R}^n \times \Omega; \mathbb{R}^{m^2}))$, for $i, j = 0, 1, \ldots, n$, ($\mathbf{C}_b^{s+1}$ replaced by $\mathbf{C}_b^{[s]+2}$ if $[s] \neq s$),

$$b \in \mathbf{Lip}([0, T_0]; \mathbf{C}_b^{s+1}(\mathbb{R}^n \times \Omega; \mathbb{R}^m)),$$

($\mathbf{C}_b^{s+1}$ replaced by $\mathbf{C}_b^{[s]+2}$ if $[s] \neq s$), $\mathbf{b}(0, \cdot, 0) \in \mathbf{H}^s(\mathbb{R}^n; \mathbb{R}^m)$;

(a2) $a_{ij} = a_{ji}^*$;

(a3) (Hyperbolicity) There is an $\varepsilon > 0$, such that the inequalities

$$a_{00}(t, \mathbf{x}, \mathbf{p}) \geqslant \varepsilon I$$

and

$$\sum_{i,j=1}^{n} \xi_i \xi_j a_{ij}(t, \mathbf{x}, \mathbf{p}) \geqslant \varepsilon \left(\sum_{i=1}^{n} \xi_i^2 \right) I$$

hold, in the sense of matrix entry comparisons, for all $(t, \mathbf{x}, \mathbf{p}) \in [0, T_0] \times \mathbb{R}^n \times \Omega$, and all $(\xi_1, \ldots, \xi_n) \in \mathbb{R}^n$.

For $s > n/2 + 1$ ($s > n/2$ if the a_{ij} do not depend on the derivatives of $\boldsymbol{\psi}$), set

$$\mathbf{X} = \mathbf{H}^s(\mathbb{R}^n; \mathbb{R}^m) \times \mathbf{H}^{s-1}(\mathbb{R}^n; \mathbb{R}^m) = \mathbf{V},$$
$$\mathbf{Y} = \mathbf{H}^{s+1}(\mathbb{R}^n; \mathbb{R}^m) \times \mathbf{H}^s(\mathbb{R}^n; \mathbb{R}^m),$$
$$\mathbf{Z} = \mathbf{H}^1(\mathbb{R}^n; \mathbb{R}^m) \times \mathbf{H}^0(\mathbb{R}^n; \mathbb{R}^m).$$

Adjoined to (7.3.1) are the initial conditions

$$\boldsymbol{\psi}(0, \cdot) = \boldsymbol{\psi}_0 \in \mathbf{H}^{s+1}(\mathbb{R}^n; \mathbb{R}^m), \tag{7.3.2a}$$

$$\frac{d\boldsymbol{\psi}(0, \cdot)}{dt} = \dot{\boldsymbol{\psi}}_0 \in \mathbf{H}^s(\mathbb{R}^n; \mathbb{R}^m), \tag{7.3.2b}$$

where it is assumed that

$$(\boldsymbol{\psi}_0(\mathbf{x}), \dot{\boldsymbol{\psi}}_0(\mathbf{x}), \nabla\boldsymbol{\psi}_0(\mathbf{x})) \in \Omega, \qquad \text{for all} \quad \mathbf{x} \in \mathbb{R}^n. \tag{7.3.3}$$

We suppose that $\mathbf{W} \subset \mathbf{Y}$ is a ball, centered at $\mathbf{u}_0 = (\boldsymbol{\psi}_0, \dot{\boldsymbol{\psi}}_0)$, with radius small enough, so that $\mathbf{u} \in \mathbf{W}$ satisfies (7.3.3). This is possible by Sobolev's inequality. We denote the general element in $\mathbf{W}$ by $\mathbf{w} = (\boldsymbol{\sigma}, \dot{\boldsymbol{\sigma}})$.

Remark 7.3.1. We may reduce (7.3.1) to a first-order system in t by the following standard device. For $\mathbf{w} \in \mathbf{W}$, define $\mathrm{A}(t, \mathbf{w})$, formally (see Proposition 7.3.1 for a precise statement), by

$$\mathrm{A}(t,\mathbf{w}) = -\begin{bmatrix} 0 & \mathrm{I} \\ a_{00}^{-1} \sum_{i,j=1}^{n} a_{ij} \dfrac{\partial^2}{\partial x_i \partial x_j} & a_{00}^{-1} \sum_{i=1}^{n} (a_{0i} + a_{i0}) \dfrac{\partial}{\partial x_i} \end{bmatrix}. \tag{7.3.4}$$

Then, for $(\psi, \dot{\psi}) \in \mathbf{Y}$(a core), $\mathrm{A}(t, w)(\psi, \dot{\psi}) \in \mathbf{X}$ is given by

$$\mathrm{A}(t,\mathbf{w})(\psi,\dot{\psi}) = -\left(\dot{\psi}, a_{00}^{-1}\left(\sum_{i,j=1}^{n} a_{ij}\frac{\partial^2 \psi}{\partial x_i \partial x_j}\right) + a_{00}^{-1}\sum_{i=1}^{n}(a_{0i}+a_{i0})\frac{\partial \dot{\psi}}{\partial x_i}\right), \tag{7.3.5}$$

so that

$$\frac{d}{dt}(\psi, \dot{\psi}) + \mathrm{A}(t, (\psi, \dot{\psi}))(\psi, \dot{\psi}) = (\mathbf{0}, \mathbf{b}(t, \cdot, \psi, \dot{\psi}, \nabla\psi)), \tag{7.3.6}$$

if and only if ψ satisfies (7.3.1), $0 \leqslant t \leqslant T'_0$, $T'_0 \leqslant T_0$, *with* $\dot{\psi} = d\psi/dt$. By a solution of (7.3.1)–(7.3.2), we shall understand a pair $(\psi, \dot{\psi})$ satisfying (7.3.6) and (7.3.2).

Remark 7.3.2. Later in this section, we shall discuss the ring properties of a_{ij}. For the moment, we need only observe that a_{ij} are pointwise bounded, together with $(\partial/\partial x_k)a_{ij}$, by our assumptions.

Definition 7.3.2. Define the bilinear forms

$$B(t,\mathbf{w};\psi_1,\psi_2) = \sum_{i,j=1}^{n}\left(a_{ij}\frac{\partial \psi_1}{\partial x_i}, \frac{\partial \psi_2}{\partial x_j}\right)_{\mathbf{L}^2(\mathbb{R}^n)} \tag{7.3.7a}$$

and

$$\begin{aligned} C(t,\mathbf{w};\psi_1,\psi_2) &= \sum_{i,j=1}^{n}\left(\frac{\partial}{\partial x_i}(a_{ij}\psi_1), \frac{\partial \psi_2}{\partial x_j}\right)_{\mathbf{L}^2(\mathbb{R}^n)} - \sum_{i=1}^{n}\left(\psi_1, (a_{0i}+a_{i0})\frac{\partial \psi_2}{\partial x_i}\right)_{\mathbf{L}^2(\mathbb{R}^n)} \\ &= B(t,\mathbf{w};\psi_1,\psi_2) + \sum_{i,j=1}^{n}\left(\frac{\partial a_{ij}}{\partial x_i}\psi_1, \frac{\partial \psi_2}{\partial x_j}\right)_{\mathbf{L}^2(\mathbb{R}^n)} \\ &\quad - \sum_{i=1}^{n}\left(\psi_1, (a_{0i}+a_{i0})\frac{\partial \psi_2}{\partial x_i}\right)_{\mathbf{L}^2(\mathbb{R}^n)} \end{aligned} \tag{7.3.7b}$$

on $\mathbf{H}^1(\mathbb{R}^n; \mathbb{R}^m) \times \mathbf{H}^1(\mathbb{R}^n; \mathbb{R}^m)$. On $\mathbf{Z}$, we define the inner product

$$\begin{aligned} ((\psi_1, \dot{\psi}_1), (\psi_2, \dot{\psi}_2))_{N(t,\mathbf{w})} &= B(t,\mathbf{w};\psi_1,\psi_2) + d_0(\psi_1,\psi_2)_{\mathbf{L}^2(\mathbb{R}^n)} \\ &\quad + (a_{00}\dot{\psi}_1, \dot{\psi}_2)_{\mathbf{L}^2(\mathbb{R}^n)}, \end{aligned} \tag{7.3.8}$$

where d_0 is a constant, independent of t and $\mathbf{w}$, specified by Proposition 7.3.1, to follow.

Remark 7.3.3. The equivalence of the norms, defined by $N(t, \mathbf{w})$ and $\|\cdot\|_{\mathbf{Z}}$ on $\mathbf{Z}$, follows from Proposition 7.3.1 (see (7.3.9)) and (a3). In fact, the existence of the constant λ_N of hypothesis (N) follows from (7.3.9).

Proposition 7.3.1. There are constants c, c_0, $d_0 > 0$, such that

$$|B(t, \mathbf{w}; \psi_1, \psi_2)| \leqslant c\|\psi_1\|_{\mathbf{H}^1(\mathbb{R}^n)}\|\psi_2\|_{\mathbf{H}^1(\mathbb{R}^n)}, \tag{7.3.9a}$$

$$B(t, \mathbf{w}; \psi, \psi) \geqslant c_0\|\psi\|^2_{\mathbf{H}^1(\mathbb{R}^n)} - d_0\|\psi\|^2_{\mathbf{L}^2(\mathbb{R}^n)}, \tag{7.3.9b}$$

for all $\psi, \psi_1, \psi_2 \in \mathbf{H}^1(\mathbb{R}^n; \mathbb{R}^m)$, $t \in [0, T_0]$, and $\mathbf{w} \in \mathbf{W}$. Similar constants c', c'_0, and d'_0 exist, so that $C(\cdot, \cdot; \cdot, \cdot)$ satisfies the continuity condition (7.3.9a) and the Gårding inequality (7.3.9b). If $\mathrm{C}(t, \mathbf{w})$ is the closed linear operator in $\mathbf{L}^2(\mathbb{R}^n; \mathbb{R}^m)$, specified by the Lax–Milgram lemma, according to

$$C(t, \mathbf{w}; \psi_1, \psi) = (\psi_1, \mathrm{C}(t, \mathbf{w})\psi)_{\mathbf{L}^2(\mathbb{R}^n)}, \tag{7.3.10}$$

for all $\psi \in \mathbf{D}_{\mathrm{C}(t,\mathbf{w})}$, $\psi_1 \in \mathbf{H}^1(\mathbb{R}^n; \mathbb{R}^m)$, then $\mathbf{D}_{\mathrm{C}(t,\mathbf{w})} = \mathbf{H}^2(\mathbb{R}^n; \mathbb{R}^m)$, and

$$\mathrm{C}(t, \mathbf{w})(\psi) = -\sum_{i,j=1}^{n} a_{ij}\frac{\partial^2\psi}{\partial x_i \partial x_j} - \sum_{i=1}^{n}(a_{0i} + a_{i0})\frac{\partial\psi}{\partial x_i}, \qquad \psi \in \mathbf{D}_{\mathrm{C}(t,\mathbf{w})}. \tag{7.3.11}$$

In particular, the operator $\mathrm{A}(t, \mathbf{w})$, defined formally by (7.3.4), is a closed linear operator in $\mathbf{Z}$, with domain $\mathbf{D}_{\mathrm{A}(t,\mathbf{w})} = \mathbf{H}^2(\mathbb{R}^n; \mathbb{R}^m) \times \mathbf{H}^1(\mathbb{R}^n; \mathbb{R}^m)$.

Proof: Inequality (7.3.9b) is Gårding's inequality for uniformly elliptic systems on $\mathbb{R}^n$, and may be adapted from the bounded domain case (see the bounded domain case in Morrey (Chapter 4 [25], Section 6.5), while (7.3.9a) is immediate from the Cauchy– chwarz inequality.

By (7.3.7b) and the Cauchy–Schwarz inequality, we have

$$C(t, \mathbf{w}; \psi, \psi) \geqslant B(t, \mathbf{w}; \psi, \psi) - c_1\|\psi\|_{\mathbf{H}^1(\mathbb{R}^n)}\|\psi\|_{\mathbf{L}^2(\mathbb{R}^n)}, \tag{7.3.12}$$

for some constant c_1, independent of t, $\mathbf{w}$, and ψ. Hence, the Gårding inequality for C is immediate from (7.3.9b), if we use (7.3.12) and

$$c_1\|\psi\|_{\mathbf{H}^1(\mathbb{R}^n)}\|\psi\|_{\mathbf{L}^2(\mathbb{R}^n)} \leqslant \frac{c_0}{2}\|\psi\|^2_{\mathbf{H}^1(\mathbb{R}^n)} + \frac{c_1^2}{2c_0}\|\psi\|^2_{\mathbf{L}^2(\mathbb{R}^n)}. \tag{7.3.13}$$

If $C(t, \mathbf{w})$ is defined by (7.3.10), the domain assertion $\mathbf{D}_{\mathrm{C}(t,\mathbf{w})} = \mathbf{H}^2(\mathbb{R}^n; \mathbb{R}^m)$ is the classical statement that weak and strong solutions coincide (Chapter 4 [25]) and, hence, (7.3.11) follows upon use of (a2). The fact $\mathrm{A}(t, \mathbf{w})$ is closed in $\mathbf{Z}$

and its domain assertion follow from the identity (see (7.3.5))

$$\mathrm{A}(t,\mathbf{w})(\psi,\dot{\psi}) = \left(-\dot{\psi}, a_{00}^{-1}\mathrm{C}(t,\mathbf{w})(\psi) + a_{00}^{-1}\left(\sum_{i=1}^{n}(a_{0i}+a_{i0})\left(\frac{\partial\psi}{\partial x_i} - \frac{\partial\dot{\psi}}{\partial x_i}\right)\right)\right). \tag{7.3.14}$$

This concludes the proof. ■

Lemma 7.3.2. There exists a constant ω, such that, for $t \in [0, T_0]$, $\mathbf{w} \in \mathbf{W}$, and $\mathbf{v} \in \mathbf{Z}$, the energy inequality

$$(\mathrm{A}(t,\mathbf{w})\mathbf{v},\mathbf{v})_{N(t,\mathbf{w})} \geqslant -\omega\|\mathbf{v}\|_{N(t,\mathbf{w})}^2 \tag{7.3.15}$$

holds, where $\omega \in \mathbb{R}^1$. In particular, $\mathrm{A}(t,\mathbf{w})$ satisfies (A1).

Proof: Let $\mathbf{v} = (\psi,\dot{\psi})$, with $\psi,\dot{\psi} \in \mathbf{C}_0^\infty(\mathbb{R}^n)$. Then, by (7.3.5), (7.3.7), and (7.3.8),

$$\begin{aligned}
&(\mathrm{A}(t,\mathbf{w})\mathbf{v},\mathbf{v})_{N(t,\mathbf{w})}\\
&\quad= B(t,\mathbf{w};-\dot{\psi},\psi) + d_0(-\dot{\psi},\psi)_{\mathbf{L}^2(\mathbb{R}^n)}\\
&\qquad+\left(a_{00}\left(-a_{00}^{-1}\sum_{i,j=1}^{n}a_{ij}\frac{\partial^2\psi}{\partial x_i\partial x_j} - a_{00}^{-1}\sum_{i=1}^{n}(a_{0i}+a_{i0})\frac{\partial\dot{\psi}}{\partial x_i}\right),\dot{\psi}\right)_{\mathbf{L}^2(\mathbb{R}^n)}\\
&\quad= B(t,\mathbf{w};-\dot{\psi},\psi) + d_0(-\dot{\psi},\psi)_{\mathbf{L}^2(\mathbb{R}^n)} + C(t,\mathbf{w};\dot{\psi},\psi)\\
&\qquad+\left(\sum_{i=1}^{n}(a_{0i}+a_{i0})\left(\frac{\partial\psi}{\partial x_i}-\frac{\partial\dot{\psi}}{\partial x_i}\right),\dot{\psi}\right)_{\mathbf{L}^2(\mathbb{R}^n)}\\
&\quad= \sum_{i,j=1}^{n}\left(\frac{\partial a_{ij}}{\partial x_i}\dot{\psi},\frac{\partial\psi}{\partial x_j}\right)_{\mathbf{L}^2(\mathbb{R}^n)} - \sum_{i=1}^{n}\left((a_{0i}+a_{i0})\frac{\partial\dot{\psi}}{\partial x_i},\dot{\psi}\right)_{\mathbf{L}^2(\mathbb{R}^n)}\\
&\qquad+ d_0(-\dot{\psi},\psi)_{\mathbf{L}^2(\mathbb{R}^n)}\\
&\quad= \sum_{i,j=1}^{n}\left(\frac{\partial a_{ij}}{\partial x_i}\dot{\psi},\frac{\partial\psi}{\partial x_j}\right)_{\mathbf{L}^2(\mathbb{R}^n)} + \tfrac{1}{2}\sum_{i=1}^{n}\left(\frac{\partial}{\partial x_i}(a_{0i}+a_{i0})\dot{\psi},\dot{\psi}\right)_{\mathbf{L}^2(\mathbb{R}^n)}\\
&\qquad+ d_0(-\dot{\psi},\psi)_{\mathbf{L}^2(\mathbb{R}^n)},
\end{aligned}$$

where we have used an argument similar to that in obtaining (6.4.12). It follows that

$$(\mathrm{A}(t,\mathbf{w})\mathbf{v},\mathbf{v})_{N(t,\mathbf{w})} \geqslant -c_1\|\psi\|_{\mathbf{H}^1(\mathbb{R}^n)}\|\dot{\psi}\|_{\mathbf{L}^2(\mathbb{R}^n)} - c_2\|\dot{\psi}\|_{\mathbf{L}^2(\mathbb{R}^n)}^2,$$

for positive constants c_1 and c_2, hence,

$$(\mathrm{A}(t,\mathbf{w})\mathbf{v},\mathbf{v})_{N(t,\mathbf{w})} \geqslant -c\|\mathbf{v}\|_{\mathbf{Z}}^2,$$

from which (7.3.15) follows. The fact that $A(t, \mathbf{w}) \in G(\mathbf{Z}_{N(t,\mathbf{w})}, 1, \omega)$ follows from a spectral analysis, as in the example of symmetric linear hyperbolic systems of Section 6.4 (see Lemma 6.4.4). ■

The format (7.3.6) and (7.3.2) is now clearly related to the abstract Cauchy problem (7.2.1). The spaces **Y**, **X**, **V**, and **Z** have been identified in Definition 7.3.1, and the family $\{N(t, \mathbf{w})\}$ is contained in $\mathbf{N}(\mathbf{Z})$ (see Definition 7.2.1), by Remark 7.3.3. Moreover, by the previous lemma, $A(t, \mathbf{w}) \in G(\mathbf{Z}_{N(t,\mathbf{w})}, 1, \omega)$ for some fixed constant ω. We shall begin systematically to verify the hypotheses (N)–(F) of Definition 7.2.2. Fundamental to the various continuity, invariance, and boundedness concepts of these definitions are the ring properties, satisfied by the Sobolev and uniformly local Sobolev spaces. We develop these now.

Definition 7.3.3. For **P** a fixed Hilbert space, and $1 \leqslant p < \infty$, let $\mathbf{L}^p_{u\ell}(\mathbb{R}^n; \mathbf{P})$ denote the set of all (equivalence classes of) **P**-valued strongly measurable functions u on $\mathbb{R}^n$, such that

$$\|u\|_{\mathbf{L}^p_{u\ell}} = \sup_{\mathbf{x} \in \mathbb{R}^n} \left(\int_{|\mathbf{y}-\mathbf{x}|<1} |u(\mathbf{y})|^p \, d\mathbf{y} \right)^{1/p} < \infty. \tag{7.3.16}$$

The space $\mathbf{L}^p_{u\ell}$ is called a uniformly local $\mathbf{L}^p$–space. For each integer $s \geqslant 0$, we denote by $\mathbf{H}^s_{u\ell}$ the set of $u \in \mathbf{L}^2_{u\ell}(\mathbb{R}^n; \mathbf{P})$, such that the distribution derivatives $D^{\boldsymbol{\alpha}} u$ of order $|\boldsymbol{\alpha}| \leqslant s$ are in $\mathbf{L}^2_{u\ell}(\mathbb{R}^n; \mathbf{P})$. The norm in $\mathbf{H}^s_{u\ell}$ is given by

$$\|u\|_{\mathbf{H}^s_{u\ell}(\mathbb{R}^n)} = \sup_{|\boldsymbol{\alpha}| \leqslant s} \|D^{\boldsymbol{\alpha}} u\|_{\mathbf{L}^2_{u\ell}(\mathbb{R}^n)}.$$

Remark 7.3.4. The spaces $\mathbf{L}^p_{u\ell}$ and $\mathbf{H}^s_{u\ell}$ are Banach spaces. Interpolation yields the spaces $\mathbf{H}^s_{u\ell}$ for nonintegral s. The following properties will be necessary (see Kato [18] and Hughes, Kato, and Marsden [13]).

(1) If $s > n/2 + k$, k a nonnegative integer, then

$$\mathbf{H}^s(\mathbb{R}^n; \mathbb{R}^m) \subset \mathbf{H}^s_{u\ell}(\mathbb{R}^n; \mathbb{R}^m) \subset \mathbf{C}^k_b(\mathbb{R}^n; \mathbb{R}^m),$$

and the inclusions are continuous.

(2) If $s > n/2$, then pointwise multiplication induces continuous bilinear maps

$$\mathbf{H}^{s-\ell}(\mathbb{R}^n; \mathbb{R}^m) \times \mathbf{H}^{k+\ell}(\mathbb{R}^n; \mathbb{R}^1) \to \mathbf{H}^k(\mathbb{R}^n; \mathbb{R}^m),$$

and

$$\mathbf{H}^{s-\ell}_{u\ell}(\mathbb{R}^n; \mathbb{R}^m) \times \mathbf{H}^{k+\ell}(\mathbb{R}^n; \mathbb{R}^1) \to \mathbf{H}^k(\mathbb{R}^n; \mathbb{R}^m),$$

for $0 \leqslant \ell \leqslant s$, $0 \leqslant k \leqslant s - \ell$.

8), we obtain, by first using $\mathbf{H}^{s-1} \cdot \mathbf{H}^r \subset \mathbf{H}^r$ (if $s > n/2 + 1$),

$$\mathbf{w}) - \mathbf{A}(t', \mathbf{w}'))(\psi, \dot{\psi})\|_{\mathbf{H}^{r+1}(\mathbb{R}^n) \times \mathbf{H}^r(\mathbb{R}^n)}$$

$$\leqslant C_1 \sum_{i,j=1}^{n} \|a_{00}^{-1} a_{ij} - (a_{00}^{-1} a_{ij})'\|_{\mathbf{H}^r(\mathbb{R}^n)} \|\psi\|_{\mathbf{H}^{s+1}(\mathbb{R}^n)}$$

$$+ C_1 \sum_{i=1}^{n} \|a_{00}^{-1}(a_{0i} + a_{i0}) - [a_{00}^{-1}(a_{0i} + a_{i0})]'\|_{\mathbf{H}^r(\mathbb{R}^n)} \|\dot{\psi}\|_{\mathbf{H}^s(\mathbb{R}^n)}$$

$$\leqslant C_2(|t - t'| + \|\mathbf{w} - \mathbf{w}'\|_{\mathbf{H}^{r+1}(\mathbb{R}^n) \times \mathbf{H}^r(\mathbb{R}^n)}) \|(\psi, \dot{\psi})\|_{\mathbf{Y}}, \tag{7.3.22}$$

$< s$. If $s > n/2$, and a_{ij} does not depend on the derivatives of $\mathbf{w}$, the quality holds, by use of $\mathbf{H}^{s-1} \cdot \mathbf{H}^{r+1} \subset \mathbf{H}^r$ and a strengthened f (7.3.18), with $\mathbf{H}^r$ replaced by $\mathbf{H}^{r+1}$ on the left-hand side. We v verify the hypotheses of Definition 7.2.2 in two stages, leaving ition of S and the verification of the associated properties (S) and l the conclusion of the following lemma and its proof.

7.3.3. The hypotheses (N), (A3), and (F) hold if $s > (n/2) + 1$. esis (A4) holds for a dense subset of $\mathbf{W}$.

$\mathbf{A}(t, \mathbf{w}) \in \mathbf{B}(\mathbf{Y}, \mathbf{X})$, with $\lambda_{\mathbf{A}} = C_2$, follows from (7.3.21). The remaining continuity properties follow directly from (7.3.22), with $r = s - 1$; note here the choice $\mathbf{V} = \mathbf{X}$.

An easy calculation shows that such a dense set is given by

$$\mathbf{W} \cap (\mathbf{H}^{s+2}(\mathbb{R}^n; \mathbb{R}^m) \times \mathbf{H}^{s+1}(\mathbb{R}^n; \mathbb{R}^m)).$$

The constant λ_0 depends only on y_0.

The continuity properties of $\mathbf{F} = (\mathbf{0}, \mathbf{b})$ follow from (7.3.19). Since $\mathbf{b}$ is not obtained via a multiplier, it is necessary to conclude that $\mathbf{F}(t, \mathbf{w}) \in \mathbf{Y}$, by using (3c) of Remark 7.3.4, in conjunction with the assumption on $\mathbf{b}(0, \cdot, 0)$. The $\mathbf{Y}$-range boundedness property is now immediate from (7.3.17b).

Using $\mathbf{H}^{s-1} \cdot \mathbf{H}^0 \subset \mathbf{H}^0$ and (7.3.18), with $r = s - 1$, we obtain (for $s > n/2 + 1$)

$$|B(t, \mathbf{w}; \psi, \psi) - B(t', \mathbf{w}'; \psi, \psi)|$$
$$\leqslant C_1(|t - t'| + \|\mathbf{w} - \mathbf{w}'\|_{\mathbf{H}^s(\mathbb{R}^n) \times \mathbf{H}^{s-1}(\mathbb{R}^n)}) \cdot \|(\psi, \psi)\|_{\mathbf{Z}}^2, \tag{7.3.23}$$

which leads to

$$|\|\mathbf{u}\|_{N(t,\mathbf{w})}^2 - \|\mathbf{u}\|_{N(t',\mathbf{w}')}^2| \leqslant \mu(|t - t'| + \|\mathbf{w} - \mathbf{w}'\|_{\mathbf{X}}) \|\mathbf{u}\|_{\mathbf{Z}}^2, \tag{7.3.24}$$

for t, $t' \in [0, T_0]$ and $\mathbf{w}, \mathbf{w}' \in \mathbf{W}$, $\mathbf{u} \in \mathbf{Z}$. The same inequality (7.3.24)

(3a) Assume that $\Omega_0 \subset \mathbb{R}^N$ is open, values in Ω_0. Suppose $\mathbf{W}_0 \subset \mathbf{H}^s($ radius chosen small enough so an $K \subset \Omega_0$, and that $\mathbf{G}: \mathbb{R}^n \times \Omega_0 \to$ there is a constant C_1, such that,

$$\|\mathbf{G}(\cdot, \mathbf{u}(\cdot))\|_{\mathbf{H}^s_{u\ell}(\mathbb{R}^n)}$$

(3b) If $\mathbf{G}$ is of class $\mathbf{C}_b^{s+1}$ ($\mathbf{C}_b^{[s]+2}$ if $[s] \neq$

$$\|\mathbf{G}(\cdot, \mathbf{u}(\cdot)) - \mathbf{G}(\cdot, \mathbf{v}(\cdot))\|_{\mathbf{H}}$$

for $\mathbf{u}, \mathbf{v} \in \mathbf{W}$ and $0 \leqslant r \leqslant s$.

(3c) If, in addition, $0 \in \Omega_0$, if Ω_0 is co $\mathbf{H}^r(\mathbb{R}^n; \mathbb{R}^N)$, then $\mathbf{G}(\cdot, \mathbf{u}(\cdot)) \in \mathbf{H}^r(\mathbb{R}^n; \mathbb{R}$

Remark 7.3.5. We mention some direct c the previous remark as applied to the giv (7.3.17a,b), we conclude that

$$a_{ij} - a'_{ij} := a_{ij}(t, \cdot, \boldsymbol{\sigma}, \dot{\boldsymbol{\sigma}}, \nabla\boldsymbol{\sigma}) - a$$

and

$$\mathbf{b} - \mathbf{b}' := b(t, \cdot, \boldsymbol{\sigma}, \dot{\boldsymbol{\sigma}}, \nabla\boldsymbol{\sigma}) - \mathbf{b}(t',$$

are in $\mathbf{H}^r(\mathbb{R}^n)$ for $0 \leqslant r \leqslant s$ and $i, j = 0, 1, \ldots,$ μ_1 and μ_2,

$$\|a_{ij} - a'_{ij}\|_{\mathbf{H}^r(\mathbb{R}^n)} \leqslant \mu_1(|t - t'| + \|\mathbf{w} - \mathbf{w}$$

$$\|\mathbf{b} - \mathbf{b}'\|_{\mathbf{H}^r(\mathbb{R}^n)} \leqslant \mu_2(|t - t'| + \|\mathbf{w} - \mathbf{w}'$$

for $t, t' \in [0, T_0]$ and $\mathbf{w} = (\boldsymbol{\sigma}, \dot{\boldsymbol{\sigma}})$, $\mathbf{w}' = (\boldsymbol{\sigma}', \dot{\boldsymbol{\sigma}}')$. From

$$a_{00}^{-1} a_{ij} \in \mathbf{H}^s_{u\ell}(\mathbb{R}^n; \mathbb{R}^{m^2}) \qquad \text{is uniforml}$$

for $\mathbf{w} \in \mathbf{W}$ and $t \in [0, T_0]$. Thus, since, in this case, have

$$\|\mathrm{A}(t, \mathbf{w})(\boldsymbol{\psi}, \dot{\boldsymbol{\psi}})\|_{\mathbf{X}} \leqslant C_1(\|\dot{\boldsymbol{\psi}}\|_{\mathbf{H}^s(\mathbb{R}^n)} + \sum_{i,j=1}^{n} \|a_{00}^{-1} a_{ij}$$

$$+ \sum_{i=1}^{n} \|a_{00}^{-1}(a_{0i} + a_{i0})\|_{\mathbf{H}^s_{u\ell}(\mathbb{R}^n)}$$

$$\leqslant C_2 \|(\boldsymbol{\psi}, \dot{\boldsymbol{\psi}})\|_{\mathbf{Y}}.$$

† The second space in this inclusion consists of scalar functions This convention is maintained throughout this section.

From (7.3.1

$\|(\mathrm{A}(t$

for $0 \leqslant r$ same ine version c shall nov the defin (A2) unt

Lemma Hypoth

Proof:

(A3)

(A4)

(F)

(N)

holds for $s > n/2$, if a_{ij} does not depend on derivatives, by use of $\mathbf{H}^s \cdot \mathbf{H}^0 \subset \mathbf{H}^0$. Now write

$$1 \leqslant \frac{\|\mathbf{u}\|_{N(t,\mathbf{w})}^2}{\|\mathbf{u}\|_{N(t',\mathbf{w}')}^2} = \frac{\|\mathbf{u}\|_{N(t,\mathbf{w})}^2 - \|\mathbf{u}\|_{N(t',\mathbf{w}')}^2}{\|\mathbf{u}\|_{N(t',\mathbf{w}')}^2} + 1,$$

for $\|\mathbf{u}\|_{N(t,\mathbf{w})} \geqslant \|\mathbf{u}\|_{N(t',\mathbf{w}')}$. The other inequality is parallel. For $x \geqslant 0$, $\ln(1 + x) \leqslant x$. Thus, by (7.3.24),

$$\ln\left(\frac{\|\mathbf{u}\|_{N(t,\mathbf{w})}^2}{\|\mathbf{u}\|_{N(t',\mathbf{w}')}^2}\right) \leqslant \mu(|t - t'| + \|\mathbf{w} - \mathbf{w}'\|_{\mathbf{X}}) \frac{\|\mathbf{u}\|_{\mathbf{Z}}^2}{\|\mathbf{u}\|_{N(t',\mathbf{w}')}^2}$$
$$\leqslant \mu\lambda_N^2(|t - t'| + \|w - w'\|_{\mathbf{X}}),$$

where the existence of λ_N follows from Remark 7.3.3. Hypothesis (N) is now verified with $\mu_N = \mu\lambda_N^2/2$. ■

Lemma 7.3.4. Let $\Lambda = (\mathrm{I} - \Delta)^{1/2}$, with domain $\mathbf{H}^1(\mathbb{R}^n; \mathbb{R}^m)$ and range $\mathbf{L}^2(\mathbb{R}^n; \mathbb{R}^m)$. Then, for $a \in \mathbf{L}^\infty(\mathbb{R}^n; \mathbb{R}^1)$, with grad $a \in \mathbf{H}^{s-1}(\mathbb{R}^n; \mathbb{R}^n)$, we have, for some constant c,

$$\|[\Lambda^s, \mathrm{a}]\Lambda^{1-s}\|_{\mathbf{L}^2(\mathbb{R}^n), \mathbf{L}^2(\mathbb{R}^n)} \leqslant c\|\operatorname{grad} a\|_{\mathbf{H}^{s-1}(\mathbb{R}^n)}. \tag{7.3.25}$$

Proof: See Kato [17], p. 66. Note the diagonal operator interpretation of Λ here and throughout the section. Note also the operator connotation of a. ■

Remark 7.3.6. We select the operator

$$\mathrm{S} = \begin{bmatrix} \Lambda^s & 0 \\ 0 & \Lambda^s \end{bmatrix}.$$

A simple computation gives

$$\mathrm{S}\mathrm{A}(t, \mathbf{w})\mathrm{S}^{-1} = \mathrm{A}(t, \mathbf{w}) + \mathrm{B}(t, \mathbf{w}), \tag{7.3.26}$$

where

$$\mathrm{B}(t, \mathbf{w}) = \begin{bmatrix} 0 & 0 \\ [\Lambda^s, \mathrm{A}_1]\Lambda^{-s} & [\Lambda^s, \mathrm{A}_2]\Lambda^{-s} \end{bmatrix}, \tag{7.3.27a}$$

and

$$\mathrm{A}_1(t, w) = -\sum_{i,j=1}^{n} a_{00}^{-1} a_{ij} \frac{\partial^2}{\partial x_i \partial x_j}, \tag{7.3.27b}$$

$$\mathrm{A}_2(t, w) = -\sum_{i=1}^{n} a_{00}^{-1}(a_{0i} + a_{i0}) \frac{\partial}{\partial x_i}, \tag{7.3.27c}$$

and [,] denotes the commutator.

Lemma 7.3.5. Properties (A2) and (S) hold if $s > n/2 + 1$.

Proof: By the commuting property of $\Lambda^{-\sigma}$, $\sigma > 0$, and $\partial/\partial x_i$, we have, for $(\psi, \dot{\psi}) \in \mathbf{Z}$,

$$[\Lambda^s, \mathrm{A}_1]\Lambda^{-s}\psi = -\sum_{i,j=1}^{n} [\Lambda^s, a_{00}^{-1}a_{ij}]\Lambda^{1-s}\frac{\partial^2}{\partial x_i \partial x_j}\Lambda^{-1}\psi, \tag{7.3.28}$$

$$[\Lambda^s, \mathrm{A}_2]\Lambda^{-s}\dot{\psi} = -\sum_{i=1}^{n} [\Lambda^s, a_{00}^{-1}(a_{0i} + a_{i0})]\Lambda^{1-s}\frac{\partial}{\partial x_i}\Lambda^{-1}\dot{\psi}. \tag{7.3.29}$$

By Lemma 7.3.4, and property (2) of Remark 7.3.4, applied with (7.3.20) to the expanded gradient,

$$\|[\Lambda^s, a_{00}^{-1}a_{ij}]\Lambda^{1-s}\|_{\mathbf{L}^2(\mathbb{R}^n),\mathbf{L}^2(\mathbb{R}^n)} \leqslant c\|\mathrm{grad}\, a_{00}^{-1}a_{ij}\|_{\mathbf{H}^{s-1}(\mathbb{R}^n)} \leqslant C,$$

and

$$\|[\Lambda^s, a_{00}^{-1}(a_{0i} + a_{i0})]\Lambda^{1-s}\|_{\mathbf{L}^2(\mathbb{R}^n),\mathbf{L}^2(\mathbb{R}^n)} \leqslant \|\mathrm{grad}\, a_{00}^{-1}(a_{0i} + a_{i0})\|_{\mathbf{H}^{s-1}(\mathbb{R}^n)} \leqslant C.$$

When these inequalities are applied to the $\mathbf{L}^2$ estimations of (7.3.28) and (7.3.29), there results the estimate

$$\|\mathrm{B}(t, \mathbf{w})(\psi, \dot{\psi})\|_{\mathbf{Z}} \leqslant \lambda_{\mathrm{B}}\|(\psi, \dot{\psi})\|_{\mathbf{Z}}, \tag{7.3.30}$$

where λ_{B} does not depend on $t \in [0, T_0]$ and $\mathbf{w} \in \mathbf{W}$. Property (A2) follows directly. Property (S), of course, is immediate from the time independence of $\mathrm{S}(\cdot)$ and the continuity of the embedding $\mathbf{Z} \to \mathbf{X}$. ■

Theorem 7.3.6.[†] Assume $s > n/2 + 1$ and that (a1)–(a3) hold (if the a_{ij} do not depend on the derivatives of ψ, $s > n/2$ need only be assumed). Then, the Cauchy problem (7.3.1)–(7.3.2) is well-posed in the following sense. Given $(\psi_0, \dot{\psi}_0)$, satisfying (7.3.2)–(7.3.3), there is a neighborhood $\mathbf{Y}_0$ of $(\psi_0, \dot{\psi}_0)$ in $\mathbf{H}^{s+1}(\mathbb{R}^n) \times \mathbf{H}^s(\mathbb{R}^n)$, and a positive number $T_0' \leqslant T_0$, such that, for any initial condition in $\mathbf{Y}_0$, (7.3.1)–(7.3.2) has a unique solution $\psi(t, \cdot)$ for $t \in [0, T_0']$, satisfying (7.3.3):

$$\left(\psi(t, \mathbf{x}), \frac{d\psi}{dt}(t, \mathbf{x}), \nabla\psi(t, \mathbf{x})\right) \in \Omega, \qquad \text{for all} \quad \mathbf{x} \in \mathbb{R}^n. \tag{7.3.31}$$

Moreover, $\psi \in \mathbf{C}^r([0, T_0']; \mathbf{H}^{s+1-r}(\mathbb{R}^n; \mathbb{R}^m))$, $0 \leqslant r \leqslant s$, and the map $(\psi_0, \dot{\psi}_0) \mapsto (\psi(t, \cdot), (d\psi/dt)(t, \cdot))$ is continuous in the topology of $\mathbf{H}^1 \times \mathbf{H}^0$, uniformly in $t \in [0, T_0']$.

Proof: The $\mathbf{C}^r$ continuity of ψ for $r > 1$ follows from the cases $r = 0$ and $r = 1$ by direct use of (7.3.1). The remaining conclusions follow from Theorem

[†] This theorem and its proof are adapted from Hughes, Kato, and Marsden [13].

7.2.4. The hypotheses of this theorem are verified by Lemmas 7.3.2, 7.3.3, and 7.3.5. ∎

Remark 7.3.7. It can be shown that (see Hughes *et al.* [13])

$$(\psi_0, \dot{\psi}_0) \mapsto (\psi(t, \cdot), \dot{\psi}(t, \cdot))$$

is continuous in $\mathbf{H}^{s+1} \times \mathbf{H}^s$, uniformly in t.

7.4 THE VACUUM FIELD EQUATIONS OF GENERAL RELATIVITY

The field equations of general relativity may be viewed as an analog of Poisson's equation of the Newtonian theory. In fact, suppose we begin with the metric tensor coefficients g_{ij} of the gravitational potential, under the assumption of a well-defined semi-Riemannian, space–time manifold, with $ds^2 = \sum_{i,j=0}^{3} g_{ij}\, dx^i\, dx^j$.† Suppose a symmetric tensor expression G_{ij} in the g_{ij} is sought, satisfying the following conditions as described in Einstein [5].

(1) The tensor G_{ij} may contain no differential coefficients of the g_{ij} higher than the second order.
(2) The tensor G_{ij} must be linear and homogeneous in these second-order coefficients.

Then, the expression must be of the form

$$G_{ij} = R_{ij} + a g_{ij} R + b g_{ij}, \tag{7.4.1a}$$

where R_{ij} denotes the Ricci curvature tensor,

$$R_{ij} = R^k_{ikj}, \tag{7.4.1b}$$

of second order, R is the scalar curvature,

$$R = g^{ij} R_{ij} = R^k_k \qquad ((g^{ij}) = (g_{ij})^{-1}), \tag{7.4.1c}$$

and a and b are constants. Here, we have adopted the Einstein summation convention. The quantity $\{R^i_{jk\ell}\}$ is the Riemann curvature tensor. A coordinate-free definition of this tensor is given in differential geometry as the unique $(1, 3)$ tensor field $\bar{R}$, such that

$$\bar{R}(\omega, Z, X, Y) = \omega(\mathrm{R}_{XY} Z), \tag{7.4.2}$$

† This includes Lorentz manifolds as a special case. These are the solution manifolds of interest.

for all 1–forms ω, and for all vector fields X, Y, Z. Here, R_{XY} is the curvature operator which assigns, to each ordered pair X, Y of vector fields on the space–time manifold, the operator R_{XY} on vector fields, defined by

$$\mathrm{R}_{XY}Z = \mathrm{D}_X\mathrm{D}_YZ - \mathrm{D}_Y\mathrm{D}_XZ - \mathrm{D}_{[X,Y]}Z, \tag{7.4.3}$$

where D_X denotes the covariant derivative in the direction X, and $[\cdot,\cdot]$ denotes the Lie bracket. The quantity D is referred to as the Levi–Civita, or semi-Riemannian, connection, and, in particular satisfies the torsion-free property

$$\mathrm{D}_XY - \mathrm{D}_YX = [X, Y]. \tag{7.4.4}$$

If Γ^k_{ij} denote the connection coefficients for dual local bases $\{\omega^i\}$ and $\{X_i\}$, then

$$\mathrm{D}_{X_i}(X_j) = \sum_k \Gamma^k_{ij}X_k, \tag{7.4.5}$$

and if $\{X_i\}$ is chosen, so that $\Gamma^k_{ij} = \Gamma^k_{ji}$ (that is, $[X_i, X_j] = 0$), then we deduce from (7.4.2) and (7.4.3) the standard format, with the notation $f_{,k} = X_k(f)$,

$$R^i_{jk\ell} = \bar{R}(\omega^i, X_j, X_k, X_\ell) = \Gamma^i_{j\ell,k} - \Gamma^i_{jk,\ell} + \Gamma^i_{mk}\Gamma^m_{j\ell} - \Gamma^i_{m\ell}\Gamma^m_{jk}. \tag{7.4.6}$$

Moreover, the metric coefficients

$$g_{ij} = \langle X_i, X_j\rangle \qquad (\langle\cdot,\cdot\rangle := \text{metric tensor}) \tag{7.4.7}$$

define the connection coefficients, via the formula

$$\Gamma^k_{ij} = \tfrac{1}{2}g^{k\ell}(g_{\ell i,j} + g_{\ell j,i} - g_{ij,\ell}), \tag{7.4.8}$$

if $[X_m, X_n] = 0$, all m, n. Note that (7.4.8) is a direct consequence of the conjunction of (7.4.5) and (7.4.7), with

$$X_i\langle X_r, X_j\rangle + X_j\langle X_r, X_i\rangle - X_r\langle X_i, X_j\rangle = 2\langle \mathrm{D}_{X_i}X_j, X_r\rangle.$$

The latter is an immediate consequence of the torsion-free property (7.4.4) and the identity

$$Z\langle X, Y\rangle = \langle \mathrm{D}_ZX, Y\rangle + \langle X, \mathrm{D}_ZY\rangle,$$

both of which hold for semi-Riemannian connections.

Returning to the derivation of the field equations, we may require, on the basis of conservation principles, the following.

(3) The divergence of G_{ij} must vanish identically. This imposes the condition $a = -\frac{1}{2}$ in (7.4.1a). If, in addition, we require

(4) $G_{ij} = 0$ in flat space–time (that is, when $R^i_{jk\ell} = 0$), then $b = 0$. Thus, under the assumptions (1)–(4), the Einstein tensor G_{ij} has the form

$$G_{ij} = R_{ij} - \tfrac{1}{2}g_{ij}R, \tag{7.4.9}$$

and the generalization of Poisson's equation is

$$G_{ij} = -8\pi G T_{ij}, \tag{7.4.10}$$

where T_{ij} is the stress-energy tensor of matter, and G is Newton's gravitational constant, and units are normalized by a speed of light of unity. In this section, we discuss the general existence theory only for the vacuum case, $T_{ij} = 0$.

Remark 7.4.1. If a coordinate base field $X_i = \partial/\partial x_i$ is selected at each point of the manifold, then the Lie bracket of any two basis vectors is zero and (7.4.6) holds, where the connection coefficients are described by (7.4.8). In both equations, commas denote partial differentiation. Thus, the Einstein field equations in a vacuum are completely described by the system

$$G_{ij} = 0, \qquad 0 \leqslant i, j \leqslant 3, \tag{7.4.11}$$

of partial differential equations in the metric tensor coefficients, where G_{ij} is given by (7.4.9), (7.4.1b,c), (7.4.6), and (7.4.8). If the constraint (4) is discarded, then $b \neq 0$ is permitted in (7.4.1). This (universal) constant b is the much-heralded cosmological constant, originally introduced by Einstein to obtain a static universe solution to the "dust" field equations of the form

$$ds^2 = c^2\,dt^2 - R^2\left\{\frac{d\rho^2}{1-\rho^2} + \rho^2(d\theta^2 + \sin^2\theta\,d\phi^2)\right\}, \tag{7.4.12}$$

where $b = 1/R^2$ is the constant curvature of the associated 3-space and ρ, θ, and ϕ are spherical coordinates. The solution is static, since ρ is necessarily constant. In the case of "dust", an average density ρ_0 is assumed, leading to $T_{ij} = \operatorname{diag}(0, 0, 0, \rho_0)$ and $b = 4\pi G\rho_0$. Although Einstein later abandoned the cosmological constant in the light of red-shifted spectral evidence of an expanding universe, cosmologists have not. A nonzero value of b is, of course, consistent with expanding universe models. In fact, the Friedmann models are those of the form

$$ds^2 = c^2\,dt^2 - R^2(t)\left\{\frac{d\rho^2}{1-k\rho^2} + \rho^2(d\theta^2 + \sin^2\theta\,d\phi^2)\right\}, \tag{7.4.13}$$

for $k = 0, \pm 1$ and associated 3-space curvature $k/R^2(t)$ (cf. also the equivalent form given in [23, p. 210]). The form of $R(\cdot)$ is determined by the "dust" model field equations, and the classification schemes permit $b = 0$ and $b \neq 0$. The metric (7.4.13) describes a universe satisfying large-scale isotropy and homogeneity, and some of the special solutions are associated with various names (cf. Rindler [23] Section 9.10).

Remark 7.4.2. Before we can address the Cauchy problem for $G_{ij} = 0$, we must make several remarks related to the transformation and/or reduction of the system. First, we make the standard observation that $G_{ij} = 0 \Leftrightarrow R_{ij} = 0$ (as systems). If we set

$$\Gamma^k = g^{ij}\Gamma^k_{ij}, \tag{7.4.14a}$$

and

$$-\tilde{R}_{ij} = R_{ij} + \tfrac{1}{2}\left(g_{ik}\frac{\partial \Gamma^k}{\partial x_j} + g_{jk}\frac{\partial \Gamma^k}{\partial x_i}\right), \tag{7.4.14b}$$

then we shall solve the reduced system $\tilde{R}_{ij} = 0$, subject to the condition that the local coordinates x_i on the initial manifold be harmonic coordinates, defined by the condition $\Gamma^k(0, x_i) = 0$. The reason for the term harmonic is the general identity

$$\Box\phi = g^{ij}\frac{\partial^2 \phi}{\partial x_i \partial x_j} - \Gamma^k \frac{\partial \phi}{\partial x_k},$$

where $\Box$ is the generalized d'Alembertian, defined by

$$\Box\phi = \frac{1}{\sqrt{-g}}\frac{\partial}{\partial x_i}\left\{\sqrt{-g}g^{ij}\frac{\partial \phi}{\partial x_j}\right\}, \qquad (g = \det(g_{ij})).$$

Here we explicitly assume a Lorentzian manifold with $g < 0$. Provided the initial data satisfy the necessary conditions referred to in the next paragraph, the evolution $\tilde{R}_{ij} = 0$ preserves the harmonic property of the coordinates (see Fourès-Bruhat [3]), so that $\tilde{R}_{ij} = 0 \Leftrightarrow R_{ij} = 0$. It can be shown that $\tilde{R}_{ij} = 0$ has the form

$$-\tfrac{1}{2}g^{k\ell}\frac{\partial^2 g_{ij}}{\partial x_k \partial x_\ell} + H_{ij}\left(g_{k\ell}, \frac{\partial g_{k\ell}}{\partial x_m}\right) = \tilde{R}_{ij} = 0, \tag{7.4.15}$$

where H_{ij} is an explicit rational function of $g_{k\ell}$ and is homogeneous quadratic in its first derivatives (see Fock [8] p. 423). The reduction to (7.4.15) achieves an uncoupling of the second derivatives.

The initial data are not completely arbitrary, but must satisfy certain constraints, which are related to the second fundamental form of the embedding of the initial 3-manifold into the space–time manifold (see Fischer and Marsden [7]). The remarkable fact accrues that, given initial data satisfying the constraints in arbitrary coordinates, there exists a sufficiently regular transformation to harmonic coordinates, so that the initial data continue to satisfy the embedding constraints in the new coordinates. The evolution (7.4.15) preserves the harmonic coordinates (locally in time), as well as the embedding constraints.

Still another technical difficulty is that the metric coefficients $g_{ij}(t, \cdot)$ need not be in $\mathbf{H}^s(\mathbb{R}^3; \mathbb{R}^1)$. In the elementary case of the Minkowski metric, $g_{ij} = \text{diag}(1, -1, -1, -1)$ and the diagonal entries are clearly not $\mathbf{L}^2$ functions on $\mathbb{R}^3$, though their derivatives surely are. However, for a wide class g_{ij} of space–time metrics, including the asymptotically flat metrics, with perturbations of order $O(r^{-1})$ at spatial infinity, it is to be expected that the variables

$$\psi_{ij} = g_{ij} - g_{ij}^0 \tag{7.4.16}$$

are Sobolev class variables. Accordingly, we recast the reduced system (7.4.15) into the format of (7.3.1), with

$$a_{00} = \tfrac{1}{2}\psi^{00}I, \qquad a_{ij} = -\tfrac{1}{2}\psi^{ij}I, \qquad (i, j) \neq (0, 0), \tag{7.4.17a}$$

and

$$\mathbf{b} = \left(H_{ij} + \sum_{k,l=0}^{3} \frac{\partial^2 g_{ij}^0}{\partial x_k \partial x_l} \psi^{kl} \right)_{ij}, \tag{7.4.17b}$$

with $n = 3$ and $m = 10$. Note that we discard redundancy due to the symmetry of the g_{ij}. Here $\psi^{k\ell}$ is the entry with index (k, ℓ) in the (inverse) matrix $(\psi_{ij} + g_{ij}^0)^{-1}$. The generality of the previous section, which required only that the a_{ij} multipliers be in a uniformly local Sobolev space, is clearly necessary here.

Remark 7.4.3. The Cauchy problem to be considered, then, is defined by the quasi-linear system (7.3.1), (7.4.17) in the 10-variable ψ_{ij}, subject to initial data $\psi_{ij}^0 = 0$ and $\dot{\psi}_{ij}^0$. We shall assume that the initial data satisfy certain regularity conditions and shall prescribe the domain Ω of "hyperbolicity" as follows. The quantity s is assumed to satisfy $s > \frac{3}{2}$.

(g1)[†]
$$g_{ij}^0 \in C_b^{s+3}(\mathbb{R}^3; \mathbb{R}^1), \qquad \dot{g}_{ij}^0 \in H^s(\mathbb{R}^3; \mathbb{R}^1),$$
$$\frac{\partial g_{ij}^0}{\partial x_i} \in H^s(\mathbb{R}^3; \mathbb{R}^1), \qquad 0 \leqslant i, j \leqslant 3$$

($\mathbf{C}_b^{s+3}$ replaced by $\mathbf{C}_b^{s+4}$ if $[s] \neq s$).

(g2) $\Omega \subset \mathbb{R}^{10} \times \mathbb{R}^{10} \times \mathbb{R}^{30}$ is chosen as follows:

$$\Omega = \Omega_0 \times \mathbb{R}^{10} \times \mathbb{R}^{30}, \tag{7.4.18}$$

where Ω_0 is a ball centered at 0, such that, if $\psi_{ij}(x) = g_{ij}(x) - g_{ij}^0(x) \in \Omega_0$, then $g_{ij}(x)$ is of Lorentz signature $(+, -, -, -)$. This has the standard

[†] $\mathbf{C}_b^{s+3}$ can be replaced by $\mathbf{C}_b^{s+2}$ if one verifies (F) of Section 7.2 directly.

meaning that the matrix (g_{ij}) is unitarily similar to diag$(1, -1, -1, -1)$. Two immediate implications are that the two requirements of (a3) of Definition 7.3.1 are satisfied. It may, of course, be necessary to replace Ω_0 with an open subset with compact closure in Ω_0. It is straightforward to check that (g1) implies (a1). It follows that, when (g1) and (g2) are satisfied, the Cauchy problem, stated at the beginning of this remark, has a solution locally in time in the sense of Theorem 7.3.6. Rather than state our result in terms of the artificial variables ψ_{ij}, we present the natural formulation in terms of the g_{ij}.

Theorem 7.4.1. Let (g1) and (g2) hold.† Then, for $s > \frac{3}{2}$ and initial data in a neighborhood of $(g_{ij}^0, \dot{g}_{ij}^0)$ in $\mathbf{H}_{g_{\alpha\beta}^0}^{s+1} \times \mathbf{H}_{\dot{g}_{\alpha\beta}^0}^{s}$, equations (7.4.15) have a unique solution in the same space for a time interval $[0, T'_0]$, $T'_0 > 0$. Here, $\mathbf{H}_{u_0}^s$ is the space of u, such that $u - u_0 \in \mathbf{H}^s$, with correspondingly induced topology.

7.5 INVARIANT TIME INTERVALS FOR THE ARTIFICIAL VISCOSITY METHOD

Suppose

$$\mathrm{A}(t, \mathbf{w}) = -\nu\Delta + \mathrm{E}(t, \mathbf{w}), \qquad \nu \geqslant 0, \tag{7.5.1a}$$

where Δ has the interpretation of a diagonal operator and the first-order operator $\mathrm{E}(t, \mathbf{w})$ is given formally by

$$\mathrm{E}(t, \mathbf{w}) = \mathrm{P}\left[\sum_{j=1}^{n} a_j(t, \cdot, \mathbf{w}) \frac{\partial}{\partial x_j} + b(t, \cdot, \mathbf{w})\right], \tag{7.5.1b}$$

and P is an orthogonal projection in $\mathbf{L}^2(\mathbb{R}^n; \mathbb{R}^m)$. We shall suppose, for simplicity, that a_j and b are defined on $[0, T_0] \times \mathbb{R}^n \times \mathbb{R}^m$ with range in $\mathbb{R}^{m^2}$, and that $u_0 \in \mathbf{PH}^s(\mathbb{R}^n; \mathbb{R}^m)$ is given for $s > n/2 + 1$. The problem addressed in this section is the following. We seek to determine a pair (r', T'_0), not depending on ν, and a solution $\mathbf{u}$ of the Cauchy problem

$$\frac{d\mathbf{u}}{dt} + \mathrm{A}(t, \mathbf{u})\mathbf{u} = 0, \qquad 0 \leqslant t \leqslant T'_0, \tag{7.5.2a}$$

$$\mathbf{u}(0, \cdot) = \mathbf{u}_0, \tag{7.5.2b}$$

†Our hypotheses are stronger than those of Hughes, Kato, and Marsden [13].

such that

$$\mathbf{u} \in \mathbf{Y}_s = \mathbf{W}^{1,\infty}([0, T_0']; \mathbf{H}^{\sigma(\nu)}(\mathbb{R}^n; \mathbb{R}^m)) \cap \mathbf{L}^2((0, T_0'); \mathbf{PH}^s(\mathbb{R}^n; \mathbb{R}^m)), \quad (7.5.3)$$

with $\sigma(\nu) = s - 1 - \operatorname{sgn}(\nu)$, and

$$\|\mathbf{u}(t, \cdot)\|_{\mathbf{H}^s(\mathbb{R}^n)} \leqslant r', \qquad 0 \leqslant t \leqslant T_0'. \quad (7.5.4)$$

The solutions $\mathbf{u} = \mathbf{u}^\nu$ of the "parabolic" systems are required to converge in $\mathbf{L}^2((0, T_0') \times \mathbb{R}^n)$ as $\nu \downarrow 0$ to the solution $\mathbf{u} = \mathbf{u}^0$ of the "hyperbolic" system.

Space limitations prevent us from supplying full details of these results, which are appearing elsewhere (see Jerome [14]). However, we shall state some of the results and present the core arguments. At the basis is a stability analysis of the horizontal-line method applied to (7.5.2), in which ideas of the previous chapter play a decisive role. We shall develop the general theory for this, and return to the applications at the end of the section. However, for the reader who wishes to proceed no further, we briefly describe the choice of T_0' and r'. Suppose, for $r > 0$, such that $\mathbf{u}_0 \in \mathbf{B}(0, r) \subset \mathbf{H}^s(\mathbb{R}^n; \mathbb{R}^m)$, that $\mathrm{A}(t, \mathbf{w}) \in G(\mathbf{PL}^2, 1, \omega(r))$ for $0 \leqslant t \leqslant T_0'$, $\|\mathbf{w}\|_{\mathbf{L}^2(\mathbb{R}^n)} \leqslant r$, and a restriction of A satisfies $\tilde{\mathrm{A}}(t, \mathbf{w}) \in G(\mathbf{PH}^s, 1, \tilde{\omega}(r))$ for $\|\mathbf{w}\|_{\mathbf{H}^s(\mathbb{R}^n)} \leqslant r$ and $0 \leqslant t \leqslant T_0'$. Then, for $\gamma = 1 + \max(\omega, \tilde{\omega})$, and T_0' satisfying (see (7.5.33))

$$\|\mathbf{u}_0\|_{\mathbf{H}^s(\mathbb{R}^n)} e^{\gamma(r) T_0'} < r,$$

the interval $[0, T_0']$ is an invariant interval. If T_0' is fixed, then $r' = r$ may be chosen to maximize the function $f(r) = re^{-\gamma(r)T_0'}$ when this occurs. Explicit solutions are possible in the case of the Navier–Stokes/Euler system, for example, and the maximization may be employed to determine the critical value $T_0 = \sup(T_0')$. We now begin the basic development; we require a more stringent notion of stability than introduced in Definition 6.2.1.

Definition 7.5.1. Let $\mathbf{X}$ be a Banach space, and $\overline{\mathbf{W}}$ a closed subset of $\mathbf{X}$. Suppose that a family $\mathrm{A}(t, u) \in G(\mathbf{X})$ is given for $0 \leqslant t \leqslant T_0$ and $u \in \overline{\mathbf{W}}$. The family $\{\mathrm{A}(t, u)\}$ is said to be stable if there are (stability) constants M and ω, such that

$$\left\| \prod_{j=1}^{k} [\mathrm{A}(t_j, u_j) + \lambda]^{-1} \right\| \leqslant M(\lambda - \omega)^{-k}, \qquad \lambda > \omega, \quad (7.5.5)$$

for any finite families $\{t_j\}_{j=1}^k$ and $\{u_j\}_{j=1}^k \subset \overline{\mathbf{W}}$, with $0 \leqslant t_1 \leqslant \cdots \leqslant t_k \leqslant T_0$, $k = 1, 2, \ldots$. Moreover, $\prod$ is time-ordered.

Remark 7.5.1. It can be shown (see Proposition 6.2.1) that (7.5.5) can be replaced by the equivalent, but superficially stronger, condition

$$\left\| \prod_{j=1}^{k} [\mathrm{A}(t_j, u_j) + \lambda_j]^{-1} \right\| \leqslant M \prod_{j=1}^{k} (\lambda_j - \omega)^{-1}, \qquad \lambda_j > \omega. \quad (7.5.6)$$

Note also that a family $\mathrm{A}(t,u) \in G(\mathbf{X}, 1, \omega)$ is automatically stable, with constants $1, \omega$.

Definition 7.5.2. Let $\mathbf{Y}$ be a reflexive Banach space densely and continuously embedded in a Banach space $\mathbf{X}$, such that the norm of $\mathbf{Y}$ is determined by an isomorphism S of $\mathbf{Y}$ onto $\mathbf{X}$, i.e.,

$$\|u\|_{\mathbf{Y}} = \|\mathrm{S}u\|_{\mathbf{X}}. \tag{7.5.7}$$

Let $\bar{\mathbf{W}}$ be a closed ball in $\mathbf{Y}$, centered at 0, and let $\{\mathrm{A}(t,u)\}$, $0 \leqslant t \leqslant T_0'$, $u \in \bar{\mathbf{W}}$, be stable, with constants M and ω. Suppose that $\mathrm{A}_1(t,u) \in G(\mathbf{X})$ is defined by

$$\mathrm{A}_1(t,u) = \mathrm{S}\mathrm{A}(t,u)\mathrm{S}^{-1}, \tag{7.5.8a}$$

$$\mathbf{D}_{\mathrm{A}_1(t,u)} = \{v \in \mathbf{X} : \mathrm{A}(t,u)\mathrm{S}^{-1}v \in \mathbf{Y}\}, \tag{7.5.8b}$$

and that $\{\mathrm{A}_1(t,u)\}$ is also stable, with constants M_1 and ω_1. It is explicitly assumed that $\mathbf{Y}$ is contained in the domain of $\mathrm{A}(t,u)$ for each (t,u).

Remark 7.5.2. The restriction of $\mathrm{A}(t,u)$ to $\mathrm{S}^{-1}\mathbf{D}_{\mathrm{A}_1(t,u)}$ is in $G(\mathbf{Y})$, with stability constants ω_1 and M_1 (see Proposition 6.2.4).

Now, let $N \geqslant 1$ be given, and set $\Delta t = T_0'/N$, $T_0' \leqslant T_0$. Given $u_0 \in \bar{\mathbf{W}}$, we are interested in the recursive solution of

$$\mathrm{A}(t_k, u_k^N)u_k^N + \left(\frac{1}{\Delta t}\right)u_k^N = \left(\frac{1}{\Delta t}\right)u_{k-1}^N, \qquad k = 1, \ldots, N, \tag{7.5.9}$$

for $t_k = k\,\Delta t$, $k = 0, 1, \ldots, N$. Under the format of Definitions 7.5.1 and 7.5.2, set

$$M' = \max(M, M_1), \qquad \omega' = \max(\omega, \omega_1), \qquad \gamma = \omega' + 1, \tag{7.5.10}$$

where $M' \geqslant 1$ is assumed. Suppose that $\bar{\mathbf{W}}$ (closed in $\mathbf{X}$ and $\mathbf{Y}$) is given explicitly by

$$\bar{W} = \{w \in \mathbf{Y} : \|w\|_{\mathbf{Y}} \leqslant r, \|w\|_{\mathbf{X}} \leqslant r\}, \tag{7.5.11}$$

and that T_0' and r satisfy

$$0 < M' \max(\|u_0\|_{\mathbf{Y}}, \|u_0\|_{\mathbf{X}})e^{\gamma T_0'} < r. \tag{7.5.12}$$

Select ρ, such that

$$0 < M' \max(\|u_0\|_{\mathbf{Y}}, \|u_0\|_{\mathbf{X}}) \exp\left\{\left(1 + \frac{1}{\rho}\right)\gamma T_0'\right\} < r, \tag{7.5.13}$$

and define δ and σ by

$$\delta = \left(r \exp\left\{ -\left(1 + \frac{1}{\rho}\right)\gamma T'_0 \right\} \Big/ (M' \max(\|u_0\|_{\mathbf{Y}}, \|u_0\|_{\mathbf{X}})) \right) - 1, \quad (7.5.14a)$$

$$\sigma = (1 + \delta) M' \exp\left\{ \left(1 + \frac{1}{\rho}\right)\gamma T'_0 \right\}. \quad (7.5.14b)$$

Then, clearly, $\bar{\mathbf{W}}_0 \subset \bar{\mathbf{W}}$, where

$$\bar{\mathbf{W}}_0 = \{u \in \mathbf{Y} : \|u\|_{\mathbf{Y}} \leqslant \sigma \|u_0\|_{\mathbf{Y}}, \|u\|_{\mathbf{X}} \leqslant \sigma \|u_0\|_{\mathbf{X}}\}. \quad (7.5.15)$$

Proposition 7.5.1. Suppose that there exists a constant C, such that

$$\|\mathrm{A}(t, u) - \mathrm{A}(t, u')\|_{\mathbf{Y},\mathbf{X}} \leqslant C \|u - u'\|_{\mathbf{X}}, \quad (7.5.16)$$

for $0 \leqslant t \leqslant T'_0$ and $u, u' \in \bar{\mathbf{W}}_0$, where T'_0 and $\bar{\mathbf{W}}_0$ are described in Definition 7.5.2 and Remark 7.5.2. For

$$\mu^2 = \frac{N}{T'_0} \geqslant [(1 + \delta^{-1})M' + (1 + \rho)\gamma], \quad (7.5.17a)$$

the $\{u_k^N\}_{k=1}^N$ are characterized formally as fixed points of mappings $\mathrm{Q}_k^N : \bar{\mathbf{W}}_0 \to \bar{\mathbf{W}}_0$ (see (7.5.20)). If, in addition, μ^2 satisfies

$$1 > \frac{1}{\mu^2 - 1 - \omega} \left[M + MC\sigma \|u_0\|_{\mathbf{X}} + \frac{(\omega_1 + 1)C\sigma\delta}{(1 + \delta)^2} \|u_0\|_{\mathbf{X}} \right], \quad (7.5.17b)$$

then $\{\mathrm{Q}_k^N\}$ are strict contractions on $\mathbf{X}$, with contraction constants given by the right-hand side of (7.5.17b) as functions of N. In particular, the recursive equations (7.5.9) have unique solutions in this case.

Proof: For fixed N, assume, inductively, that (7.5.9) possesses a solution u_k^N for $k < m$, where $m \geqslant 1$, such that

$$u_{k-1}^N = \prod_{j=1}^{k-1} \mu^2 \mathrm{R}(\mu^2, -\mathrm{A}(t_j, u_j^N)) u_0, \qquad \mu^2 = \left(\frac{1}{\Delta t}\right). \quad (7.5.18)$$

Rewriting (7.5.9), we see that u_k^N may be displayed as a fixed point of the mapping

$$v \mapsto \mathrm{Q}v = -\mathrm{R}(\mu^2 - 1, -\mathrm{A}(t_k, v))v + \mu^2 \mathrm{R}(\mu^2 - 1, -\mathrm{A}(t_k, v)) u_{k-1}^N. \quad (7.5.19)$$

To prove that $\mathrm{Q} = \mathrm{Q}_k^N$ maps $\bar{\mathbf{W}}_0$ into itself, we combine (7.5.18) and (7.5.19) to obtain, for $v \in \bar{\mathbf{W}}_0$,

$$\begin{aligned} \mathrm{Q}v = &-\mathrm{R}(\mu^2 - 1, -\mathrm{A}(t_k, v))v \\ &+ \mu^2 \mathrm{R}(\mu^2 - 1, -\mathrm{A}(t_k, v)) \prod_{j=1}^{k-1} \mu^2 \mathrm{R}(\mu^2, -\mathrm{A}(t_j, u_j^N))u_0, \end{aligned} \tag{7.5.20}$$

and application of (7.5.5) yields

$$\|\mathrm{Q}v\|_{\mathbf{X}} \leqslant \frac{M}{\mu^2 - 1 - \omega}\sigma\|u_0\|_{\mathbf{X}} + \frac{M\mu^2}{\mu^2 - 1 - \omega}\left(\frac{\mu^2}{\mu^2 - \omega}\right)^{k-1}\|u_0\|_{\mathbf{X}}.$$

By the choice of N,

$$\frac{M}{\mu^2 - 1 - \omega} \leqslant \delta/(1 + \delta),$$

and, by choice of σ and N,

$$\left(\frac{\mu^2}{\mu^2 - 1 - \omega}\right)\left(\frac{\mu^2}{\mu^2 - \omega}\right)^{k-1} \leqslant \left(\frac{\mu^2}{\mu^2 - 1 - \omega}\right)^N \leqslant e^{(1+1/\rho)(1+\omega)T_0'} \leqslant \sigma/[(1+\delta)M'].$$

In particular, $\|\mathrm{Q}v\|_{\mathbf{X}} \leqslant \sigma\|u_0\|_{\mathbf{X}}$. Now, by (7.5.8) and (6.2.7),

$$\mathrm{R}(\lambda, -\mathrm{A}_1(t, u)) = \mathrm{S}\mathrm{R}(\lambda, -\mathrm{A}(t, u))\mathrm{S}^{-1}. \tag{7.5.21}$$

Applying S to (7.5.20), and using (7.5.21) yields

$$\begin{aligned} \mathrm{SQ}v = &-\mathrm{R}(\mu^2 - 1, -\mathrm{A}_1(t_k, v))\mathrm{S}v \\ &+ \mu^2 \mathrm{R}(\mu^2 - 1, -\mathrm{A}_1(t_k, v)) \prod_{j=1}^{k-1} \mu^2 \mathrm{R}(\mu^2, -\mathrm{A}_1(t_j, u_j^N))Su_0, \end{aligned}$$

and the estimation of $\|\mathrm{SQ}v\|_{\mathbf{X}}$ proceeds much like that of $\|\mathrm{Q}v\|_{\mathbf{X}}$. In particular,

$$\|\mathrm{SQ}v\|_{\mathbf{X}} \leqslant \frac{M_1}{\mu^2 - 1 - \omega_1}\sigma\|\mathrm{S}u_0\|_{\mathbf{X}} + \frac{M_1\mu^2}{\mu^2 - 1 - \omega_1}\left(\frac{\mu^2}{\mu^2 - \omega_1}\right)^{k-1}\|\mathrm{S}u_0\|_{\mathbf{X}},$$

so that $\|\mathrm{SQ}v\|_{\mathbf{X}} \leqslant \sigma\|\mathrm{S}u_0\|_{\mathbf{X}}$. We conclude that $\mathrm{Q}\bar{\mathbf{W}}_0 \subset \bar{\mathbf{W}}_0$.

The contractive property of Q reduces to proving the estimate

$$\|\mathrm{R}(\lambda, -\mathrm{A}(t, v)) - \mathrm{R}(\lambda, -\mathrm{A}(t, w))\|_{\mathbf{X}} \leqslant C_1\|v - w\|_{\mathbf{X}}/[(\lambda - \omega)(\lambda - \omega_1)], \tag{7.5.22}$$

where $C_1 = MM_1C$. Indeed, an application of (7.5.22) to $\mathrm{Q}v - \mathrm{Q}u$, where Q is given by (7.5.20), leads to

$$\begin{aligned} \|\mathrm{Q}v - \mathrm{Q}w\|_{\mathbf{X}} \leqslant \frac{1}{\mu^2 - 1 - \omega}\Bigg[M &+ \frac{C_1\sigma}{\mu^2 - 1 - \omega_1}\|u_0\|_{\mathbf{X}} \\ &+ C_1\|u_0\|_{\mathbf{X}}\left(\frac{\mu^2}{\mu^2 - 1 - \omega_1}\right)\left(\frac{\mu^2}{\mu^2 - \omega}\right)^N\Bigg]\|v - w\|_{\mathbf{X}}, \end{aligned}$$

and the right-hand side of this expression is bounded, via the right-hand side of (7.5.17b), since $M_1/(\mu^2 - 1 - \omega_1) \leqslant \delta/(1 + \delta)$, and since

$$\frac{\mu^2}{\mu^2 - 1 - \omega_1} \leqslant 1 + \frac{\omega_1 + 1}{(1 + \delta^{-1})M_1}, \qquad \left(\frac{\mu^2}{\mu^2 - \omega}\right)^N \leqslant \frac{\sigma}{(1 + \delta)M'}.$$

To verify (7.5.22), we note that, by a perturbation theorem for resolvents,

$$\begin{aligned} &\mathrm{R}(\lambda, -\mathrm{A}(t, w)) - \mathrm{R}(\lambda, -\mathrm{A}(t, v)) \\ &\quad = \mathrm{R}(\lambda, -\mathrm{A}(t, w))[\mathrm{A}(t, v) - \mathrm{A}(t, w)]\mathrm{R}(\lambda, -\mathrm{A}(t, v)), \end{aligned} \tag{7.5.23}$$

so that, by (7.5.16),

$$\begin{aligned} &\|\mathrm{R}(\lambda, -\mathrm{A}(t, w)) - \mathrm{R}(\lambda, -\mathrm{A}(t, v))\| \\ &\quad \leqslant \|\mathrm{R}(\lambda, -\mathrm{A}(t, w))\|_{\mathbf{X}} C\|v - w\|_{\mathbf{X}} \|\mathrm{R}(\lambda, -\mathrm{A}(t, v))\|_{\mathbf{Y}} \\ &\quad \leqslant \left(\frac{M}{\lambda - \omega}\right)\left(\frac{M_1}{\lambda - \omega_1}\right) C\|v - w\|_X. \end{aligned}$$

The final statement follows from the contraction mapping principle applied to the complete metric subspace $\bar{\mathbf{W}}_0$ of $\mathbf{X}$; note that, here, we use the fact that $\mathbf{Y}$ is reflexive. ■

Corollary 7.5.2. Suppose the hypotheses of Proposition 7.5.1 are satisfied, including (7.5.16) and (7.5.17). Suppose that there is a uniform bound C_1 for $\{\mathrm{A}(t, w)\}$ as bounded linear mappings from $\mathbf{Y}$ into $\mathbf{X}$, for $0 \leqslant t \leqslant T_0'$ and $w \in \bar{\mathbf{W}}_0$. Then, the solutions of (7.5.9) uniquely exist and satisfy

$$\max(\|u_k^N\|_{\mathbf{Y}}, \|u_k^N\|_{\mathbf{X}}) \leqslant r, \qquad k = 0, \ldots, N, \quad N \geqslant N_0, \tag{7.5.24a}$$

$$\|u_k^N - u_{k-1}^N\|_{\mathbf{X}} \leqslant C_1 r\,\Delta t, \qquad N \geqslant N_0, \tag{7.5.24b}$$

where N_0 is defined via (7.5.17).

Proof: The first of these inequalities has been proved already. The second results from a simple estimation of (7.5.9). ■

Remark 7.5.3. It is possible to use the previous corollary as a starting point for a parallel development of quasi-linear equations of evolution in Banach spaces. However, this would needlessly repeat the results of Section 7.2. We shall, instead, return to the ideas introduced at the beginning of this section. In order to emphasize that the convergence of the viscosity method is routine, once the invariant time interval has been established, we prove the former result first under a dissipation hypothesis. We now define this concept and present the hypotheses upon a_j, b and P.

Definition 7.5.3. We shall call the system (7.5.1)–(7.5.3) dissipative on $(0, T_0')$ if, for $0 < t < T_0'$ and $\mathbf{w} \in \mathbf{PH}^s(\mathbb{R}^n; \mathbb{R}^m)$,

$$(\mathrm{E}(t, \mathbf{w})\mathbf{v}, \mathbf{v})_{\mathbf{L}^2(\mathbb{R}^n)} \geqslant 0, \qquad \text{for all} \quad \mathbf{v} \in \mathbf{Y}_s. \tag{7.5.25}$$

The coefficients a_j and b are assumed to belong to the class

$$\mathbf{Lip}([0, T_0]; \mathbf{C}_b^{s+1}(\mathbb{R}^n \times \Omega; \mathbb{R}^{m^2})) \qquad (\mathbf{C}_b^{s+1} \text{ replaced by } \mathbf{C}_b^{[s]+2} \text{ if } s \neq [s])$$

for *every* relatively compact open subset Ω of $\mathbb{R}^m$. The $\mathbf{L}^2$ orthogonal projection P is explicitly assumed to satisfy

$$\mathbf{PH}^\sigma(\mathbb{R}^n; \mathbb{R}^m) \subset \mathbf{H}^\sigma(\mathbb{R}^n; \mathbb{R}^m), \qquad \sigma \geqslant 1, \tag{7.5.26a}$$

$$\Delta\mathbf{PH}^2(\mathbb{R}^n; \mathbb{R}^m) \subset \mathbf{PL}^2(\mathbb{R}^n; \mathbb{R}^m), \tag{7.5.26b}$$

$$\mathrm{P}(\mathrm{I} - \Delta)^{\sigma/2} = (\mathrm{I} - \Delta)^{\sigma/2}\mathrm{P}, \qquad \sigma \geqslant 1. \tag{7.5.26c}$$

Finally, we assume that $\mathrm{E}(t, w) \in G(\mathbf{PL}^2, 1, \omega(t, \mathbf{w}))$ for all $\mathbf{w} \in \mathbf{PH}^s(\mathbb{R}^n)$.

Remark 7.5.4. The Navier–Stokes/Euler system projection satisfies (7.5.26) and the generator property on $\mathrm{E}(t, \mathbf{w})$ (see Kato [17]). The first hypothesis implies that PH^σ is closed in $\mathbf{H}^\sigma$, and the third hypothesis permits a resolvent estimation on $\mathbf{H}^s$. Note that $(\mathrm{I} - \Delta)^{\sigma/2}$, here, has a diagonal operator interpretation, as does Δ in (7.5.26b).

Proposition 7.5.3. Consider the system (7.5.1)–(7.5.2), with the associated properties (7.5.3) and (7.5.4) and the dissipation property (7.5.25). There exist constants C_1 and C_2, such that, if $\mathbf{u}_1$ and $\mathbf{u}_2$ are solutions of (7.5.2), with $v = v_1$ and $v = v_2$, respectively, the estimate

$$\|(\mathbf{u}_1 - \mathbf{u}_2)(t)\|^2_{\mathbf{L}^2(\mathbb{R}^n)} \leqslant C_1(v_1 - v_2)^2 e^{C_2 T_0'} \tag{7.5.27}$$

holds for $0 \leqslant t \leqslant T_0'$.

Proof: Subtract the respective relations (7.5.2a), involving $\mathbf{u} = \mathbf{u}_1$, $v = v_1$ and $\mathbf{u} = \mathbf{u}_2$, $v = v_2$, and dot multiply by $\mathbf{u}_1 - \mathbf{u}_2$ to obtain, after integration over $\mathbb{R}^n$,

$$\begin{aligned}
&\tfrac{1}{2}\frac{d}{dt}\|\mathbf{u}_1 - \mathbf{u}_2\|^2_{\mathbf{L}^2(\mathbb{R}^n)} - v_2(\Delta(\mathbf{u}_1 - \mathbf{u}_2), \mathbf{u}_1 - \mathbf{u}_2)_{\mathbf{L}^2(\mathbb{R}^n)} \\
&\quad = (v_1 - v_2)(\Delta\mathbf{u}_1, \mathbf{u}_1 - \mathbf{u}_2)_{\mathbf{L}^2(\mathbb{R}^n)} - (\mathrm{E}(\mathbf{u}_1)\mathbf{u}_1 - E(\mathbf{u}_2)\mathbf{u}_2, \mathbf{u}_1 - \mathbf{u}_2)_{\mathbf{L}^2(\mathbb{R}^n)} \\
&\quad = (v_1 - v_2)(\Delta\mathbf{u}_1, \mathbf{u}_1 - \mathbf{u}_2)_{\mathbf{L}^2(\mathbb{R}^n)} - (E(\mathbf{u}_1)(\mathbf{u}_1 - \mathbf{u}_2), \mathbf{u}_1 - \mathbf{u}_2)_{\mathbf{L}^2(\mathbb{R}^n)} \\
&\qquad - ([\mathrm{E}(\mathbf{u}_1) - \mathrm{E}(\mathbf{u}_2)]\mathbf{u}_2, \mathbf{u}_1 - \mathbf{u}_2)_{\mathbf{L}^2(\mathbb{R}^n)} \\
&\quad \leqslant \tfrac{1}{2}(v_1 - v_2)^2\|\Delta\mathbf{u}_1\|^2_{\mathbf{L}^2(\mathbb{R}^n)} + \tfrac{1}{2}\|\mathbf{u}_1 - \mathbf{u}_2\|^2_{\mathbf{L}^2(\mathbb{R}^n)} \\
&\qquad + C\|\mathbf{u}_2\|_{\mathbf{L}^2(\mathbb{R}^n)}\|\mathbf{u}_1 - \mathbf{u}_2\|^2_{\mathbf{L}^2(\mathbb{R}^n)},
\end{aligned}$$

where we have used the dissipation property (7.5.25) and the Lipschitz property

$$\|\mathrm{E}(t, \mathbf{w}) - \mathrm{E}(t, \mathbf{w}')\|_{\mathbf{H}^s(\mathbb{R}^n), \mathbf{L}^2(\mathbb{R}^n)} \leqslant C\|\mathbf{w} - \mathbf{w}'\|_{\mathbf{L}^2(\mathbb{R}^n)}, \tag{7.5.28}$$

for $\mathbf{w}, \mathbf{w}' \in \overline{\mathbf{B}(0, r')} \subset \mathbf{H}^s(\mathbb{R}^n; \mathbb{R}^m)$ (see Remark 7.3.4). By the classical Gronwall inequality applied, after integration in t, we have (7.5.27), where $C_1 \geqslant \|\Delta \mathbf{u}_1\|^2_{\mathbf{L}^2(\mathbb{R}^n)}$, $C_2 \geqslant 2C\|\mathbf{u}_2\|_{\mathbf{L}^2(\mathbb{R}^n)} + 1$ are constants independent of v_1 and v_2, since $\mathbf{u}_1$ and $\mathbf{u}_2$ are assumed to satisfy (7.5.4). ■

Remark 7.5.5. The method of selecting the pair (T'_0, r') may now be implemented, as described prior to Definition 7.5.1. We note that $-\nu\Delta \in G(\mathbf{PL}^2, 1, 0)$ for $\nu > 0$. The perturbation $\mathrm{E}(t, \mathbf{w})$ is relatively bounded, with respect to $-\nu\Delta$, by a standard interpolation, with bound arbitrarily small, and by assumption is in $G(\mathbf{PL}^2, 1, \omega)$, where an upper bound for ω can frequently be determined. Thus, $\mathrm{A}(t, w) \in G(\mathbf{PL}^2, 1, \omega)$ (see Kato [15] pp. 499 and 500). In the particular case of the Navier–Stokes/Euler system, the operator

$$\mathrm{A}(t, \mathbf{w}) = -\nu\Delta + \mathrm{P}(\mathbf{w} \cdot \nabla) \tag{7.5.29}$$

is unconditionally stable on $\mathbf{L}^2(\mathbb{R}^n)$, in the sense that $\mathrm{A}(t, \mathbf{w}) \in G(\mathbf{L}^2, 1, 0)$ (see Kato [17]). In this case, the choice $\omega(r) \equiv 0$ may be made.

The estimation of $\omega_1(r)$ uses the similarity relation (7.5.8a), with $\mathrm{A}_1 = \mathrm{A} + \mathrm{B}$. Earlier results (see Proposition 6.3.1) reveal that $\|\mathbf{B}\|_{\mathbf{L}^2(\mathbb{R}^n), \mathbf{L}^2(\mathbb{R}^n)} \leqslant \beta \Rightarrow \omega_1 = \omega + \beta$ is an acceptable choice. We have already discussed the choice of β in the context of second-order quasi-linear hyperbolic systems. The present analysis is similar. Thus, a simple computation gives

$$\mathrm{SA}(t, \mathbf{w})\mathrm{S}^{-1} = \mathrm{A}(t, \mathbf{w}) + \mathrm{P}\left[\sum_{j=1}^{n} [\mathrm{S}, \mathrm{a}_j]\Lambda^{1-s}\left(\frac{\partial}{\partial x_j}\right)\Lambda^{-1} + [\mathrm{S}, \mathrm{b}]\Lambda^{1-s}\Lambda^{-1}\right],$$

where $\Lambda = (\mathrm{I} - \Delta)^{1/2}$, $\mathrm{S} = \Lambda^s$, and $[\cdot, \cdot]$ denotes the commutator. According to the commutator result of Kato (see Lemma 7.3.4), the estimates

$$\|[\mathrm{S}, \mathrm{a}_j]\Lambda^{1-s}\|_{\mathbf{L}^2(\mathbb{R}^n), \mathbf{L}^2(\mathbb{R}^n)} \leqslant c\|\operatorname{grad} a_j\|_{\mathbf{H}^{s-1}(\mathbb{R}^n)} \leqslant \beta(r)/(2n), \tag{7.5.30a}$$

$$\|[\mathrm{S}, \mathrm{b}]\Lambda^{1-s}\|_{\mathbf{L}^2(\mathbb{R}^n), \mathbf{L}^2(\mathbb{R}^n)} \leqslant c\|\operatorname{grad} b\|_{\mathbf{H}^{s-1}(\mathbb{R}^n)} \leqslant \beta(r)/2 \tag{7.5.30b}$$

hold for $s > n/2 + 1$. Since $\partial/\partial x_j \Lambda^{-1}$ and Λ^{-1} are bounded on $\mathbf{L}^2(\mathbb{R}^n; \mathbb{R}^m)$ by 1, the choice of β and, hence, ω_1 is possible.

Remark 7.5.6. In the case of the operator given by (7.5.29), a careful study of Kato's proof reveals that $\beta(r)$ in (7.5.30) may be chosen as the linear function

$$\beta(r) = n^{3/2}s(c_1 + c_2)r = cr, \tag{7.5.31a}$$

where c_1 is the least constant in the ring inequality

$$\|fg\|_{\mathbf{H}^{s-1}(\mathbb{R}^n)} \leqslant c_1\|f\|_{\mathbf{H}^{s-1}(\mathbb{R}^n)}\|g\|_{\mathbf{H}^{s-1}(\mathbb{R}^n)}, \tag{7.5.31b}$$

for real-valued functions f and g, and c_2 is the Sobolev constant in the inequality

$$\|f\|_{\mathbf{L}^\infty(\mathbb{R}^n)} \leqslant c_2\|f\|_{\mathbf{H}^{s-1}(\mathbb{R}^n)}, \tag{7.5.31c}$$

for real-valued functions f. Here we have adjusted $\beta(r)$ for the fact that $\mathrm{b} = 0$ for (7.5.29). Now, let $T_0 = \sup T'_0$ represent an as yet undetermined critical value (actual or spurious). Maximize the function

$$f(r) = re^{-\gamma(r)T_0}, \qquad r > 0,$$

where $\gamma(r) = \beta(r) + 1$, and β is given by (7.5.31a). The solution is rendered by elementary calculus as

$$r' = \frac{1}{cT_0}, \tag{7.5.32a}$$

giving the relation

$$\|\mathbf{u}_0\|_{\mathbf{H}^s(\mathbb{R}^n)} = \frac{1}{cT_0}\, e^{-(1+T_0)} \tag{7.5.32b}$$

as the defining relation for T_0, when taken in conjunction with (7.5.33) to follow. The quantity T'_0 may be chosen as any value satisfying $T'_0 < T_0$.

It is time to bring all of this to a close. We end with a summarizing statement, which may be proved with the aid of Corollary 7.5.2 and the standard convergence techniques for the method of horizontal lines.

Theorem 7.5.4. Let $\mathrm{A}(t, \mathbf{w})$ be given as in (7.5.1). Suppose $v \geqslant 0$ and $r' > 0$ under the hypotheses on P and $\mathrm{E}(t, \mathbf{w})$ of Definition 7.5.3. Then there exists a positive constant γ, not depending on v, such that, if $\mathbf{u}_0 \in \mathbf{B}(0, r') \subset \mathbf{H}^s(\mathbb{R}^n; \mathbb{R}^m)$, $s > n/2 + 1$, and $\mathbf{u}_0$ and T'_0 satisfy the relation

$$\|\mathbf{u}_0\|_{\mathbf{H}^s(\mathbb{R}^n)} e^{\gamma T'_0} < r', \tag{7.5.33}$$

there exists a solution of (7.5.1)–(7.5.2), satisfying the properties (7.5.3)–(7.5.4). If the system is dissipative, the convergence result (7.5.27) holds and all solutions are unique. In the particular case of the (dissipative) Navier–Stokes/Euler system, r' and T'_0 are given explicitly via (7.5.31) and (7.5.32), and a unique solution of the Cauchy problem (7.5.2), satisfying (7.5.3) and (7.5.4), exists for the operator $\mathrm{A}(\cdot,\cdot)$ of (7.5.29) on $[0, T'_0]$.

7.6 BIBLIOGRAPHICAL REMARKS

The development of Sections 7.2 and 7.3 largely follows that of Hughes, Kato, and Marsden [13] in conceptual content, although the hypotheses here differ in certain respects. Not only is the linear theory, developed in the previous chapter, indispensable for this development, but certain associated ideas, developed by Kato in earlier papers (see Kato [17, 18]), some of which are presented in Section 7.1, are also necessary. Other ideas which prove decisive, and which were developed earlier by Kato, include the role of the uniformly local Sobolev spaces in a generalized ring theory (see Remark 7.3.4), and the commutator estimate of Lemma 7.3.4, which implies the uniform boundedness and, hence the upper Lebesgue integrability of the family $\{B^v(\cdot)\}$. The hypotheses of Sections 7.2 and 7.3 are sufficiently strong to guarantee that $\{\mathrm{A}^v(t)\}$ is stable both on the ground space and the smooth space **Y**, so that the full strength of the linear theory of Chapter 6 is not required.

Many sources have been used for the example of Section 7.4. Of course, the basic source is [13] for the existence theorem. For the physics, we have been guided by the eminently readable monograph of Einstein [5] and the ambitious treatise of Hawking and Ellis [11], where existence theorems are also proved. Other valuable ideas were derived from the expository book of Rindler [23]. The mathematical development of the section was assisted by reference to Bishop and Goldberg [1], Hicks [12], and Sachs and Wu [24]. A clear account of the role of harmonic coordinates in the analytic reduction is presented by Fock [9]. This reduction was employed by Choquet-Bruhat [3] in the first existence proof for the field equations, where it is noted that the evolution preserves the harmonic coordinates. An informative mathematical analysis of the elliptic constraint equations satisfied by the initial data has been given by Cantor [2] in the general context of tensor field decomposition. An alternative approach, making use of the Hamiltonian formalism for general relativity, via the lapse and shift functions, has been given by Fischer and Marsden [8] (see also Marsden [21]).

The final section follows Jerome [14]. The viscosity method in fluid dynamics was examined by Golovkin [10] for planar flow, and by Swann [25] and Kato [16], among others, for three-dimensional flows. The estimate for the time interval given by (7.5.31)–(7.5.32) in this case appears more explicit than that given in the two previous references. However, it depends directly upon the commutator estimates obtained by Kato and referred to above. The reader may have already conjectured that the semidiscrete analysis can probably spawn a semidiscrete viscosity convergence result.

This proves correct, and details may be found in Jerome [14]. A study of the viscosity method, employing global analysis, has been carried out by Ebin and Marsden [4].

There are, of course, other approaches to the theory of nonlinear Cauchy problems in Banach spaces. An approach which constructs the evolution operators directly for the nonlinear system has been given by Evans [6] for accretive systems. These systems, though more restricted, permit global solutions in time.

Though the hypotheses that **Y**, **X**, **V**, and **Z** are separable and reflexive are more stringent than necessary, we have used these hypotheses to invoke the Dunford–Pettis theorem and the fundamental theorem of calculus in Banach spaces in our analysis of the similarity relations embodied, say, in (A2). Stronger hypotheses here would yield results valid in general Banach spaces.

We note, finally, that Theorem 7.5.4 serves as a useful counterpoint to the global weak existence theory of Chapter 5, particularly for models such as the Navier–Stokes model, where uniqueness has not been demonstrated for the general class of weak solutions. Other applications of the theory of this chapter are possible, for example, to elastodynamics. We refer to the monograph of Marsden and Hughes [22] for an account of this. We also refer the reader to Kato [19] for a discussion of the redundancy of condition (A4) of Definition 7.2.2.

REFERENCES

[1] R. L. Bishop and S. I. Goldberg, "Tensor Analysis on Manifolds." MacMillan, New York, 1968.

[2] M. Cantor, Elliptic operators and the decomposition of tensor fields, *Bull. Amer. Math. Soc.* **5**, 235–262 (1982).

[3] Y. Choquet-Bruhat (Y. Fourès-Bruhat), Théorème d'existence pour certain systèmes d'équations aux dérivées partielles nonlinéaires, *Acta Math.* **88**, 141–225 (1952).

[4] D. Ebin and J. Marsden, Groups of diffeomorphisms and the motion of an incompressible fluid, *Ann. of Math.* **92**, 102–163 (1970).

[5] A. Einstein, "The Meaning of Relativity," 3rd. ed. Princeton Univ. Press, Princeton, New Jersey, 1950.

[6] L. C. Evans, Nonlinear evolution equations in an arbitrary Banach Space, *Israel J. Math.* **26**, 1–42 (1977).

[7] A. Fischer and J. Marsden, The Einstein evolution equations as a first order quasi-linear symmetric hyperbolic system, I, *Comm. Math. Phys.* **28**, 1–38 (1972).

[8] A. Fischer and J. Marsden, The initial value problem and the dynamical formulation of general relativity, in, "General Relativity and Einstein's Centenary Survey" (S. Hawking and W. Israel, eds.), Chapter 4. Cambridge Univ. Press, London and New York, 1979.

[9] V. Fock, "Theory of Space, Time and Gravitation." MacMillan, New York, 1964.

[10] K. Golovkin, Vanishing viscosity in Cauchy's problem for hydromechanics, *Trudy Mat. Inst. Steklov* **92**, 31–49 (1966); and *Proc. Steklov Inst. Math.* **92**, 33–53 (1966).

[11] S. W. Hawking and G. F. R. Ellis, "The Large Scale Structure of Space-Time," Cambridge Univ. Press, London and New York, 1973.

[12] N. J. Hicks, "Notes on Differential Geometry." Van Nostrand, Princeton, New Jersey, 1965.

[13] T. Hughes, T. Kato, and J. Marsden, Well-posed, quasi-linear, second-order hyperbolic systems with applications to nonlinear elastodynamics and general relativity, *Arch. Rational Mech. Anal.* **63**, 273–294, (1977).

[14] J. W. Jerome, Quasilinear hyperbolic and parabolic systems: contractive semidiscretizations and convergence of the discrete viscosity method, *J. Math. Anal. Appl.* **90**, 185–206 (1982).

[15] T. Kato, "Perturbation Theory for Linear Operators." Springer–Verlag, Berlin and New York, 1966.

[16] T. Kato, Nonstationary flows of viscous and ideal fluids in $\mathbb{R}^3$, *J. Functional Anal.* **9**, 296–305 (1972).

[17] T. Kato, Quasi-linear equations of evolution, with applications to partial differential equations. Springer Lecture Notes in Mathematics 448, pp. 25–70. Springer–Verlag, Berlin and New York, 1975.

[18] T. Kato, The Cauchy problem for quasi-linear symmetric hyperbolic systems, *Arch. Rational Mech. Anal.* **58**, 181–205 (1975).

[19] T. Kato, "Linear and Quasilinear Equations of Hyperbolic Type," Bressanone Lectures. Centro Inter. Mat. Estivo, Rome, 1977.

[20] Y. Komura, Nonlinear semi-groups in Hilbert space, *J. Math. Soc. Japan* **19**, 473–507 (1967).

[21] J. E. Marsden, "Lectures on Geometric Methods in Mathematical Physics." SIAM, Philadelphia, Pennsylvania, 1981.

[22] J. E. Marsden and T. J. R. Hughes, "Topics in the Mathematical Foundations of Elasticity." Prentice–Hall, Englewood Cliffs, New Jersey, 1982.

[23] W. Rindler, "Essential Relativity," 2nd ed. Springer–Verlag, Berlin and New York, 1977.

[24] R. K. Sachs and H. Wu, "General Relativity for Mathematicians." Springer–Verlag, Berlin and New York, 1977.

[25] H. Swann, The convergence with vanishing viscosity of nonstationary Navier–Stokes flow to ideal flow in $\mathbb{R}^3$, *Trans. Amer. Math. Soc.* **157**, 373–397 (1971).

SUBJECT INDEX

H

I

K

L

M

N

O

P

Q